Ecosystem Services from Forest Landscapes

Ajith H. Perera • Urmas Peterson
Guillermo Martínez Pastur • Louis R. Iverson
Editors

Ecosystem Services from Forest Landscapes

Broadscale Considerations

 Springer

Editors
Ajith H. Perera
Ontario Forest Research Institute
Ministry of Natural Resources and Forestry
Sault Ste. Marie, ON, Canada

Urmas Peterson
Institute of Forestry and Rural Engineering
Estonian University of Life Sciences
Tartu, Estonia

Guillermo Martínez Pastur
Centro Austral de Investigaciones
Científicas (CADIC)
Consejo Nacional de Investigaciones
Científicas y Técnicas (CONICET)
Ushuaia, Tierra del Fuego, Argentina

Louis R. Iverson
Northern Institute of Applied Climate
Science
Northern Research Station
US Forest Service
Delaware, OH, USA

ISBN 978-3-319-74514-5   ISBN 978-3-319-74515-2   (eBook)
https://doi.org/10.1007/978-3-319-74515-2

Library of Congress Control Number: 2018935250

Printed on acid-free paper

This Springer imprint is published by the registered company Springer International Publishing AG part of Springer Nature.
The registered company address is: Gewerbestrasse 11, 6330 Cham, Switzerland

Preface

Over the last two decades, the topic of ecosystem services has attracted the attention of researchers, land managers, and policy makers around the globe. The ecosystems addressed thus include an array of aquatic and terrestrial systems including oceans, lakes, rivers, wetlands, grasslands, forests, croplands, and urban areas. The services rendered by these ecosystems are also a long list, ranging from intrinsic to anthropocentric benefits that are typically grouped as provisioning, regulating, supporting, and cultural. The research efforts, assessments, and attempts to manage ecosystems for their sustained services are now widely published in scientific literature.

Nearly 200 researchers gathered in Tartu, Estonia, from August 23 to 30, 2015, under the sponsorship of the IUFRO working party on landscape ecology to discuss the topic "sustaining ecosystem services from forest landscapes." A major theme that emerged from the proceedings was the necessity to broaden the scope of land use planning through adopting a landscape-scale approach. Even though this approach is complex and involves multiple ecological, social, cultural, economic, and political dimensions, the landscape perspective appears to offer the best opportunity for a sustained provision of forest ecosystem services.

This book is a compilation of keynote presentations and syntheses of symposia of the Tartu meeting, focusing on broadscale aspects of forest ecosystem services, beyond individual stands to large landscapes. In doing so, our goal is to create an awareness of the conceptual and practical opportunities as well as challenges involved with planning for forest ecosystem services across landscapes, regions, and nations. However, we must remind the reader that our goal here is not to offer an exhaustive literature review or a comprehensive assessment of the state of knowledge in forest ecosystem services. For that purpose, many general reviews and syntheses can be easily found in scientific literature.

This volume is composed of nine chapters. It begins with a brief introduction to ecosystem services from forest landscapes to provide a topical overview and describe the terminology. The next two chapters draw attention to two relatively lesser known regulatory services from forest ecosystems that have broadscale connotations.

Chapters "Towards Functional Green Infrastructure in the Baltic Sea Region: Knowledge Production and Learning Across Borders", "Sustainable Planning for Peri-urban Landscapes" and "Barriers and Bridges for Landscape Stewardship and Knowledge Production to Sustain Functional Green Infrastructures" address the complexities and multiple issues that are associated with attempts to sustain forest ecosystem services across large landscapes and multiple administrative and political boundaries, whether local or international. Chapters "Solving Conflicts Among Conservation, Economic and Social Objectives in Boreal Production Forest Landscapes: Fennoscandian Perspectives" and "Natural Disturbances and Forest Management: Interacting Patterns on the Landscape" focus on both practical and conceptual aspects deriving ecosystem services from forest landscapes. The concluding chapter summarizes the overall contents and emergent messages of the book and offers some thoughts for future research and applications.

We hope that both developers of scientific knowledge and those who apply that knowledge through policy development and land management will benefit from this discourse. The geographical scope of this book is primarily focused on temperate forest landscapes, and the array of case studies and topics discussed here is by no means globally exhaustive. We anticipate, however, that this volume will offer useful insights to readers in different geographic contexts and also to those who focus on services from non-forested ecosystems. We believe that the various concepts, questions, issues, and solutions presented here, which transcend individual ecosystems and narrower scales around the globe, are valuable contributions to the collective endeavor of expanding our knowledge of this important topic.

Finally, we are indebted to the colleagues who critically reviewed chapter manuscripts and offered suggestions for improvements: Mariano Amoroso, Peter Besseau, Juan Manuel Cellini, Guy Chiasson, Trevor F. Keenan, Timo Kuuluvainen, Lars Laestadius, Silvia Matteucci, Sergio Menéndez, Josep Peñuelas, Chris J. Peterson, Sanna-Riikka Saarela, Andreas Schindlbacher, Ayanda Sigwela, and Susan Smith. Their critiques helped us to greatly improve the veracity and clarity of the messages in this book. We also acknowledge the assistance of Andrea Sandell and Janet Slobodien of Springer New York, who guided us through the publication process.

Sault Ste. Marie, ON, Canada	Ajith H. Perera
Tartu, Estonia	Urmas Peterson
Ushuaia, Argentina	Guillermo Martínez Pastur
Delaware, OH, USA	Louis R. Iverson

Contents

Contributors

Christian Albert Institute of Environmental Planning, Leibniz Universität Hannover, Hannover, Germany

Per Angelstam School for Forest Management, Swedish University of Agricultural Sciences, Skinnskatteberg, Sweden

Endijs Baders Latvian State Forest Research Institute, "Silava", Salaspils, Latvia

Daniel Burgas Department of Biological and Environmental Sciences, University of Jyvaskyla, Jyvaskyla, FL, Finland

Klaus Butterbach-Bahl Institute of Meteorology and Climate Research, Atmospheric Environmental Research (IMK-IFU), Karlsruhe Institute of Technology, Garmisch-Partenkirchen, Germany

Lucas Dawson Department of Physical Geography and Quaternary Geology, Stockholm University, Stockholm, Sweden

Eugenio Díaz-Pinés Institute of Soil Research, University of Natural Resources and Life Sciences (BOKU), Vienna, Austria

Institute of Meteorology and Climate Research, Atmospheric Environmental Research (IMK-IFU), Karlsruhe Institute of Technology, Garmisch-Partenkirchen, Germany

Marine Elbakidze School for Forest Management, Swedish University of Agricultural Sciences, Skinnskatteberg, Sweden

Kyle Eyvindson Department of Biological and Environmental Sciences, University of Jyvaskyla, Jyvaskyla, FL, Finland

Lee E. Frelich University of Minnesota, Center for Forest Ecology, Saint Paul, MN, USA

Christine Fürst Institute for Geosciences and Geography, Dept. Sustainable Landscape Development, Martin Luther University Halle, Halle, Germany

Davide Geneletti Department of Civil, Environmental and Mechanical Engineering, University of Trento, Trento, Italy

Aleksandra Grebenzhikova Pskovlesproekt Company, Pskov, Russian Federation

Austra Irbe Zemgale Planning Region Administration, Jelgava, Latvia

Louis R. Iverson Northern Institute of Applied Climate Science, Northern Research Station, US Forest Service, Delaware, OH, USA

Kalev Jõgiste Institute of Forestry and Rural Engineering, Estonian University of Life Sciences, Tartu, Estonia

Daniele La Rosa Department Civil Engineering and Architecture, University of Catania, Catania, Italy

Anna Lawrence Scottish School of Forestry, University of the Highlands and Islands, Inverness, UK

Eric Le Tortorec Department of Biological and Environmental Sciences, University of Jyvaskyla, Jyvaskyla, FL, Finland

Liga Liepa Faculty of Forestry, Latvia University of Agriculture, Jelgava, Latvia

Michael Manton Faculty of Forest Sciences and Ecology, Aleksandras Stulginskis University, Kaunas, Lithuania

School for Forest Management, Swedish University of Agricultural Sciences, Skinnskatteberg, Sweden

State Environmental Institution National Park "Braslavskie Ozera", Braslav, Belarus

Guillermo Martínez Pastur Centro Austral de Investigaciones Científicas (CADIC), Consejo Nacional de Investigaciones Científicas y Técnicas (CONICET), Ushuaia, Tierra del Fuego, Argentina

Viesturs Melecis Institute of Biology, University of Latvia, Salaspils, Latvia

Mikko Mönkkönen Department of Biological and Environmental Sciences, University of Jyvaskyla, Jyvaskyla, FL, Finland

Vladimir Naumov School for Forest Management, Swedish University of Agricultural Sciences, Skinnskatteberg, Sweden

Marharyta Nestsiarenka State Environmental Institution, National Park "Braslavskie Ozera", Braslav, Belarus

Ülo Niinemets Institute of Agricultural and Environmental Sciences, Estonian University of Life Sciences, Tartu, Estonia

Kristi Parro Institute of Forestry and Rural Engineering, Estonian University of Life Sciences, Tartu, Estonia

Ajith H. Perera Ontario Forest Research Institute, Ontario Ministry of Natural Resources, Sault Ste. Marie, ON, Canada

Urmas Peterson Institute of Forestry and Rural Engineering, Estonian University of Life Sciences, Tartu, Estonia

Maiju Peura Department of Biological and Environmental Sciences, University of Jyvaskyla, Jyvaskyla, FL, Finland

Tähti Pohjanmies Department of Biological and Environmental Sciences, University of Jyvaskyla, Jyvaskyla, FL, Finland

Zigmārs Rendenieks Faculty of Geography and Earth Sciences, University of Latvia, Riga, Latvia

Anna Repo Department of Biological and Environmental Sciences, University of Jyvaskyla, Jyvaskyla, FL, Finland

Alena Shushkova SSPA "The Scientific and Practical Center of the National Academy of Sciences of Belarus for Biological Resources", Minsk, Belarus

Marcin Spyra Institute for Geosciences and Geography, Dept. Sustainable Landscape Development, Martin Luther University Halle, Halle, Germany

John A. Stanturf Center for Forest Disturbance Science, Southern Research Station, US Forest Service, Athens, Georgia, USA

Laura Trasūne Faculty of Geography and Earth Sciences, University of Latvia, Riga, Latvia

María Triviño Department of Biological and Environmental Sciences, University of Jyvaskyla, Jyvaskyla, FL, Finland

Siarhei Uhlianets State Scientific Institution, "The Institute of Experimental Botany named after V.F. Kuprevich of the National Academy of Sciences of Belarus", Minsk, Belarus

Uladzimir Ustsin SSPA "The Scientific and Practical Center of the National Academy of Sciences of Belarus for Biological Resources", Minsk, Belarus

Christian Werner Senckenberg Biodiversity and Climate Research Centre (BiK-F), Frankfurt, Germany

Maxim Yermokhin State Scientific Institution, "The Institute of Experimental Botany named after V.F. Kuprevich of the National Academy of Sciences of Belarus", Minsk, Belarus

Natalia Yurhenson SSPA "The Scientific and Practical Center of the National Academy of Sciences of Belarus for Biological Resources", Minsk, Belarus

Anton Zhivotov State Scientific Institution, "The Institute of Experimental Botany named after V.F. Kuprevich of the National Academy of Sciences of Belarus", Minsk, Belarus

Ecosystem Services from Forest Landscapes: An Overview

Guillermo Martínez Pastur, Ajith H. Perera, Urmas Peterson, and Louis R. Iverson

1 What are Ecosystem Services?

Human beings derive direct benefit from an array of ecosystem goods as well as from the activities and products of organisms, in both wild and human-dominated ecosystems (Daily et al. 1997; Levin and Lubchenco 2008). These benefits from nature have been readily available throughout most of human history. To this day, societies take many of these natural services for granted (Daily 1997, MEA 2005), even while the support systems that provide them are being severely degraded (Vitousek et al. 1997; Levin and Lubchenco 2008; Seppelt et al. 2011). The central challenge of this century is to develop economic and social systems and supporting systems of governance from local to global scales that will achieve sustainable levels of human population and consumption while also maintaining the ecosystem life-support services that underpin human well-being (Guerry et al. 2015).

The full range of ecosystem benefits to human life is grouped under the concept "ecosystem services" (ES). Since this concept was first introduced (Ehrlich and Mooney 1983), it has evolved (Daily 1997; MEA 2005) into a global phenomenon (e.g., Kubiszewski et al. 2017). ES can be briefly defined as the benefits that humans

G. Martínez Pastur (✉)
Centro Austral de Investigaciones Científicas (CADIC), Consejo Nacional de Investigaciones Científicas y Técnicas (CONICET), Ushuaia, Tierra del Fuego, Argentina
e-mail: gpastur@conicet.gov.ar

A. H. Perera
Ontario Forest Research Institute, Sault Ste. Marie, ON, Canada

U. Peterson
Institute of Forestry and Rural Engineering, Estonian University of Life Sciences, Tartu, Estonia

L. R. Iverson
Northern Institute of Applied Climate Science, Northern Research Station, US Forest Service, Delaware, OH, USA

© Springer International Publishing AG, part of Springer Nature 2018
A. H. Perera et al. (eds.), *Ecosystem Services from Forest Landscapes*,
https://doi.org/10.1007/978-3-319-74515-2_1

1

Fig. 1 Four categories of
ecosystem services defined
by The Millennium
Ecosystem Assessment
(2005)

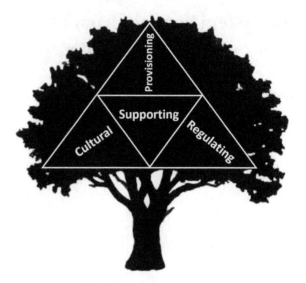

obtain from ecological systems (Levin and Lubchenco 2008), consisting of flows of
materials, energy, and information from natural capital stocks which, when combined with services derived from human capital to produce human welfare (Costanza
et al. 1997). ES comprise ecosystem functions, which refer to the habitat, biological
or system properties or processes of ecosystems, and also the ecosystem goods
(such as food) and services (such as waste assimilation) which human populations
derive, directly or indirectly, from ecosystem functions (Costanza et al. 1997, 2014).

It is possible to recognize four categories of ES (Fig. 1): (i) provisioning services
or the provision of food or habitat; (ii) regulating services, such as the regulation of
erosion or climate; (iii) supporting services, such as primary production or nutrient
cycling; and (iv) cultural services, such as aesthetic enjoyment or recreation (MEA
2005). This classification gave rise to wider understanding of the potential uses of
ES and also provided a framework for analyzing the various influences, active and
passive, by which ecosystem services enhance human well-being (Boyd and
Banzhaf 2007; Fisher et al. 2009). Nevertheless, most of the functions and services
included under any one of the four ES categories are interdependent and support
human welfare through their contribution to the joint products of the ecosystem
(Costanza et al. 1997).

2   What are Forest Landscapes?

Here we define a forest landscape as either a natural or built-up area, at any scale, in
which trees dominate the main ecosystems. We include in this definition all of the
natural components that are present, together with their spatial heterogeneity, but
also the human activities which create and affect patterns and processes within the

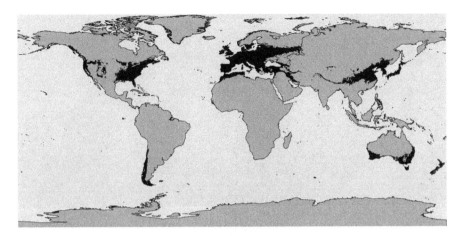

Fig. 2 Distribution of world's temperate forest biome that include broad-leaved, coniferous, and mixed forests (based on www.worldwildlife.org/biomes)

landscape. Forest landscapes cover more than four billion hectares, close to 30% of the Earth's land area, and account for 75% of terrestrial gross primary production and 80% of total plant biomass. They contain more carbon (in biomass and soils) than the total stored in the atmosphere (Beer et al. 2010; Pan et al. 2011). Forest landscapes also harbor most of the species on Earth and provide the most valuable goods and services to humanity (Costanza et al. 1997; Daily et al. 1997).

Temperate forests (Fig. 2), defined here as those forests located between 25° and 55° N and S latitudes, are highly diverse in species and soils and in the carbon pool of their ecosystems (Lal and Lorenz 2012). Temperate forest types vary among broad-leaved evergreen, broad-leaved deciduous, and coniferous, both pure or in combination. These forests are located primarily in the northern hemisphere across all continents but also in southern South America, Africa, and Oceania. Temperate regions of the world have been the most extensively altered by human activities, with significant impacts on the provision of goods and services, as well as the loss of biodiversity (Franklin 1988; Lindenmayer et al. 2012). These forests are the primary focus of our discourse because the need for improved strategies of management and conservation is particularly important there.

3 Ecosystem Services from Forests

The Millennium Ecosystem Assessment (MEA 2005) concluded that since about 1950, 60% of all ES had declined as a direct result of the growth of agriculture, forestry, fisheries, industries, and urban settlement, mainly through the increase in markets for provisioning services, but that similar declines did not occur in the other categories of benefit that ES provide (Kinzig et al. 2011). Forest ecosystems, in

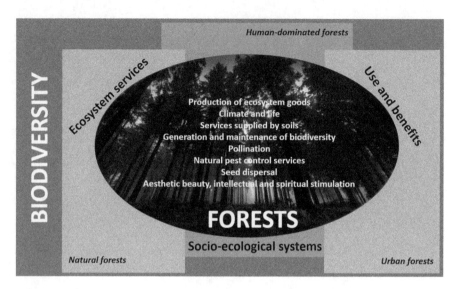

Fig. 3 Importance of forest ecosystem services in natural and anthropogenic landscapes

particular, provide critical ES to humanity (FAO 2010) and harbor most of the global terrestrial biodiversity (Gustafsson et al. 2012). Forests play a multifunctional role in which attempts are made to balance human commodity needs with the production of other goods and services, including the habitat needs of forest-dependent organisms (Thompson et al. 2011; Lindenmayer and Franklin 2002). More than 2 billion hectares of the world's forests (55%) are managed as production forests to supply ES and, at the same time, revenue from timber products to help pay for forest management (Gustafsson et al. 2012; Lindenmayer et al. 2012). When management strategies are developed, however, consideration is seldom given to the full range of ES that forest landscapes provide (Myers 1996; Daily et al. 1997; Nahuelhual et al. 2007) (Fig. 3). Some examples of ES that need to be taken into account are as follows:

- Production of ecosystem goods: The range of products obtained from forests includes food (e.g., fruits, nuts, mushrooms, honey, or spices), fuelwood, fiber, pharmaceuticals, and industrial products (Alamgir et al. 2016; Quintas-Soriano et al. 2016). In addition, animals such as cattle, goats, and sheep are raised in forests' silvopastoral systems (Peri et al. 2016), and these animals are the source of many trade products (e.g., meat, milk, wool, and leather). Hunting is also important in the forests of many countries, both for food and for sport, and can be critical to the survival of low-income people in developing countries (Golden et al. 2014).
- Climate and life: Climate plays a major role in the evolution and distribution of life over the planet, and forests are one of the main factors in the regulation of global climate. Forests help stabilize the climate, lessening extreme events (e.g., by slowing down water runoff) and removing greenhouse gases and other

pollutants from the atmosphere (Beer et al. 2010; Pan et al. 2011; Lal and Lorenz 2012).

- Services supplied by soils: Forests provide a critical role in forming soils, as well as in retaining them through reducing soil erosion. Forest soils moderate the water and carbon cycles, they retain and deliver nutrients to other organisms, and they provide a consistent and high quality source of water within forested basins (Kreye et al. 2014; Panagos et al. 2015; Sun and Vose 2016).
- Generation and maintenance of biodiversity: Forests support most of the terrestrial biological diversity, which benefits humanity through the direct delivery of goods (genetic and biochemical resources) used by humans or through the interaction of complex ecological systems (Daily and Ehrlich 1995).
- Pollination: About one-third of the human diet depends on insect-pollinated vegetables, legumes, and fruits. These pollinators, most of which live only in forested lands, allow for the successful reproduction of innumerable economic and noneconomic flowering plants (Karp et al. 2015; Martins et al. 2015; Quintas-Soriano et al. 2016).
- Natural pest control services: Several species compete with humans for goods and for other provisioning services. One approach to pest control is to use biotechnology or chemical compounds. Another option is to take advantage of biological control species that occur in nature, as many species (e.g., insects such as wasps and other species such as owls and bats) help humans live in forested landscapes (González et al. 2015; Karp et al. 2015; Quintas-Soriano et al. 2016).
- Seed dispersal: Many species of plants need animals as their dispersal agents or require passage through the gut of a bird or mammal before they can germinate. Many of these animals live only in forested lands, and several of the dispersing plant species (e.g., the fruit tree and shrubs species of temperate forests) have a long tradition of bringing goods to humans (Bregman et al. 2015; Karp et al. 2015; Peres et al. 2016).
- Aesthetic beauty, together with intellectual and spiritual stimulation: Human beings have a deep appreciation of natural ecosystems, especially forests, as evidenced by enjoyment of such pursuits as nature photography, bird watching, ecotourism, hiking, and camping. In forests, humans find an unparalleled source of wonderment and inspiration, peace and beauty, fulfillment, and rejuvenation (Daily 1997; Martínez Pastur et al. 2016).

4 Managing for Forest Ecosystem Services

The differences among human-dominated ecosystems, natural ecosystems, and ecosystems built-up through human activity have increased in recent years. Some ES provided by human-dominated ecosystems are traded on formal markets, and society tends to set a higher value on these than is actually due. The other two types of ecosystem are undervalued because their ES are not traded on formal markets, so they do not send price signals that warn of changes in their supply or condition

(Daily et al. 1997). However, the provisioning ES that flow from built-up and natural ecosystems have greatly increased. In response, it is essential to incorporate natural capital and ES into decision-making (Guerry et al. 2015). Costanza et al. (2014) estimated that ecosystems provide at least US$33 trillion dollars' worth of services annually, where about 38% of the estimated value comes from terrestrial systems, mainly from forests (US$4.7 trillion yr.$^{-1}$) and wetlands (US$4.9 trillion. yr.$^{-1}$). Our current economic, political, and social systems are not well suited to the challenge of representing the real value of ecosystems not dominated by human population and activity. There is a fundamental asymmetry at the heart of economic systems that rewards short-term production and consumption of marketed commodities, at the expense of stewardship of natural capital necessary for human well-being in the long term. Conservation and economic development have been considered as separate spheres for too long. Sustainable development requires explicit recognition that social and economic development are part of a stable and resilient biosphere (Guerry et al. 2015).

The Millennium Ecosystem Assessment (MEA 2005) combined both the applied and basic motives of sustainability science. It challenged the research community to synthesize what is known about sustainability science in policy-relevant ways, exposing both the strengths and the gaps in the underlying science (Carpenter et al. 2009). As human populations grow, and increasingly disconnect from nature, sustainability requires increasing focus and effort. For this, Guerry et al. (2015) proposed the following strategies to achieve sustainable development: (i) developing solid evidence linking decisions to impacts on natural capital and ES and then to human well-being; (ii) working closely with leaders in governments, businesses, and civil society to develop and make accessible the knowledge, tools, and practices necessary to integrate natural capital and ecosystem services into everyday decision-making; and (iii) reforming policies and institutions and building capacity to better align private short-term goals with societal long-term goals.

Conservation and development come from two distinct agendas: (i) conservationists who seek to increase public support for biodiversity protection by integrating economic development into protection initiatives and (ii) development agencies that seek to provide for the stewardship of nature under the mantra of sustainable development (Tallis et al. 2008). However, to achieve sustainability in ecosystem management, it is not enough to create partial reserves protecting some percentage of nature: the objectives of maintaining ES and biodiversity must be incorporated into intensively managed temperate landscapes at the landscape level (Franklin 1988; Lindenmayer et al. 2012; Gustafsson et al. 2012).

For ecosystem management, which aims to provide sustainable ES to society while also preserving and fostering biodiversity, the divergent disturbance impacts of these goals present a paradox, as they are at the same time risk factors and facilitators of management objectives (Thom and Seidl 2016). Therefore, it is necessary to develop management strategies for forestry which also incorporate broader protection and maintenance of ES and species diversity. It is probable that such new strategies will lead to reduced production of commodities but will increase the provision of ES for the whole of society.

In addition, many of the ES provided by forests are closely associated with ecosystem resilience, the ability of ecosystems to resist stresses and shocks, to absorb disturbance, and to recover from disruptive change (Myers 1996; Levin and Lubchenco 2008). If resilience declines, ES can generally be expected to decline, too. In this framework, proposals for managing forest landscapes which use ES to advance both conservation and human agendas, simultaneously, would benefit from improved scientific understanding of four key issues: sustainable use of ES, trade-offs among different types of ES, the spatial flows of ES, and economic feedbacks in ES markets (Tallis et al. 2008). The role of the market economy in developing this new management process lies in helping to design institutions which will provide incentives for the conservation of important natural systems and will also mediate human impacts on the biosphere so that these natural systems are sustainable (Heal 2000).

MEA (2005) did not, however, deliver a fully operational method for implementing the ES concept, including tools to assist policy-makers and policy-oriented researchers in taking the provisioning of natural goods and services into account (Armsworth et al. 2007). As a result, the ES label is currently used in a range of studies with widely differing aims. This divergence presents a problem for policy-makers as well as researchers because it makes it difficult to assess the credibility of assessment results and reduces the comparability of studies (Seppelt et al. 2011). Yet it is clear that, to strengthen the political relevance of the concept of ES, the scientific basis for its practical implementation must likewise be solidified (Ash et al. 2010).

5 Broader-Scale Consideration of Forest Ecosystem Services and their Sustenance

Even though much has been written on ES in forests, few examples exist in which the concept was effectively included in the planning, conservation, and management of the temperate forest ecosystems around the world. A great many of the studies and land management plans have focused on local scales, especially with respect to the types of ES addressed but also with respect to the land management policies and practices designed to sustain them.

To realize the full potential of the concept, broader-scale analyses of ES are required. We expect that the scale of focus will shift, for both the scientific community and the land managers, toward addressing broader-scale ecosystem services and design plans to sustain them. This paradigm shift to the adoption of a broader-scale consideration of forest ecosystem services will likely be made less daunting by advances in landscape ecological concepts, in remote sensing and GIS technologies and in simulation modeling methodologies.

Adoption of the concept of ES creates will create a significant change in the point of view of scientists, managers, and policy-makers, and studies on land and resource management will inevitably turn to the broader tools of ES types and landscape ecology. Landscape management with multiple objectives is a better solution for most of the urgent problems of our modern society, in which provision services cannot be divorced from either regulation or cultural services. The foundation for this shift is a better understanding of ES on a broad, even global, scale. Such a perspective is required for designing landscapes that serve human well-being while preserving the ecosystems and biodiversity on which that well-being depends.

References

Alamgir M, Turton SM, Macgregor CJ, Pert PL (2016) Assessing regulating and provisioning ecosystem services in a contrasting tropical forest landscape. Ecol Indic 64:319–334

Armsworth PR, Chan KM, Daily GC, Ehrlich PR, Kremen C, Ricketts TH, Sanjayan MA (2007) Ecosystem-service science and the way forward for conservation. Conserv Biol 21:1383–1384

Ash N, Blanco H, García K, Tomich T, Vira B, Brown C, Zurek M (2010) Ecosystems and human well-being: a manual for assessment practitioners. Island Press, Washington, DC

Beer C, Reichstein M, Tomelleri E, Ciais P, Jung M, Carvalhais N, Rödenbeck C, Arain A, Baldocchi D, Bonan G, Bondeau A, Cescatti A, Lasslop G, Lindroth A, Lomas M, Luyssaert S, Margolis H, Oleson K, Roupsard O, Veenendaal E, Viovy N, Williams C, Woodward I, Papale D (2010) Terrestrial gross carbon dioxide uptake: global distribution and covariation with climate. Science 329:834–838

Boyd J, Banzhaf S (2007) What are ecosystem services? The need for standardized environmental accounting units. Ecol Econ 63:616–626

Bregman TP, Lees AC, Seddon N, Macgregor HEA, Darski B, Aleixo A, Bonsall MB, Tobias JA (2015) Species interactions regulate the collapse of biodiversity and ecosystem function in tropical forest fragments. Ecology 96(10):2692–2704

Carpenter SR, Mooney HA, Agard J, Capistrano D, DeFries RS, Díaz S, Dietz T, Duraiappah AK, Oteng-Yeboah A, Miguel Pereira H, Perrings C, Reid W, Sarukhan J, Scholes RJ, Whyte A (2009) Science for managing ecosystem services: beyond the millennium ecosystem assessment. PNAS 106(5):1305–1312

Costanza R, Arge R, de Groot R, Farber S, Grasso M, Hannon B, Limburg K, Naeem S, Oneill RV, Paruelo J, Raskin RG, Sutton P, van den Belt M (1997) The value of the world's ecosystem services and natural capital. Nature 387:253–260

Costanza R, de Groot R, Sutton P, van der Ploeg S, Anderson SJ, Kubiszewski I, Farber S, Turner RK (2014) Changes in the global value of ecosystem services. Glob Environ Chang 26:152–158

Daily GC, Ehrlich PR (1995) Population diversity and the biodiversity crisis. In: Perrings C, Maler K, Folke C, Holling C, Jansson B (eds) Biodiversity conservation: problems and policies. Kluwer Academic Press, Dordrecht, pp 41–51

Daily GC (1997) Nature's services: societal dependence on natural ecosystems. Island Press, Washington, DC

Daily GC, Alexander SE, Ehrlich PR, Goulder LH, Lubchenco J, Matson PA, Mooney HA, Postel S, Schneider SH, Tilman D, Woodwell GM (1997) Ecosystem services: benefits supplied to human societies by natural ecosystems. Issues in Ecology 2:1–18

Ehrlich PR, Mooney HA (1983) Extinction, substitution, and ecosystem services. Bioscience 33(4):248–254

Fisher B, Turner R, Morling P (2009) Defining and classifying ecosystem services for decision making. Ecol Econ 68(3):643–653

Food and Agricultural Organization of the United Nations (FAO) (2010) Global forest resources assessment 2010: Main Report. FAO Forestry Paper 163

Franklin J (1988) Structural and functional diversity in temperate forests. In: Wilson EO, Peter FM (eds) Biodiversity. National Academies Press, Washington, DC

Golden CD, Bonds MH, Brashares JS, Rodolph Rasolofoniaina BJ, Kremen C (2014) Economic valuation of subsistence harvest of wildlife in Madagascar. Conserv Biol 28(1):234–243

González E, Salvo A, Valladares G (2015) Sharing enemies: evidence of forest contribution to natural enemy communities in crops, at different spatial scales. Insect conservation and. Diversity 8(4):359–366

Guerry AD, Polasky S, Lubchenco J, Chaplin-Kramer R, Daily G, Griffin R, Ruckelshaus M, Bateman I, Duraiappah A, Elmqvist T, Feldman M, Folke C, Hoekstra J, Kareiva P, Keeler B, Li S, McKenzie E, Ouyang Z, Reyers B, Ricketts T, Rockström J, Tallis H, Vira B (2015) Natural capital and ecosystem services informing decisions: from promise to practice. PNAS 112(24):7348–7355

Gustafsson L, Baker S, Bauhus J, Beese W, Brodie A, Kouki J, Lindenmayer D, Lõhmus A, Martínez Pastur G, Messier C, Neyland M, Palik B, Sverdrup-Thygeson A, Volney J, Wayne A, Franklin J (2012) Retention forestry to maintain multifunctional forests: a world perspective. Bioscience 62(7):633–645

Heal G (2000) Valuing ecosystem services. Ecosystems 3:24–30

Karp DS, Mendenhall CD, Callaway E, Frishkoff LO, Kareiva PM, Ehrlich PR, Daily GC (2015) Confronting and resolving competing values behind conservation objectives. PNAS 112(35):11132–11137

Kinzig A, Perrings C, Chapin F III, Polasky S, Smith V, Tilman D, Turner IIB (2011) Paying for ecosystem services: promise and peril. Science 334:603–604

Kreye MM, Adams DC, Escobedo FJ (2014) The value of forest conservation for water quality protection. Forests 5(5):862–884

Kubiszewski I, Costanza R, Anderson S, Sutton P (2017) The future value of ecosystem services: global scenarios and national implications. Ecosyst Serv 26:289–301

Lal R, Lorenz K (2012) Carbon sequestration in temperate forests. In: Lal R, Lorenz K, Hüttl R, Schneider B, von Braun J (eds) Recarbonization of the biosphere. Springer, Amsterdam

Levin SA, Lubchenco J (2008) Resilience, robustness, and marine ecosystem based management. Bioscience 58:27–32

Lindenmayer D, Franklin J (2002) Conserving forest biodiversity: a comprehensive multiscaled approach. Island Press, Washington, DC

Lindenmayer D, Franklin J, Lõhmus A, Baker S, Bauhus J, Beese W, Brodie A, Kiehl B, Kouki J, Martínez Pastur G, Messier C, Neyland M, Palik B, Sverdrup-Thygeson A, Volney J, Wayne A, Gustafsson L (2012) A major shift to the retention approach for forestry can help resolve some global forest sustainability issues. Conserv Lett 5(6):421–431

Martínez Pastur G, Peri PL, Lencinas MV, García Llorente M, Martín López B (2016) Spatial patterns of cultural ecosystem services provision in southern Patagonia. Landsc Ecol 31:383–399

Martins KT, Gonzalez A, Lechowicz MJ (2015) Pollination services are mediated by bee functional diversity and landscape context. Agric Ecosyst Environ 200:12–20

Millennium Ecosystem Assessment Panel (MEA) (2005) Island press. Washington, USA

Myers N (1996) Environmental services of biodiversity. PNAS 93(7):2764–2769

Nahuelhual L, Donoso P, Lara A, Núñez D, Oyarzún C, Neira E (2007) Valuing ecosystem services of Chilean temperate rainforests. Environ Dev Sustain 9:481–499

Quintas-Soriano C, Martín-López B, Santos-Martín F, Loureiro M, Montes C, Benayas J, García-Llorente M (2016) Ecosystem services values in Spain: a meta-analysis. Environ Sci Policy 55(01):186–195

Pan Y, Birdsey RA, Fang J, Houghton R, Kauppi PE, Kurz W, Phillips O, Shvidenko A, Lewis S, Canadell J, Ciais P, Jackson R, Pacala S, McGuire D, Piao S, Rautiainen A, Sitch S, Hayes D (2011) A large and persistent carbon sink in the world's forests. Science 333:988–993

Panagos P, Borrelli P, Poesen J, Ballabio C, Lugato E, Meusburger K, Montanarella L, Alewell C (2015) The new assessment of soil loss by water erosion in Europe. Environ Sci Policy 54:438–447

Peres CA, Emilio T, Schietti J, Desmoulière SJM, Levi T (2016) Dispersal limitation induces long-term biomass collapse in overhunted Amazonian forests. PNAS 113(4):892–897

Peri, P, Dube F, Varella A (2016) Silvopastoral systems in southern South America. Springer, Series: Advances in agroforestry 11, Amsterdam

Seppelt R, Dormann CF, Eppink FV, Lautenbach S, Schmidt S (2011) A quantitative review of eco-system service studies: approaches, shortcomings and the road ahead. J Appl Ecol 48:630–636

Sun G, Vose JM (2016) Forest management challenges for sustaining water resources in the Anthropocene. Forests 7(3):e68

Tallis H, Kareiva P, Marvier M, Chang A (2008) An ecosystem services framework to support both practical conservation and economic development. PNAS 105(28):9457–9464

Thom D, Seidl R (2016) Natural disturbance impacts on ecosystem services and biodiversity in temperate and boreal forests. Biol Rev Camb Philos Soc 91(3):760–81

Thompson ID, Okabe K, Tylianakis JM, Kumar P, Brockerhoff EG, Schellhorn NA, Parrotta JA, Nasi R (2011) Forest biodiversity and the delivery of ecosystem goods and services: translating science into policy. Bioscience 61:972–981

Vitousek P, Aber J, Howarth R, Likens G, Matson P, Schindler D, Schlesinger W, Tilman D (1997) Human alteration of the global nitrogen cycle: sources and consequences. Ecol Appl 7(3):737–750

Effects of Climate Change on CH_4 and N_2O Fluxes from Temperate and Boreal Forest Soils

Eugenio Díaz-Pinés, Christian Werner, and Klaus Butterbach-Bahl

1 Introduction

Boreal and temperate forests cover 1210 and 680 million ha, respectively (Keenan et al. 2015). In contrast to tropical forests, whose extent is decreasing due to current deforestation activities resulting in huge emissions of greenhouse gases (Roman-Cuesta et al. 2016), the area of boreal forests remained constant, while the area of temperate forests slightly increased in the last 25 years at an average rate of 2.7 million ha a^{-1} (Keenan et al. 2015). In total, boreal and temperate forests cover approximately 13% of the global terrestrial land surface.

Temperate and boreal forests are known to provide a wide range of ecosystem services (e.g., Gamfeldt et al. 2013), including timber production, water regulation, soil protection and erosion control, support of biodiversity, or recreation. The role of forests in regulating the climate has been also well acknowledged, due to their strong potential for sequestering atmospheric CO_2 in its biomass and soils (De Vries et al. 2003; Vesterdal et al. 2008). In contrast to CO_2, the role of forests as both significant sinks and sources of other powerful greenhouse gases, i.e., CH_4 and N_2O has received comparatively little attention. The crucial role played by forests in

E. Díaz-Pinés (✉)
Institute of Soil Research, University of Natural Resources and Life Sciences (BOKU), Vienna, Austria

Institute of Meteorology and Climate Research, Atmospheric Environmental Research (IMK-IFU), Karlsruhe Institute of Technology, Garmisch-Partenkirchen, Germany
e-mail: eugenio.diaz-pines@boku.ac.at

C. Werner
Senckenberg Biodiversity and Climate Research Centre (BiK-F), Frankfurt, Germany

K. Butterbach-Bahl
Institute of Meteorology and Climate Research, Atmospheric Environmental Research (IMK-IFU), Karlsruhe Institute of Technology, Garmisch-Partenkirchen, Germany

© Springer International Publishing AG, part of Springer Nature 2018
A. H. Perera et al. (eds.), *Ecosystem Services from Forest Landscapes*,
https://doi.org/10.1007/978-3-319-74515-2_2

11

regulating nutrient cycling is only possible due to the microbial-mediated transformation processes of the soil organic matter, which make nutrients available again for plant metabolism while also resulting in a substantial release of CO_2, as well as CH_4 and N_2O, to the atmosphere.

Temperate and boreal forests represent one of the major global pools of carbon (C) and nitrogen (N), with more than half of the C being stored in soils (Batjes 1996). Pan et al. (2011) estimated that boreal forest ecosystems store approximately 271 ± 22 Pg C, distributed in the living (54 Pg C) and dead biomass (43 Pg C) and in the soils (down to 1 m) (175 Pg C). This estimate excludes some deep organic boreal forest soils, which explains the significant difference with a recent estimate by Bradshaw and Warkentin (2015) (average: 1096 Pg C, range: 367–1715 Pg C), who included peats, or assessments from IPCC (2007) (471 Pg C). However, all estimates agree that 2/3 to ¾ of all C is stored in soils and peats. For temperate forests, Pan et al. (2011) estimated that the living and dead biomass pool is 62 Pg C, approximately equal to the amount of C stored in soils down to 1 m (57 Pg). Global amounts of N in soils down to 1 m are estimated to be 133–140 Pg (Batjes 1996), while only 10 Pg of N is held in the global plant biomass (Davidson 1994). Figure 1 shows the distribution of forests (Fig. 1A), the soil N stocks down to 1 m (Fig. 1B), and the C:N ratio of these soils (Fig. 1C) for the temperate and boreal zones of the

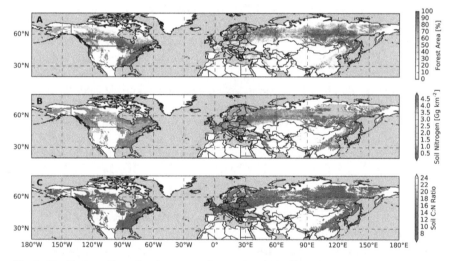

Fig. 1 Distribution of boreal and temperate forests (Panel A), the total nitrogen in soils (Panel B), and C:N ratio of soils (Panel C). The areal extent considered is based on the Olson ecoregions "temperate broadleaf and mixed forests" and "temperate coniferous forest" (Olson et al. 2001). The relative forest cover percentage is based on GlobCover 2009 v2.3 (Bontemps et al. 2011) including the following classes: "mosaic vegetation," "closed to open broadleaf (B) evergreen (E) forest," "closed B deciduous (D) forest," "open BD forest," "closed needleleaf (N) E forest," "open NE/ or BE forest," "closed to open mixed BD/ND forest," "mosaic forest," "closed to open B regularly flooded," "closed B forest permanently flooded." Spatially explicit soil C and N stocks were derived from the ISRIC-WISE soil map (Batjes 2012)

northern hemisphere. The areal extent considered is based on the Olson ecoregions (Olson et al. 2001), and the relative forest cover percentage (Fig. 1A) is based on the GlobCover 2009 (Bontemps et al. 2011). Spatially explicit soil C and N stocks were derived from the ISRIC-WISE soil map (Batjes 2012). According to this approach, the total soil N stocks in boreal and temperate forest soils are 10.4 Pg N and 7.2 Pg N, respectively.

Globally, both boreal and temperate forest soils have been identified as a source of atmospheric N$_2$O and as a net sink for atmospheric CH$_4$. The IPCC 2001 report listed the source strength of temperate forests for atmospheric N$_2$O with 1.0 Tg N$_2$O-N a^{-1} (0.1–2.0 Tg N$_2$O-N a^{-1}), while for boreal forest soils, an estimate was missing. More recently, Dalal and Allen (2008) estimated that boreal forests are a weak source for N$_2$O (0.33 ± 0.27 Tg N$_2$O-N a^{-1}) and confirmed earlier estimates for temperate forests (1.05 ± 0.37 Tg N$_2$O-N a^{-1}). With regard to atmospheric CH$_4$, Dutaur and Verchot (2007) estimated the sink strength of boreal and temperate forests to be 3.4 ± 5.0 Tg CH$_4$-C a^{-1} and 2.5 ± 2.6 Tg CH$_4$-C a^{-1}, respectively. However, since wetland forests were excluded from the study, this estimate is likely biased, because wetlands show net CH$_4$ emissions at the annual scale. Dalal and Allen (2008) estimated a smaller, more variable sink strength of boreal forest soils (2.0 ± 4.0 Tg CH$_4$-C a^{-1}), whereas the estimated CH$_4$ sink strength of temperate forests was with 3.7 ± 0.5 Tg CH$_4$-C a^{-1} higher and highly constrained.

Climate change refers here to the human-induced alteration of weather patterns, such as temperature and rainfall (amount, frequency, seasonal distribution). Climate change affects soil environmental conditions, as well as landscape hydrology, vegetation cover, and substrate supply. Indirect effects of climate change on land use (e.g., it is expected an agricultural expansion further north (Kicklighter et al. 2014)) are not covered here. Different climate models indicate that the temperate and boreal zones will experience warming in the range of 1.4–5.8 °C by 2100 (Hanewinkel et al. 2013), accompanied by an increase in extreme weather events, which will provoke the shrinkage of permafrost, and the reduction of the snow cover period (IPCC 2013). This will result in changing environmental conditions in both forest canopies and soils, along with shifts of vegetation zones, i.e., upward and northward expansion of the temperate and boreal forest biomes. Alterations of forest species composition, forest growth, and vitality of natural and managed forest landscapes will result in modification of the ecosystem services sustained by forests. While synthesis on the contribution of forests to several ecosystem services is already available (e.g., Millennium Ecosystem Assessment 2005), comprehensive studies linking forests and the exchange of non-CO$_2$ greenhouse gases between forests and the atmosphere are much more scarce. This chapter aims at evaluating the role of temperate and boreal forests as providers of climate regulation services. Special emphasis is given to the production and consumption of non-CO$_2$ greenhouse gases by forest soils under changing environmental conditions. Specifically, this chapter assesses how the changes in climate and associated effects may affect temperate and boreal forest soils N$_2$O and CH$_4$ fluxes, thereby summarizing existing knowledge and identifying research gaps.

2 Governing Processes and Mechanisms of Forest Soil-Atmosphere CH_4 and N_2O Exchange

Nitrous oxide is mainly produced by the microbial processes of nitrification and denitrification, i.e., an oxidative process converting ammonia/ammonium (NH_3/ NH_4^+, classical nitrification) or organic N (heterotrophic nitrification) to nitrate (NO_3^-) and a reductive process, which uses NO_3^- as an electron acceptor for C oxidation to finally convert it to N_2 (denitrification) (Butterbach-Bahl et al. 2013). In these key microbial processes, N_2O is either a facultative (nitrification) or obligate (denitrification) intermediate, which can be released to the soil air, consumed in other parts of the soil profile or finally be emitted to the atmosphere. Although denitrification is considered the most important source of N_2O in forest soils at the European level, nitrification activity also drives total soil N_2O emissions (Ambus et al. 2006). Other microbial processes such as NO_3^- ammonification or physico-chemical processes, e.g., chemical decomposition of reactive inorganic N species such as hydroxylamine (NH_2OH) or nitrite (NO_2^-), can lead to N_2O formation too (Butterbach-Bahl et al. 2013). The major controls for N_2O production in forests soils are substrate availability, i.e., NH_3/NH_4^+ and/ or NO_3^- as well as easily degradable C availability (Butterbach-Bahl et al. 2012), temperature (with sensitivity for N_2O emission varying widely) (Brumme 1995; Butterbach-Bahl et al. 1997; Díaz-Pinés et al. 2014; Sitaula and Bakken 1993; Zhang et al. 2016), and soil moisture and soil aeration, as both affect the soil redox potential and thus the preference of reductive processes such as denitrification (Butterbach-Bahl et al. 2013). In addition, soil N_2O emissions are indirectly controlled by tree and associated plant species, forest stand characteristics, and their effects on the abovementioned parameters, soil C:N ratios (Klemedtsson et al. 2005) and soil microbial community composition (Philippot et al. 2009). Finally, the occurrence of extreme events, such as wildfires and pronounced freeze-thaw and soil drying-rewetting cycles (Borken and Matzner 2009; Butterbach-Bahl et al. 2013), strongly affects microbial activity and availability of substrates for N_2O-producing processes.

Methane is predominantly produced in anaerobic, organic-matter-rich microsites of forest soils as a final step of the anaerobic decomposition of organic matter (Conrad 1996). CH_4 production has been observed in both the forest floor and the mineral soil (Butterbach-Bahl and Papen 2002). Forest soils can predominantly function as weak sources of CH_4 (0–20 kg CH_4-C ha^{-1} a^{-1}) if soils are poorly drained or seasonally flooded due to their topographic position in the landscape, such as many aspen or alder stands (Mander et al. 2015; Matson et al. 2009). In upland soils, CH_4 produced at anaerobic microsites or in deeper soil layers is likely to be oxidized while passing through aerobic soil layers. This implies that the observed CH_4 flux at the forest soil-atmosphere interface is the net result of simultaneously occurring production and consumption processes (Conrad 1996). Most of the temperate and boreal forest soils are upland soils, which predominantly function at annual scales as weak sinks for atmospheric CH_4 (0–5 kg CH_4-C ha^{-1} a^{-1}) (Dutaur and Verchot 2007). However, topographically complex ecosystems may lead to spatial fragmentation at the landscape level, with specific locations being net CH$_4$ sinks while others being strong "hotspot" CH$_4$ emitters (Nykänen et al. 2003). The CH$_4$-oxidizing microbial communities are mostly using O$_2$ – but under certain circumstances also use sulfate or NO$_3^-$ – as electron acceptors (Conrad 2009). High-affinity methanotrophic bacteria found in most forest soils are capable to gain energy from soil atmosphere CH$_4$ concentrations lower than 1.7 ppmv. Climate change interacts in several ways with CH$_4$ production and consumption processes in soils. On the one hand, climate change directly affects soil environmental conditions, namely, moisture and temperature, and by this the balance between oxidative and reductive processes, e.g., temperature increases, will – as long as water availability is not limiting – likely result in an increase in aerobic respiration, thus decreasing soil oxygen (O$_2$) availability and the CH$_4$ oxidizing capacity of upland soils. On the other hand, global change and increases in atmospheric CO$_2$ concentration affect plant biomass production and its aboveground-to-belowground ratio, root exudation, and litter quality. All these changes finally modify ecosystem CH$_4$ exchange, with results being different across different ecosystem types and climatic zones. Finally, climate change also affects regional water balances and thus landscape groundwater levels. This will ultimately control the future distribution of wetlands and emission magnitudes of CH$_4$ at the landscape scale (Jungkunst and Fiedler 2007).

3   Forest Composition and N$_2$O and CH$_4$ Fluxes

Forest tree species composition and tree species richness are of high significance with regard to the provision of economic and ecological services by forests (Gamfeldt et al. 2013). While extensive research has been conducted to elucidate the effect of tree species on biomass production (De Vries et al. 2003), biodiversity (Barbier et al. 2008), water regulation (Ewers et al. 2002), or soil C sequestration (Vesterdal et al. 2008; Díaz-Pinés et al. 2014), our knowledge is rather limited with regard to the relationship between forest composition and its importance for the function of forests as climate regulators, specifically in view of the importance of forest soils as sink or sources of non-CO$_2$ greenhouse gases.

Individual trees strongly interact with the surrounding environment by, e.g., reducing the amount of light reaching the soil surface, intercepting water in their canopies, taking up water and nutrients from the soil, and returning organic matter back to the soil. Specific tree species usually behave differently (due to, e.g., different growth rates, water or nutritional requirements, or canopy and root system structure) and therefore create distinct ecological conditions and biogeochemical characteristics in both the canopy (radiation levels, microclimate) and the soil (moisture, pH value, or availability of nutrients). Consequently, microbial processes responsible for production and consumption of N$_2$O and CH$_4$ in both the forest floor and the mineral soil are usually tree-species-dependent (e.g., Borken et al. 2003; Butterbach-Bahl et al. 2002; Díaz-Pinés et al. 2014).

Litter is an inherent part of nutrient and C cycling in forest ecosystems. Aboveground litter regulates microclimatic conditions by forming a protective layer on the soil surface (Sayer 2006). Litter material from conifers contains high amounts of lignin and tannins, which are mainly decomposed by fungi (Dix and Webster 1995), as opposed to litter originated from deciduous trees (e.g., beech). The latter has simpler chemical structures and can be decomposed by broader spectra of soil microorganisms. This usually provokes that coniferous forests develop a thicker forest floor, which both produces and consumes CH_4 (Butterbach-Bahl and Papen 2002) and probably limits the transport of atmospheric CH_4 into the mineral soil (Borken and Beese 2006; Borken et al. 2003). At the same time, the usually compact and moist litter layer developed under deciduous forest can lead to high N_2O production rates (Pilegaard et al. 2006).

Belowground, rhizodeposition and root decay supply soil microorganisms with C to sustain further microbial decomposition (Cheng and Kuzyakov 2005). This, along with root respiration and water and nutrient uptake, significantly alters important biochemical properties (i.e., soil moisture, pH, O_2 and CO_2 concentrations, and labile C and N concentrations) in the rhizosphere. In a rhizotron experiment, it has been recently found that roots from different tree species affect soil microorganisms and C dynamics in different ways, with *Fraxinus excelsior* showing a higher CH_4 sink and a lower N_2O source strength compared with *Fagus sylvatica* or root-free soil (Fender et al. 2013), underpinning the possible tree-species-dependent root effects on trace gas production in soils.

In addition to the inherent variation of greenhouse gas fluxes along the landscape due to changing environmental conditions, trees can also pose a strong effect on the spatial pattern of N_2O and CH_4 exchange between the soil and the atmosphere (Butterbach-Bahl et al. 2002). It has been observed that fundamental soil properties (e.g., C and N contents, O_2 availability, microbial activity, moisture) strongly vary with distance from the stem (Chang and Matzner 2000) or from the canopy edge (Simón et al. 2013), and this pattern has been found to be tree-species-dependent (Butterbach-Bahl et al. 2002; Van Haren et al. 2010). Further, tree stems can be major conduits for soil-produced CH_4 and N_2O into the atmosphere. The transport may take place through aerenchymous tissues (extra-large intercellular spaces intended to facilitate aeration in the root system) as has been described for alder trees (Rusch and Rennenberg 1998) but also as dissolved gases in the water stream of the xylem. The contribution of tree trunks and tree leaves to the total ecosystem release of N_2O has been estimated to range from 1% to 3% in temperate beech forests (Díaz-Pinés et al. 2015) to 8% in boreal pine forests (Machacova et al. 2016). With regard to CH_4, *Alnus glutinosa* and *Betula pubescens* trees were found to contribute up to 27% of the ecosystem flux of temperate forested wetlands (Pangala et al. 2015). To our knowledge, information on the contribution of trees to the release of N_2O and CH_4 in boreal forested peatlands is not available.

Coniferous forests are predominant in temperate and boreal biomes (Douglas et al. 2014). In the frame of a changing climate, tree species better adapted to warmer temperatures and more tolerant to summer drought are supposed to have an adaptive advantage compared with other more water-sensitive tree species.

Drought-induced decline of Scots pine (*Pinus sylvestris*) stands has been observed already in the Alps (Rebetez and Dobbertin 2004) or in the Pyrenees (Galiano et al. 2010), at the extent of *Quercus* species. In the Rocky Mountains, drought-induced mortality of *Abies* and *Picea* species has also been observed (Bigler et al. 2007). In addition to the vegetation succession in view of changing environmental conditions, forest managers have promoted mixed forests or broadleaf species in the last decades, under the belief of having higher stability against disturbances (Jandl et al. 2007). It has been predicted the areal contribution of coniferous forests will shrink at the extent of broadleaf forests (Hanewinkel et al. 2013), even if we lack a clear understanding of how drought- and heat-induced tree mortality will impact the composition of most forests (Anderegg et al. 2013).

Following the change of forest composition, environmental conditions and soil microbial communities are expected to change, along with the organic matter transformation processes, ultimately leading to different end-, co-, and by-products during microbial N turnover (e.g., N$_2$O, NO, N$_2$). There is a substantial number of publications showing higher N$_2$O emissions in soils under deciduous forests than under coniferous ones (Ambus et al. 2006; Brumme et al. 1999; Butterbach-Bahl et al. 2002; Díaz-Pinés et al. 2014). This has been associated to larger NO losses in conifer forests (Butterbach-Bahl et al. 1997), resulting in higher N$_2$O:NO ratios (Papen et al. 2003). Recent results from forest floor incubations in European forests support higher NO emissions from coniferous forests but also a N$_2$O sink potential in the forest floor of deciduous species (Gritsch et al. 2016). Other authors have associated the lower N$_2$O emissions under conifer forests to a decoupling between N$_2$O production and reduction processes, resulting in decreased N$_2$O:N$_2$ ratios (Menyailo and Hungate 2005). With regard to CH$_4$, broadleaf forests usually show higher atmospheric CH$_4$ oxidation rates compared with coniferous ones (Butterbach-Bahl et al. 2002; Maurer et al. 2008), probably due to the distinct CH$_4$ diffusivity of the forest floor developed under each type of forest (Borken and Beese 2006; Borken et al. 2003). However, direct tree species effects on CH$_4$ fluxes can interact with soil moisture effects (Menyailo and Hungate 2005).

Tree species composition may change naturally in the course of ecological succession, but human interventions often also actively modulate stand composition and structures. Thus, forest management plays an active role for determining forest composition, which affects the benefits provided by forests in terms of ecosystem services, including its importance as climate regulators. However, the processes responsible for emitting or taking up N$_2$O or CH$_4$ are highly dynamic, and they are the result of complex biogeochemical processes and feedbacks and usually show a high temporal and spatial variability (e.g., Brumme et al. 1999; Butterbach-Bahl et al. 2002). Further, the forest composition is strongly influenced by the topography and landscape configuration, which in turn impacts the net exchange of CH$_4$ and N$_2$O. Finally, the relevant role of other parameters such as soil texture, precipitation (Borken and Beese 2005), or N limitation (Pilegaard et al. 2006), which may largely overwhelm the direct tree species effects on the net soil-atmosphere N$_2$O and CH$_4$ exchange, appeals for more comprehensive studies including not only different tree species but also soil types, N deposition rates, and climatic regions.

4 Effects of Tree Line and Forest-Tundra Ecotone Shifts on N₂O and CH₄ Fluxes

Effects of climate change (increasing temperatures, changing precipitation patterns) are particularly intense in temperate and boreal ecosystems compared with global averages, and it has been anticipated that the trend will continue in the coming years (Callaghan et al. 2005). Climatic changes are already affecting the location of the contact line between forest and grasslands or shrublands formation, the so-called tree line ecotone (Wieser 2010). Furthermore, the tree species composition of the forest is changing because of the natural adaptation to new environmental conditions, in addition to the forest management efforts to increase the forest resilience in the frame of global change and the succession after forest disturbances (e.g., wildfires, windthrow). Finally, the spatial configuration and boundaries of temperate and boreal landscape units will probably change (Fig. 2), and the areal extent of the main land cover types (e.g., evergreen and deciduous forests, shrublands, woody wetlands) will vary following changing environmental conditions.

Both an upward shift of the tree line in mountainous areas (e.g., Kammer et al. 2009) and a northward migration of the forest-tundra ecotone in boreal latitudes (Serreze et al. 2000) have been detected, indicating an encroachment of forest areas into herbaceous and shrub communities. To our knowledge, there is no information available on the consequences of the movement of the mountainous tree line on the soil N_2O and CH_4 emissions. With regard to the forest encroachment in boreal latitudes, available studies are limited to individual case studies. When investigating the arctic tree line in Canada, Rouse et al. (2002) found significant releases of CH_4 in a fen compared with the negligible CH_4 fluxes in the forest (ca. 50 vs 8 kg CH_4-C ha^{-1} a^{-1}). Tupek et al. (2015) found higher CH_4 uptake in the upland forest than in the forest-mire transition, whereas no N_2O fluxes were observed in any of the ecosystems. In another tundra-forest comparison (Takakai et al. 2008), the forest was a small CH_4 sink compared with neutral or CH_4 sources from different grasslands. In the same study, the forest was a modest emitter of N_2O, whereas the magnitude of N_2O emitted by grasslands was highly dependent on water content. On the other hand, others have found no support for the hypothesis that conversion of tundra to forest or vice versa would result in a systematic change of net CH_4 fluxes in well-drained soils (Sjögersten and Wookey 2002). As it is usual that both CH_4 and N_2O emissions from tundra soils largely vary in space and time depending on local hydrological regimes (Zhu et al. 2014), with CH_4 annual fluxes from the same site ranging from 10 to 250 kg CH_4 ha^{-1} a^{-1} (Nykänen et al. 2003), prediction and assessment of CH_4 and N_2O emissions at the landscape or regional scale are highly challenging.

Due to the lack of a standardized terminology and methodology to locate, characterize, and observe changes, a general shift of the tree-grassland boundary was so far not detected at regional scales for northern latitudes (Callaghan et al. 2005). In addition, while climate change may promote forest encroachment through higher temperatures, human interventions lead to forest degradation and loss of forest-covered

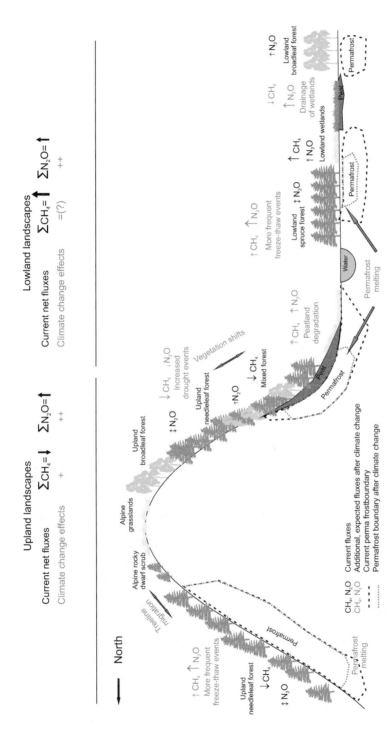

Fig. 2 Cross section representing the predominant upland and lowland landscape forms in boreal ecosystems. Size of vertical arrows is proportional to the magnitude of the fluxes (Adapted after Douglas et al. (2014) and Jorgenson et al. (2010)). Vegetation vectors designed by Natasha Sinegina and Bob Comix (creative commons license)

area. However, even in the scenario of no alteration of boundaries between ecosystem types, the soil-atmosphere balance of CH_4 and N_2O in the tree line ecotone is very sensitive to changing environmental conditions. Warming across an extensive northern gradient along the tundra-forest ecotone in Scandinavia consistently enhanced the soil CH_4 sink strength (Sjögersten and Wookey 2002), and in northern China, soil warming experiments in tundra resulted in increased uptake rates of both CH_4 and N_2O (Zhou et al. 2016). On the contrary, Karbin et al. (2015) have recently shown that soil warming does not affect soil CH_4 fluxes at the Alpine forest tree line, likely due to the negligible effect of warming on soil water contents.

5 Extreme Events and Forest Soil N_2O and CH_4 Fluxes

Most climate projections show that temperature change will be highest in northern latitudes, accompanied by significant increases in precipitation (IPCC 2013). As much of the precipitation increase will be in the form of rain, including rain-on-snow, an accelerating rate of snowmelt can be expected (Callaghan et al. 2005), and a higher frequency of freeze-thaw events is forecasted in temperate and boreal regions (Fig. 2). Freeze-thaw events have been shown to result in major pulses of soil N_2O emissions, not only from agricultural but specifically from forest soils (De Bruijn et al. 2009). Luo et al. (2012) reported for the Höglwald Forest, a temperate forest in Southern Germany, freeze-thaw events in 5 out of 14 observational years, and that in those years, annual soil N_2O emissions were at least a factor of two higher than in other years due to the strong pulse of N_2O emissions associated with freeze-thaw events. In contrast, comparable effects on soil CH_4 fluxes were not observed.

Permafrost is perennially frozen ground, which underlies 20–25% of the exposed land surface in cold climates, and many boreal forests grow on permafrost (Serreze et al. 2000). Strong arctic warming forecasted in climate scenarios will provoke a strong collapse and shrinkage of permafrost (Callaghan et al. 2005). There is a growing body of research showing that permafrost thawing will yield increased C emissions due to the enhancement of both CO_2 and CH_4 emissions (e.g., Flessa et al. 2008; Schaefer et al. 2014; Schuur et al. 2015), as well as accelerated N losses following dramatic increases in soil N_2O fluxes (Abbott and Jones 2015).

Prolonged drought and following soil-rewetting events are also likely to exert a significant effect on annual soil N_2O and CH_4 fluxes of forest soils. For example, Borken et al. (2000) showed for a Norway spruce stand in Germany that a simulated drought period of 3–5 months resulted in an increase of the net sink capacity of the soil for atmospheric CH_4 by 40–100%. On the other hand, rewetting of soils after drought has been shown to result in pulses of soil N mineralization and soil N_2O emissions as a result of the release of accessible substrate from accumulated microbial and plant necromass, the lysis of live microbial cells, and the disruption of previously protected organic matter, allowing for accelerated microbial N turnover processes (Borken and Matzner 2009). However, pulse emissions due to soil rewetting may compensate or not for reductions in soil N$_2$O emissions during the drought period compared with constantly wet scenarios, depending on the frequency of the drying-wetting cycles and the length of the drought periods and the magnitude of the priming effects following the wetting pulse (Borken and Matzner 2009). Moreover, it has been found that under drought conditions, soils might turn from a source to a sink of N$_2$O (Goldberg and Gebauer 2009), so that the expected future increases in drought periods may finally result in a decreasing atmospheric N$_2$O source strength of temperate and boreal forests.

The most important natural disturbances impacting the functioning of temperate and boreal forests are wildfires, windthrows, and insect and disease outbreaks. In a changing climate, the intensity and duration of extreme weather events will increase, with unexpected consequences for disaster, i.e., widespread damage and severe alteration in the normal functioning of ecosystems (IPCC 2012). As a result, ecosystem services provided by forests, including climate regulation, will be also affected. Studies have found a climate signal in the increased wildfire activity in North America (Gillett et al. 2004; Westerling et al. 2006). Enhanced drought in boreal forests is associated with higher frequency of defoliator outbreaks, while late-spring frost has a role in terminating outbreaks (Volney and Fleming 2000). Forest damages due to windthrows have also increased in Europe during the last century (Schelhaas et al. 2003). All these disturbances result in significant reductions or even complete losses of the canopy cover and an episodic, over-proportionated incorporation of plant litter and residues (e.g., ashes) on the forest floor. Reduced water interception and transpiration after forest disturbance has been shown to increase soil moisture levels (Huber et al. 2004), whereas reduced canopy cover leads to enhanced soil temperature amplitudes. Similar to clear-cutting effects, such disturbances usually result in strongly reduced CH$_4$ uptake activities (Strömgren et al. 2016; Wu et al. 2011) and in pulses of N$_2$O emissions during the growing season (e.g., Mcvicar and Kellman 2014; Zerva and Mencuccini 2005) as well as during freeze-thaw periods, due to accelerated organic matter mineralization, which provides substrate for soil microbial processes (Rosenkranz et al. 2010). However, the excess availability of N and the associated pulses seem to be limited to at most a few years, as N losses also increase along hydrological (NO$_3^-$ leaching; e.g., Huber et al. 2004, 2010) and gaseous pathways (N$_2$ emissions due to denitrification) and as the regrowing vegetation becomes quickly a strong competitor for the available soil N (Rosenkranz et al. 2010).

6   Synthesis

Temperate and boreal forests play a pivotal role in climate regulation, a critical ecosystem service. Dynamic changes in environmental conditions in the coming decades will likely affect the net forest-atmosphere balance of CH$_4$ and N$_2$O fluxes at different temporal and spatial scales, and these changes will feedback on chemical composition of the atmosphere and, thus, on the global climate.

Increasing temperature and changes in precipitation patterns are specifically expected for northern latitudes, with changes being more pronounced compared with tropical or temperate regions (IPCC 2013). This will affect the production and consumption of N_2O and CH_4 directly by changing the environmental conditions for microbial processes and indirectly through modifications of the landscape configuration, territory land use, and forest vegetation. A conversion of coniferous to broadleaf and mixed forests in response to climate change could lead to enhanced CH_4 uptake rates, whereas the overall effect on N_2O emissions is unclear if the role of increasing likeliness of summer droughts is taken into account. There is still insufficient empirical evidence of a direct and consistent link between tree species and the soil-atmosphere CH_4 and N_2O exchange rates, due to the complex interactions between tree species composition and site/soil characteristics. Soil drainage and more pronounced summer drought will increase the CH_4 oxidation strength of upland mineral soils, but this effect could be highly overwhelmed by the anticipated massive collapse of permafrost (Douglas et al. 2014), which is likely to result in the mobilization of currently frozen C and N stocks and years of high CH_4 and N_2O emissions. Furthermore, the lowering of the water table level will likely enhance soil N_2O emissions. Since shorter periods of snow-covered soil are expected, the occurrence and intensity of freeze-thaw events will increase, potentially leading to more frequent pulses of N_2O. Finally, acceleration of natural disturbances such as wildfires and pests will decrease the soil's capacity to uptake atmospheric CH_4 and will likely provoke further releases of N_2O as ecosystem N stocks are mobilized.

We should mention that there is already a reasonable understanding on how single climate change parameters such as temperature or moisture affect soil environmental conditions, C and N cycling, and associated soil N_2O and CH_4 emissions. However, we still do not understand how interacting effects of environmental condition changes, i.e., combined changes in tree species composition, atmospheric chemistry (i.e., CO_2, ozone, or reactive N concentrations), or enhanced and more severe extreme weather events will jointly affect the soil-atmosphere CH_4 and N_2O flux. Only a few experiments have tried to tackle this problem comprehensively (e.g., Carter et al. 2012). Even though the fact that gaining this knowledge is fundamental for predicting future feedbacks of soil N_2O and CH_4 fluxes from temperate and boreal forest soils in response to environmental changes, we are lacking suitable experimental approaches. Moreover, research has been mainly focused on plot to ecosystem scale, without confronting the issue of how landscape fluxes may change due to specific changes in environmental drivers. For example, changes in groundwater table levels, lateral water flows, and flooding events will occur at landscape scales and provoke huge changes in hydrological regimes. Those alterations may exert for many landscapes a more pronounced effect specifically on forest soil N_2O fluxes as, e.g., changes in temperature (Butterbach-Bahl and Dannenmann 2011). Here, manipulation experiments of groundwater table levels in combination with other climate change factors and a better fine-scale mapping of the affected territory offer a way forward to get a better insight in forest soil greenhouse gas fluxes in the frame of global change.

References

Abbott BW, Jones JB (2015) Permafrost collapse alters soil carbon stocks, respiration, CH$_4$, and N$_2$O in upland tundra. Glob Chang Biol 21:4570–4587

Ambus P, Zechmeister-Boltenstern S, Butterbach-Bahl K (2006) Sources of nitrous oxide emitted from European forest soils. Biogeosciences 3:135–145

Anderegg WRL, Kane JM, Anderegg LDL (2013) Consequences of widespread tree mortality triggered by drought and temperature stress. Nat Clim Chang 3:30–36

Barbier S, Gosselin F, Balandier P (2008) Influence of tree species on understory vegetation diversity and mechanisms involved—a critical review for temperate and boreal forests. For Ecol Manag 254:1–15

Batjes NH (1996) Total carbon and nitrogen in the soils of the world. Eur J Soil Sci 47:151–163

Batjes NH (2012) ISRIC-WISE derived soil properties on a 5 by 5 arc-minutes global grid (ver 1.2). Report 2012/01. ISRIC – World Soil Information, Wageningen

Bigler C, Gavin DG, Gunning C, Veblen TT (2007) Drought induces lagged tree mortality in a subalpine forest in the Rocky Mountains. Oikos 116:1983–1994

Bontemps S, Defourny P, Van Bogaert E, Arino O, Kalogirou V, Ramos Perez J (2011) GLOBCOVER 2009. Products description and validation report. Université Catholique de Louvain & ESA Team

Borken W, Beese F (2005) Control of nitrous oxide emissions in European beech, Norway spruce and scots pine forests. Biogeochemistry 76:141–159

Borken W, Beese F (2006) Methane and nitrous oxide fluxes of soils in pure and mixed stands of European beech and Norway spruce. Eur J Soil Sci 57:617–625

Borken W, Matzner E (2009) Reappraisal of drying and wetting effects on C and N mineralization and fluxes in soils. Glob Chang Biol 15:808–824

Borken W, Xu YJ, Beese F (2003) Conversion of hardwood forests to spruce and pine plantations strongly reduced soil methane sink in Germany. Glob Chang Biol 9:956–966

Borken W, Xu YJ, Brumme R (2000) Effects of prolonged soil drought on CH$_4$ oxidation in a temperate spruce forest. J Geophys Res Atmos 105:7079–7088

Bradshaw CJA, Warkentin IG (2015) Global estimates of boreal forest carbon stocks and flux. Glob Planet. Change 128:24–30

Brumme R (1995) Mechanisms of carbon and nutrient release and retention in beech forest gaps. Plant Soil 168:593–600

Brumme R, Borken W, Finke S (1999) Hierarchical control on nitrous oxide emission in forest ecosystems. Glob Biogeochem Cycles 13:1137–1148

Butterbach-Bahl K, Baggs EM, Dannenmann M, Kiese R, Zechmeister-Boltenstern S (2013) Nitrous oxide emissions from soils: how well do we understand the processes and their controls? Phil Trans R Soc B Biol Sci 368:20130122

Butterbach-Bahl K, Dannenmann M (2011) Denitrification and associated soil N$_2$O emissions due to agricultural activities in a changing climate. Curr Opin Environ Sustain 3:389–395

Butterbach-Bahl K, Díaz-Pinés E, Dannenmann M (2012) Soil trace gas emissions and climate change. In: Freedman B (ed) Global environmental change. Berlin/Heidelberg, Springer, pp 325–334

Butterbach-Bahl K, Gasche R, Breuer L, Papen H (1997) Fluxes of NO and N$_2$O from temperate forest soils: impact of forest type, N deposition and of liming on the NO and N$_2$O emissions. Nutr Cycl Agroecosyst 48:79–90

Butterbach-Bahl K, Papen H (2002) Four years continuous record of CH$_4$-exchange between the atmosphere and untreated and limed soil of a N-saturated spruce and beech forest ecosystem in Germany. Plant Soil 240:77–90

Butterbach-Bahl K, Rothe A, Papen H (2002) Effect of tree distance on N$_2$O and CH$_4$-fluxes from soils in temperate forest ecosystems. Plant Soil 240:91–103

Callaghan TV, Björn LO, Chapin FS III et al (2005) Arctic tundra and polar desert ecosystems. In: Symon C, Arris L, Heal B (eds) Arctic climate impact assessment. Cambridge University Press, New York, pp 243–352

Carter MS, Larsen KS, Emmett B et al (2012) Synthesizing greenhouse gas fluxes across nine European peatlands and shrublands – responses to climatic and environmental changes. Biogeosciences 9:3739–3755

Chang S-C, Matzner E (2000) Soil nitrogen turnover in proximal and distal stem areas of European beech trees. Plant Soil 218:117–125

Cheng W, Kuzyakov Y (2005) Root effects on soil organic matter decomposition. In: Zobel R, Wright S (eds) Roots and soil management: interactions between roots and the soil. ASA-SSSA, Madison, pp 119–144

Conrad R (1996) Soil microorganisms as controllers of atmospheric trace gases (H_2, CO, CH_4, OCS, N_2O, and NO). Microbiol Rev 60:609–640

Conrad R (2009) The global methane cycle: recent advances in understanding the microbial processes involved. Environ Microbiol Rep 1:285–292

Dalal RC, Allen DE (2008) Greenhouse gas fluxes from natural ecosystems. Aust J Bot 56:369–407

Davidson EA (1994) Climate change and soil microbial processes: secondary effects are hypothesised from better known interacting primary effects. In: Rounsevell MDA, Loveland PJ (eds) Soil responses to climate change. Springer, Berlin/Heidelberg, pp 155–168

De Bruijn AMG, Butterbach-Bahl K, Blagodatsky S, Grote R (2009) Model evaluation of different mechanisms driving freeze–thaw N_2O emissions. Agric Ecosyst Environ 133:196–207

De Vries W, Reinds GJ, Posch M et al (2003) Intensive monitoring of forest ecosystems in europe, 2003 technical report. EC, UN/ECE, Brussels

Díaz-Pinés E, Heras P, Gasche R, Rubio A, Rennenberg H, Butterbach-Bahl K, Kiese R (2015) Nitrous oxide emissions from stems of ash (Fraxinus angustifolia Vahl) and European beech (Fagus sylvatica L.) Plant Soil 398:35–45

Díaz-Pinés E, Schindlbacher A, Godino M, Kitzler B, Jandl R, Zechmeister-Boltenstern S, Rubio A (2014) Effects of tree species composition on the CO_2 and N_2O efflux of a Mediterranean mountain forest soil. Plant Soil 384:243–257

Dix NJ, Webster J (1995) Fungal ecology. Chapman & Hall, London

Douglas TA, Jones MC, Hiemstra CA, Arnold JR (2014) Sources and sins of carbon in boreal ecosystems of interior Alaska: a review. Elementa: Science of the Anthropocene 2:000032

Dutaur L, Verchot LV (2007) A global inventory of the soil CH_4 sink. Glob Biogeochem Cycles 21:GB4013

Ewers BE, Mackay DS, Gower ST, Ahl DE, Burrows SN, Samanta SS (2002) Tree species effects on stand transpiration in northern Wisconsin. Water Resour Res 38:8-1-8-11

Fender A-C, Gansert D, Jungkunst HF et al (2013) Root-induced tree species effects on the source/sink strength for greenhouse gases (CH_4, N_2O and CO_2) of a temperate deciduous forest soil. Soil Biol Biochem 57:587–597

Flessa H, Rodionov A, Guggenberger G et al (2008) Landscape controls of CH_4 fluxes in a catchment of the forest tundra ecotone in northern Siberia. Glob Chang Biol 14:2040–2056

Galiano L, Martinez-Vilalta J, Lloret F (2010) Drought-induced multifactor decline of scots pine in the pyrenees and potential vegetation change by the expansion of co-occurring oak species. Ecosystems 13:978–991

Gamfeldt L, Snäll T, Bagchi R et al (2013) Higher levels of multiple ecosystem services are found in forests with more tree species. Nat Commun 4:1340

Gillett NP, Weaver AJ, Zwiers FW, Flannigan MD (2004) Detecting the effect of climate change on Canadian forest fires. Geophys Res Lett 31:L18211

Goldberg SD, Gebauer G (2009) Drought turns a Central European Norway spruce forest soil from an N_2O source to a transient N_2O sink. Glob Chang Biol 15:850–860

Gritsch C, Egger F, Zehetner F, Zechmeister-Boltenstern S (2016) The effect of temperature and moisture on trace gas emissions from deciduous and coniferous leaf litter. J Geophys Res Biogeo 121:1339–1351

Hanewinkel M, Cullmann DA, Schelhaas M-J, Nabuurs G-J, Zimmermann NE (2013) Climate change may cause severe loss in the economic value of European forest land. Nat Clim Chang 3:203–207

Huber C, Aherne J, Weis W, Farrell EP, Göttlein A, Cummins T (2010) Ion concentrations and fluxes of seepage water before and after clear cutting of Norway spruce stands at Ballyhooly, Ireland, and Höglwald, Germany. Biogeochemistry 101:7–26

Huber C, Weis W, Baumgarten M, Göttlein A (2004) Spatial and temporal variation of seepage water chemistry after femel and small scale clear-cutting in a N-saturated Norway spruce stand. Plant Soil 267:23–40

IPCC (2001) In: Houghton JT, Ding Y, Griggs DJ, Noguer M, van der Linden PJ, Dai X, Maskell K, Johnson CA (eds) Climate change 2001: the scientific basis. Contribution of Working Group I to the third assessment report of the Intergovernmental Panel on Climate Change. Cambridge University Press, Cambridge, UK/New York

IPCC (2007) Climate change 2007: synthesis report. Contribution of Working Groups I, II and III to the fourth assessment report of the Intergovernmental Panel on Climate Change. Intergovernmental Panel on Climate Change, Geneva

IPCC (2012) In: Field CB, Barros V, Stocker TF, Qin D, Dokken DJ, Ebi KL, Mastrandrea MD, Mach KJ, Plattner GK, Allen SK, Tignor M, Midgley PM (eds) Managing the risks of extreme events and disasters to advance climate change adaptation. A special report of Working Groups I and II of the Intergovernmental Panel on Climate Change. Cambridge University Press, Cambridge, UK/New York

IPCC (2013) Climate change 2013: the physical science basis. Contribution of Working Group I to the fifth assessment of the Intergovernmental Panel on Climate Change. Cambridge University Press, Cambridge, UK/New York

Jandl R, Lindner M, Vesterdal L et al (2007) How strongly can forest management influence soil carbon sequestration? Geoderma 137:253–268

Jorgenson MT, Romanovsky V, Harden J et al (2010) Resilience and vulnerability of permafrost to climate change. Can J For Res 40:1219–1236

Jungkunst HF, Fiedler S (2007) Latitudinal differentiated water table control of carbon dioxide, methane and nitrous oxide fluxes from hydromorphic soils: feedbacks to climate change. Glob Chang Biol 13:2668–2683

Kammer A, Hagedorn F, Shevchenko I et al (2009) Treeline shifts in the Ural mountains affect soil organic matter dynamics. Glob Chang Biol 15:1570–1583

Karbin S, Hagedorn F, Dawes MA, Niklaus PA (2015) Treeline soil warming does not affect soil methane fluxes and the spatial micro-distribution of methanotrophic bacteria. Soil Biol Biochem 86:164–171

Keenan RJ, Reams GA, Achard F, De Freitas JV, Grainger A, Lindquist E (2015) Dynamics of global forest area: results from the FAO global forest resources assessment 2015. For Ecol Manag 352:9–20

Kicklighter DW, Cai Y, Zhuang Q et al (2014) Potential influence of climate-induced vegetation shifts on future land use and associated land carbon fluxes in Northern Eurasia. Environ Res Lett 9:035004

Klemedtsson L, Von Arnold K, Weslien P, Gundersen P (2005) Soil CN ratio as a scalar parameter to predict nitrous oxide emissions. Glob Chang Biol 11:1142–1147

Luo GJ, Brüggemann N, Wolf B, Gasche R, Grote R, Butterbach-Bahl K (2012) Decadal variability of soil CO₂, NO, N₂O, and CH₄ fluxes at the Höglwald Forest, Germany. Biogeosciences 9:1741–1763

Machacova K, Bäck J, Vanhatalo A et al (2016) Pinus sylvestris as a missing source of nitrous oxide and methane in boreal forest. Sci Rep 6:23410

Mander Ü, Maddison M, Soosaar K, Teemusk A, Kanal A, Uri V, Truu J (2015) The impact of a pulsing groundwater table on greenhouse gas emissions in riparian grey alder stands. Environ Sci Pollut Res 22:2360–2371

Matson A, Pennock D, Bedard-Haughn A (2009) Methane and nitrous oxide emissions from mature forest stands in the boreal forest, Saskatchewan, Canada. For Ecol Manag 258:1073–1083

Maurer D, Kolb S, Haumaier L, Borken W (2008) Inhibition of atmospheric methane oxidation by monoterpenes in Norway spruce and European beech soils. Soil Biol Biochem 40:3014–3020

Mcvicar K, Kellman L (2014) Growing season nitrous oxide fluxes across a 125+ year harvested red spruce forest chronosequence. Biogeochemistry 120:225–238

Menyailo OV, Hungate B (2005) Tree species effects on potential production and consumption of carbon dioxide, methane and nitrous oxide: the Siberian afforestation experiment. In: Binkley D, Menyailo OV (eds) Tree species effects on soils: implications for global change. Springer, Dordrecht, pp 293–305

Millennium Ecosystem Assessment (2005) Ecosystems and human well-being: biodiversity synthesis. World Resources Institute, Washington, DC

Nykänen H, Heikkinen JEP, Pirinen L, Tiilikainen K, Martikainen P (2003) Annual CO_2 exchange and CH_4 fluxes on a subarctic palsa mire during climatically different years. Glob Biogeochem Cycles 17:18–11

Olson DM, Dinerstein E, Wikramanayake ED et al (2001) Terrestrial ecoregions of the world: a new map of life on earth: a new global map of terrestrial ecoregions provides an innovative tool for conserving biodiversity. Bioscience 51:933–938

Pan Y, Birdsey RA, Fang J et al (2011) A large and persistent carbon sink in the world's forests. Science 333:988–993

Pangala SR, Hornibrook ERC, Gowing DJ, Gauci V (2015) The contribution of trees to ecosystem methane emissions in a temperate forested wetland. Glob Chang Biol 21:2642–2654

Papen H, Rosenkranz P, Butterbach-Bahl K, Gasche R, Willibald G, Brüggemann N (2003) Effects of tree species on C- and N- cycling and biosphere-atmosphere exchange of trace gases in forests. In: Binkley D, Menyailo O (eds) Tree species effects on soils: implications for global change. Springer, Dordrecht, pp 165–172

Philippot L, Cuhel J, Saby NBA et al (2009) Mapping field-scale spatial patterns of size and activity of the denitrifier community. Environ Microbiol 11:1518–1526

Pilegaard K, Skiba U, Ambus P et al (2006) Factors controlling regional differences in forest soil emissions of nitrogen oxides (NO and N_2O). Biogeosciences 3:651–661

Rebetez M, Dobbertin M (2004) Climate change may already threaten Scots pine stands in the Swiss Alps. Theor Appl Climatol 79:1–9

Roman-Cuesta RM, Rufino MC, Herold M et al (2016) Hotspots of gross emissions from the land use sector: patterns, uncertainties, and leading emission sources for the period 2000–2005 in the tropics. Biogeosciences 13:4253–4269

Rosenkranz P, Dannenmann M, Brüggemann N, Papen H, Berger U, Zumbusch E, Butterbach-Bahl K (2010) Gross rates of ammonification and nitrification at a nitrogen-saturated spruce (Picea abies (L.) Karst.) stand in southern Germany. Eur J Soil Sci 61:745–758

Rouse WR, Bello RL, Souza A, Griffis TJ, Lafleur PM (2002) The annual carbon budget for fen and forest in a wetland at Arctic treeline. Arctic 55:229–237

Rusch H, Rennenberg H (1998) Black alder (Alnus Glutinosa (L.) Gaertn.) trees mediate methane and nitrous oxide emission from the soil to the atmosphere. Plant Soil 201:1–7

Sayer EJ (2006) Using experimental manipulation to assess the roles of leaf litter in the functioning of forest ecosystems. Biol Rev 81:1–31

Schaefer K, Lantuit H, Romanovsky VE, Schuur EG, Witt R (2014) The impact of the permafrost carbon feedback on global climate. Environ Res Lett 9:085003

Schelhaas M-J, Nabuurs G-J, Schuck A (2003) Natural disturbances in the European forests in the 19th and 20th centuries. Glob Chang Biol 9:1620–1633

Schuur EG, Mcguire AD, Schädel C et al (2015) Climate change and the permafrost carbon feedback. Nature 520:171–179

Serreze MC, Walsh JE, Chapin FS et al (2000) Observational evidence of recent change in the northern high-latitude environment. Clim Chang 46:159–207

Simón N, Montes F, Díaz-Pinés E, Benavides R, Roig S, Rubio A (2013) Spatial distribution of the soil organic carbon pool in a Holm oak dehesa in Spain. Plant Soil 366:537–549

Sitaula BK, Bakken LR (1993) Nitrous oxide release from spruce forest soil: relationships with nitrification, methane uptake, temperature, moisture and fertilization. Soil Biol Biochem 25:1415–1421

Sjögersten S, Wookey PA (2002) Spatio-temporal variability and environmental controls of methane fluxes at the forest–tundra ecotone in the Fennoscandian mountains. Glob Chang Biol 8:885–894

Strömgren M, Hedwall PO, Olsson BA (2016) Effects of stump harvest and site preparation on N$_2$O and CH$_4$ emissions from boreal forest soils after clear-cutting. For Ecol Manag 371:15–22

Takakai F, Desyatkin AR, Lopez CML, Fedorov AN, Desyatkin RV, Hatano R (2008) CH$_4$ and N$_2$O emissions from a forest-alas ecosystem in the permafrost taiga forest region, eastern Siberia, Russia. J Geophys Res Biogeosci 113:G02002

Tupek B, Minkkinen K, Pumpanen J, Vesala T, Nikinmaa E (2015) CH$_4$ and N$_2$O dynamics in the boreal forest–mire ecotone. Biogeosciences 12:281–297

Van Haren JLM, De Oliveira RC, Restrepo-Coupe N, Hutyra L, De Camargo PB, Keller M, Saleska SR (2010) Do plant species influence soil CO$_2$ and N$_2$O fluxes in a diverse tropical forest? J Geophys Res Biogeosci 115:G03010

Vesterdal L, Schmidt IK, Callesen I, Nilsson LO, Gundersen P (2008) Carbon and nitrogen in forest floor and mineral soil under six common European tree species. For Ecol Manag 255:35–48

Volney WJA, Fleming RA (2000) Climate change and impacts of boreal forest insects. Agric Ecosyst Environ 82:283–294

Westerling AL, Hidalgo HG, Cayan DR, Swetnam TW (2006) Warming and earlier spring increase western U.S. forest wildfire activity. Science 313:940–943

Wieser G (2010) Lessons from the timberline ecotone in the Central Tyrolean Alps: a review. Plant Ecol Div 5:127–139

Wu X, Brüggemann N, Gasche R, Papen H, Willibald G, Butterbach-Bahl K (2011) Long-term effects of clear-cutting and selective cutting on soil methane fluxes in a temperate spruce forest in southern Germany. Environ Pollut 159:2467–2475

Zerva A, Mencuccini M (2005) Short-term effects of clearfelling on soil CO$_2$, CH$_4$, and N$_2$O fluxes in a Sitka spruce plantation. Soil Biol Biochem 37:2025–2036

Zhang J, Peng C, Zhu Q et al (2016) Temperature sensitivity of soil carbon dioxide and nitrous oxide emissions in mountain forest and meadow ecosystems in China. Atmos Environ 142:340–350

Zhou Y, Hagedorn F, Zhou C, Jiang X, Wang X, Li M-H (2016) Experimental warming of a mountain tundra increases soil CO$_2$ effluxes and enhances CH$_4$ and N$_2$O uptake at Changbai Mountain, China. Sci Rep 6:21108

Zhu R, Ma D, Xu H (2014) Summertime N$_2$O, CH$_4$ and CO$_2$ exchanges from a tundra marsh and an upland tundra in maritime Antarctica. Atmos Environ 83:269–281

What Are Plant-Released Biogenic Volatiles and How They Participate in Landscape- to Global-Level Processes?

Ülo Niinemets

1 Introduction What Are Plant Volatiles?

Plant-released organic volatiles constitute a vast spectrum of compounds, more than 30,000 different compounds with a certain capacity to escape into the gas phase from a liquid or solid (Niinemets et al. 2004). In common with the compounds characteristically called volatiles is that they have normal pressure boiling points between ca. 30 and 250 °C and, thus, support a relatively high vapor partial pressure at ambient temperatures (between ca. 10^1 and 10^5 Pa at 25 °C) (Fuentes et al. 2000; Copolovici and Niinemets 2005; Kosina et al. 2013). In addition, studies on plant volatiles also often consider semivolatiles that support a much lower vapor pressure (partial pressure ca. 10^{-6} and 10^0 at 25 °C) (Helmig et al. 2003; Widegren and Bruno 2010; Kosina et al. 2013). Semivolatiles have a large capacity for partitioning into liquid and solid phases and, once released from plants, play a major role in atmospheric particle formation (Ehn et al. 2014).

All plants emit a plethora of volatiles that are synthesized in different subcellular compartments involving multiple biochemical pathways (Fig. 1), and the emissions can be further tissue- and organ-specific. The volatiles emitted can be intermediates of normal plant metabolic activity and are released from plant tissues because the metabolic pathways are "leaky." Emissions of such compounds can be enhanced under certain periods of plant life. For instance, plants emit methanol as the result of demethylation of cell wall pectins (Nemecek-Marshall et al. 1995; Fall and Benson 1996). Methanol emissions occur at low level from all physiologically active plant tissues, but the emissions are strongly enhanced in growing tissues due to relaxation and rigidification of cell walls during tissue expansion growth (Harley

Ü. Niinemets (✉)
Institute of Agricultural and Environmental Sciences,
Estonian University of Life Sciences, Tartu, Estonia
e-mail: ylo.niinemets@emu.ee

© Springer International Publishing AG, part of Springer Nature 2018
A. H. Perera et al. (eds.), *Ecosystem Services from Forest Landscapes*,
https://doi.org/10.1007/978-3-319-74515-2_3

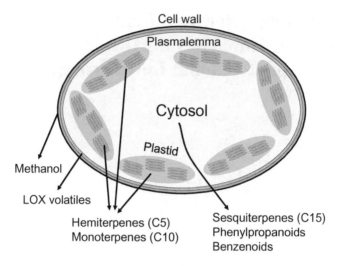

Fig. 1 Main volatiles emitted by plants are formed in different subcellular compartments and their synthesis involves a variety of biochemical pathways. Methanol and lipoxygenase (LOX) pathway volatiles (also called green leaf volatiles, dominated by various C6 aldehydes and alcohols) are ubiquitous volatiles that can be released from all plant tissues as the result of constitutive activity of key enzymes of their synthesis pathways, while volatile isoprenoids (hemiterpenes like isoprene, monoterpenes, and sesquiterpenes) and phenylpropanoids and benzenoids are specialized volatiles and are emitted as the result of induction of expression of genes coding for specific rate-limiting synthases, although in several species, certain specialized volatiles can be emitted constitutively. Methanol is released as the result of demethylation of pectins in cell walls in growing tissues or upon different biotic and abiotic stresses, whereas green leaf volatiles are formed from free polyunsaturated fatty acids released from membrane lipids upon membrane damage characteristic to exposure to severe stresses (Liavonchanka and Feussner 2006; Andreou and Feussner 2009). Emissions of phenolic compounds typically originate from cytosol, but isoprenoid emissions can originate from plastids or cytosol, depending on the compound emitted (Niinemets et al. 2013; Rosenkranz and Schnitzler 2013; Pazouki and Niinemets 2016). There is evidence that some terpenoids can be potentially also released from mitochondria (not shown in the figure, see Tholl and Lee 2011; Dong et al. 2016), but the possible mitochondrial release of volatiles is much less studied

et al. 2007; Hüve et al. 2007). Analogously, root zone hypoxia during flooding leads to ethanol formation in the roots and its transportation to the leaves with the transpiration stream (Bracho Nunez et al. 2009; Kreuzwieser and Rennenberg 2013). Ethanol that reaches the leaves can be further enzymatically oxidized to acetic acid via acetaldehyde, and enter into the primary metabolism, but some ethanol and acetaldehyde escape leaves due to limited alcohol and aldehyde dehydrogenase activities (Kreuzwieser et al. 2000, 2001; Rottenberger et al. 2008; Kreuzwieser and Rennenberg 2013).

These examples demonstrate how normal physiological processes in plant life, and the ecosystem services they provide, can be associated with major leakage of organic compounds due to relatively high vapor pressure of these compounds. Apart from metabolic intermediates, a large number of organic compounds are specifically made to be volatile, including several volatile benzenoids and phenylpropanoids and volatile isoprenoids such as the hemiterpene isoprene (C5), monoterpenes (C10), and sesquiterpenes (C15) (Fig. 1, Peñuelas and Llusià 2004; Fineschi et al. 2013; Guenther 2013a; Portillo-Estrada et al. 2015). These specialized volatiles constitute the plant "talk," fulfilling a plethora of biological and ecological functions from communication among plant organs, other plants, and other organisms to altering the plant stress resistance. Furthermore, all volatiles participate in multiple landscape- to global-scale processes, altering ambient air oxidative status, atmospheric particle condensation, and cloud cover (Peñuelas and Staudt 2010; Kulmala et al. 2013). Both the specific biological and broad-scale atmospheric roles of volatiles result in a number of key ecosystem services beneficial to humans. As discussed in this chapter, these services can be local to regional scale services such as preservation of ecosystem integrity under biotic and abiotic stresses and, thus, the preservation of the capacity to provide the "standard," well-perceived, ecosystem services to humans, e.g., wood production of forest stands. These services also include regional to global-scale services such as environmental cooling and dampening the global climate change.

In the current chapter, I first provide a short overview of key types of biological volatile emission and of the environmental controls on volatile emission and modification of emissions by abiotic and biotic stresses. Then I focus on the biological roles of volatiles, on the roles of volatiles in large-scale biosphere-atmosphere processes, and on ecological services provided by plant volatiles. I demonstrate that the trace gas release is a key vegetation characteristic that contributes a number of unique ecosystem services that alter the performance of ecosystems in current and future climates with major implications for human life. Quantitative significance of plant volatiles in Earth system processes is still poorly understood, and the role of plant volatiles in climate has been largely neglected in the last report of the Intergovernmental Panel on Climate Change (IPCC, Field et al. 2014; Stocker et al. 2014). In this chapter, I show that the evidence is accumulating that plant trace gas exchange participates in multiple feedback loops that can potentially play major roles in Earth system processes and argue that biosphere-atmosphere interactions mediated by plant volatiles need to be included in Earth system models intending to predict future climate.

2   Plant Volatile Diversity, Environmental Controls on Emission and Emission Capacities

2.1   Sites of Volatile Synthesis and Diversity

Synthesis of volatile phenolics typically occurs in cytosol, while volatile isoprenoids can be synthesized in plastids or cytosol, depending on the compound emitted (Niinemets et al. 2013; Pazouki and Niinemets 2016). Hemiterpene and monoterpene synthesis is considered to occur in the plastids where corresponding C5 and

C10 terminal enzymes responsible for terpene synthesis, terpene synthases, are located, while sesquiterpene synthases are located in the cytosol (Fig. 1, Chen et al. 2011). However, there is also recent evidence of mixed substrate specificity of some terpene synthases (Pazouki and Niinemets 2016), suggesting that product profiles can potentially vary depending on substrate availability in different subcellular compartments. Furthermore, there is evidence that some mono- and sesquiterpenes can be synthesized in mitochondria (Tholl and Lee 2011; Dong et al. 2016), further complicating the picture.

The diversity of emitted volatiles varies for different volatile compound classes. In the case of hemiterpenes, in addition to isoprene, plants also emit the oxygenated hemiterpene 2-methyl-3-buten-2-ol (Gray et al. 2006, 2011), but mono- and sesquiterpene-emitters typically release a wide spectrum of compounds. Often more than 20 different terpenoids are observed in the emissions from a single species (e.g., Niinemets et al. 2002a; Winters et al. 2009). Such a high diversity reflects the presence of multiple terpene synthases in the given emitting species (e.g., Falara et al. 2011; Jiang et al. 2016a) but also the specific reaction mechanism of terpene synthesis. In particular, terpene synthesis involves formation of a highly reactive carbocation intermediate, and depending on the extent to which the carbocation can be stabilized, the product specificity of terpene synthases strongly varies (e.g., Christianson 2008). In fact, most terpene synthases catalyze formation of multiple products, and only some terpene synthases form single terpenes, explaining the huge chemical diversity of emitted volatile terpenoids.

2.2 Constitutive and Stress-Induced Volatile Emissions

2.2.1 Constitutive Emissions of Specialized Volatiles

While all plant species can emit metabolic intermediates, only some species emit specialized volatiles, in particular, volatile isoprenoids under typical non-stressed physiological conditions, being, thus, constitutive emitters (Peñuelas and Llusià 2004; Fineschi et al. 2013). The capacity for constitutive emission of certain volatiles requires that the specific synthesis pathways are constitutively active, although the degree of activation can vary with environmental conditions, sometimes several-fold (e.g., Niinemets et al. 2010b). The constitutive emissions can result from emission of compounds stored in specialized storage structures such as oil glands in *Citrus* species, resin ducts in conifers, and glandular trichomes in species from Labiatae or Solanaceae. Typically, volatiles stored in these structures are mono- and sesquiterpenes, but sometimes benzenoids are also stored (Loreto et al. 2000; Jiang et al. 2016a). Filling up the storage structures takes typically multiple days to months, and thus, the release of the volatiles from the storage is uncoupled from the synthesis of these compounds. Thus, the rate of constitutive emissions from storage structures depends on the rate of compound evaporation and diffusion and therefore scales exponentially with temperature (Niinemets et al. 2010b; Grote et al. 2013).

The hemiterpenes, isoprene and 2-methyl-3-buten-2-ol, cannot be stored due to high volatility. In addition, several constitutive monoterpene emitters such as the Mediterranean evergreen oaks *Quercus ilex* and *Q. suber* and the broad-leaved deciduous temperate tree *Fagus sylvatica* lack specialized storage structures (Staudt et al. 2004; Dindorf et al. 2006). In hemiterpene and non-storage monoterpene emitters, the volatile emissions result from de novo compound synthesis, and the emissions respond to environmental drivers similarly to photosynthesis, increasing asymptotically with light intensity and scaling positively with temperature up to an optimum temperature and decreasing thereafter (Niinemets et al. 2010b; Monson et al. 2012). Emissions of de novo synthesized volatiles also respond to immediate changes in ambient CO_2 concentration. In particular, isoprene emissions decrease with increasing CO_2 concentration (e.g., Wilkinson et al. 2009), but the CO_2 response is less clear for monoterpene emissions (Sun et al. 2012 for a discussion). However, the CO_2 sensitivity of isoprene is gradually lost at higher temperatures (Rasulov et al. 2010; Li and Sharkey 2013), and the emission response to longer-term changes in ambient CO_2 concentration can be different from the immediate response due to acclimation of isoprene synthesis pathway to long-term ambient CO_2 concentration (Sun et al. 2012).

2.2.2   Induction of Volatiles Upon Abiotic and Biotic Stresses

In field environments, plants are often exposed to various abiotic and biotic stresses. Although only some plant species can emit volatiles constitutively, all species can be triggered to release volatiles upon biotic and abiotic stresses (Fig. 2). Among the volatiles triggered are ubiquitous stress compounds such as lipoxygenase (LOX)

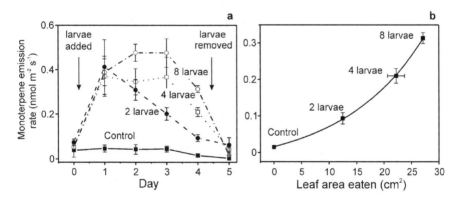

Fig. 2 Induction of monoterpene emissions in broad-leaved temperate deciduous tree *Alnus incana* upon feeding by larvae of the moth *Cabera pusaria* (**a**), and correlation of the degree of elicitation of emissions with the amount of leaf area consumed on the fourth day since the start of feeding (**b**). *Alnus incana* does not significantly emit volatiles in non-stressed conditions and is therefore considered a non-emitter species, but exposure to different stresses results in significant emissions of green leaf volatiles, mono-, sesqui-, and homoterpenes (Modified from Copolovici et al. 2011)

pathway volatiles (also called green leaf volatiles). Green leaf volatiles, typically dominated by various C6 aldehydes and alcohols, are rapidly formed when free polyunsaturated fatty acids are released from membrane lipids (Beauchamp et al. 2005; Copolovici et al. 2011). This typically occurs upon membrane-level damage characteristic to exposure to severe stresses such as mechanic wounding during biotic stresses but also upon exposure to severe heat, frost, and ozone stress (Beauchamp et al. 2005; Copolovici et al. 2011). Methanol, which can be released from non-stressed tissues constitutively, also serves as another ubiquitous stress volatile; major rapid methanol bursts are associated with both severe biotic and abiotic stresses (Beauchamp et al. 2005; von Dahl et al. 2006).

Apart from the ubiquitous compounds emitted from virtually all plant species, induction of synthesis of specialized compounds requires the presence of given stress-elicited synthase genes in plant genomes. For instance, glucosinolate pathway volatiles (short-chained S- and/or N-containing volatiles resulting from breakdown of glucosylated amino acids and their derivatives) are only released upon stress from the species in the order Brassicales (Kask et al. 2016). On the other hand, all plants do include terpene synthase genes in their genomes, and expression of these genes is typically activated upon stress. Studies have demonstrated elicitation of emissions of mono- and sesquiterpenes upon biotic stresses such as herbivory (Fig. 2, Blande et al. 2007; Copolovici et al. 2014a, 2011; Farré-Armengol et al. 2015); pathogen infections such as powdery mildew (Fig. 3), leaf rust (Fig. 4), and canker fungus (Achotegui-Castells et al. 2015) infections; and upon abiotic stresses such as ozone stress (Beauchamp et al. 2005) and heat stress (Copolovici et al. 2012; Kask et al. 2016). However, the number of terpenoid synthase genes strongly varies among species from as few as only one synthase gene to more than 80 genes (Rajabi Memari et al. 2013), implying that the diversity of induced terpenoid emission responses can also be variable. Furthermore, different stresses can trigger emissions of different volatiles (Dicke et al. 2009; Zhang et al. 2009), but so far, understanding of the overall stress-dependent emission diversity and variation of emission profiles under different stresses is very limited.

Differently from LOX volatiles and methanol, emissions of which are triggered rapidly due to the presence of a certain constitutive activity of lipoxygenases (Liavonchanka and Feussner 2006) and pectin methylesterases (Micheli 2001), elicitation of terpenoid emissions is more time-consuming because it requires gene expression and protein synthesis to reach a certain terpene synthase activity. Typically, emissions of terpenoids can be detected hours after the stress impact and the emissions peak 24–48 h after the impact (Fig. 2, Pazouki et al. 2016). On the other hand, when the stress is relieved, the emissions gradually decrease, reaching the initial non-induced level in a few days after the stress relief (Fig. 2a). This reduction is a characteristic feature to induced emissions and contrasts to constitutive emissions. Although the rate of constitutive emissions can also be affected by stress, typically negatively the level of constitutive emissions almost never reaches zero.

2.3 Relationships Among Constitutive and Stress-Induced Emissions

As noted above, constitutive emissions are present only in several plant species, while induced emissions can be elicited in all species, including the constitutive emitters. Typically, the composition of induced emissions is different from constitutive emissions. For instance, in constitutively isoprene-emitting deciduous oak *Quercus robur*, infection by oak powdery mildew (*Erysiphe alphitoides*) results in emissions of mono- and sesquiterpenes (Fig. 3). Analogously, infection of the constitutively isoprene-emitting poplar (*Populus* spp.) with the rust fungus *Melampsora larici-populina*, results in mono- and sesquiterpene emissions (Fig. 4).

In the case of constitutive monoterpene emitters, an environmental or a biotic stress often results in elicitation of emissions of monoterpenes different from constitutively emitted monoterpenes. In particular, typical stress-elicited monoterpenes are ocimenes, linalool, and 1,8-cineole, while constitutive emissions are characteristically dominated by limonene and pinenes (Staudt and Bertin 1998; Niinemets et al. 2002b). Importantly, in constitutive storage emitters, stress-induced monoterpene emissions reflect de novo synthesis of volatiles and scale similarly with light and temperature as the emissions in non-storage emitters (Niinemets et al. 2010a, b). In addition to monoterpenes, stress often results in elicitation of emissions of sesquiterpenes and homoterpenes 4,8-dimethylnona-1,3,7-triene (DMNT) and 4,8,12-trimethyltrideca-1,3,7,11-tetraene (TMTT) that are not observed in constitutive emissions (e.g., Niinemets et al. 2010b; Staudt and Lhoutellier 2011; Tholl et al. 2011).

In both of the biotic stress case studies highlighted here, the constitutive isoprene emissions decreased in pathogen-infected leaves (Figs. 3 and 4), and analogous negative relationships between induced and constitutive emissions have been demonstrated in the case of other stresses as well (e.g., Kleist et al. 2012 for heat stress). Overall, positive stress dose vs. induced emission relationships have been observed for several abiotic stresses such as frost and heat stress (Copolovici et al. 2012) and ozone stress (Beauchamp et al. 2005). Although biotic impacts have been considered to be hard to quantify (Niinemets et al. 2013), quantitative stress dose vs. induced emission responses have been observed for several biotic stresses such as herbivory (Fig. 2) and fungal pathogen infections (Figs. 3 and 4). The key issue with the biotic stresses seems to be how to quantify the severity of biotic stress (Copolovici and Niinemets 2016), but once the biotic stress severity has been properly characterized, it becomes clear that the rate of emission of volatile organic compounds scales with stress severity similarly to abiotic stresses (Niinemets et al. 2013; Copolovici and Niinemets 2016).

Fig. 3 Oak powdery mildew (*Erysiphe alphitoides*) is a major pathogen infesting pedunculate oak (*Quercus robur*) all over Europe. The visual damage symptoms can be detected through the growing season in young oak trees (**a**), and in late summer and autumn in old plants (**b**). *Erysiphe alphitoides* infections are associated with reduction of constitutive emissions of isoprene (**c**) and elicitation of emissions of monoterpenes (**d**), sesqui- and homoterpenes (data not shown) and green leaf volatiles (**e**) ((**c**–**e**) Modified from Copolovici et al. 2014b). Open symbols denote non-infected leaves

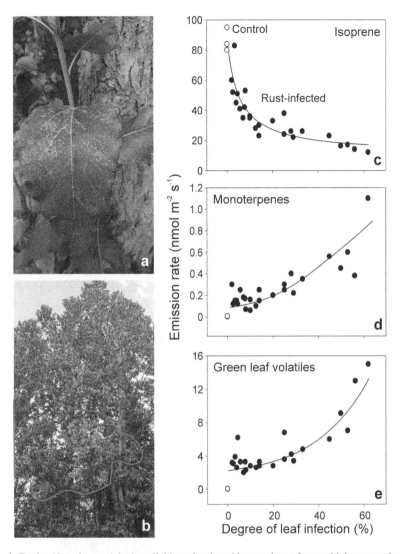

Fig. 4 Poplar (*Populus* spp.) is the telial host for the widespread rust fungus *Melampsora larici-populina*, infections of which are initially associated with diffusely spread yellow spots (**a**), followed by extensive leaf yellowing and necrosis and premature leaf senescence encompassing large parts of the canopy (encircled area in **b**). *Melampsora larici-populina* infection reduces constitutive emissions of isoprene (**c**) and induces emissions of monoterpenes (**d**), sesqui- and homoterpenes (data not shown) and green leaf volatiles (**e**) in an infection-dependent manner (Jiang et al. 2016b). Open symbols refer to emissions from leaves without visible signs of infection. Pictures (**a**) and (**b**) are for *P. laurifolia* and data (**c–e**) for *P. balsamifera* var. *suaveolens*

3 Ecosystem Services and Impacts of Plant Volatiles

3.1 Biological Role of Plant Volatiles in Ecological Processes

The biological role of constitutive emissions of isoprene and monoterpenes is not yet fully clear, but there is increasing evidence that these compounds have antioxidative and/or membrane-stabilizing properties (Sharkey et al. 2001; Loreto et al. 2004; Vickers et al. 2009; Peñuelas and Staudt 2010), and it has been postulated that they play an important role in enhancing the abiotic stress resistance in species emitting them (Vickers et al. 2009; Possell and Loreto 2013). Furthermore, volatiles released from constitutively emitting species can be taken up from the ambient air by neighboring non-emitting species (Copolovici et al. 2005; Noe et al. 2008; Himanen et al. 2010). This can enhance the stress resistance of non-emitters, resulting in an overall increase in ecosystem stress resistance. Such an increase of ecosystem stress resistance is the key ecological service which allows for maintenance of all other well-recognized ecosystem services provided by healthy ecosystems.

While the constitutive emissions can provide certain protection against chronic mild abiotic stresses that do not lead to induction of stress volatiles, once induced, chemically similar induced volatiles can also directly participate in stress protection during short-term severe stresses that trigger their emission. However, more importantly, stress-elicited volatiles play various functions in communication among plants and plants and other organisms (Dicke and Baldwin 2010; Holopainen et al. 2013; Blande et al. 2014). In plants, stress-induced volatiles serve as infochemicals eliciting stress response pathways, leading to plant acclimation to the altered environmental conditions and priming defenses against herbivore and pathogen attacks in leaves and neighboring not yet stressed plants (Dicke and Baldwin 2010; Peng et al. 2011). Such a defense priming can again augment the whole ecosystem resistance to both abiotic and biotic stresses. Furthermore, herbivory-induced volatiles serve as infochemicals for the enemies of herbivores, and thus, the release of attractants to predatory and parasitic insects can importantly reduce the spread of herbivores (D'Alessandro et al. 2006; Dicke and Baldwin 2010) and, thus, provides a further important means for enhancing the resilience of ecosystems.

3.2 Plant Volatiles in Broad-Scale Ecological Processes

Apart from the biological role of plant-emitted volatiles, plant-generated volatiles play important roles in large-scale regional and global processes. The global amount of emitted biogenic volatiles has been estimated to be roughly 1.1 Pg yr^{-1} (equivalent to ca. 0.84 Pg C yr^{-1}) (Guenther et al. 2012). Thus, the biogenic release of trace gases exceeds the anthropogenic release by more than a factor of ten (Guenther et al. 2012). Despite only certain species are capable of constitutive isoprenoid emissions, many of these species are widely distributed, often being the dominating plant species in given ecosystems. In fact, global plant emissions are currently derived using species-specific emission potentials obtained by extensive screening studies (e.g., Karlik and Winer 2001; Simon et al. 2006; Keenan et al. 2009; Llusià et al. 2010, 2014). The emphasis in these screening studies has been on constitutive emissions, and thus, global emission estimates are mainly based on constitutive emitters (Guenther et al. 2012).

Indeed, constitutively emitted isoprene is the most important plant-generated volatile compound with the global source strength predicted to be ca. 550 Tg yr^{-1} by different models (Arneth et al. 2008; Guenther et al. 2012; Guenther 2013b). While different global emission models based on profoundly different emission algorithms converge to a similar value of global isoprene emission, due to uncertainties in the share of storage vs. de novo emissions and constitutive vs. stress-induced emissions, the model estimates are more variable for mono- and sesquiterpenes than for isoprene (Arneth et al. 2008). Based on empirical model approaches, total emissions of ca. 160 Tg yr^{-1} for mono- and ca. 30 Tg yr^{-1} for sesquiterpenes have been estimated (Guenther et al. 2012). The rest of global BVOC emission source strength of ca. 0.4 Pg yr^{-1} is mainly made up of oxygenated compounds, dominated by methanol, ethanol, acetone, and acetaldehyde (Guenther et al. 2012).

The largest uncertainty in the global volatile emission estimates seems to be the lack of proper consideration of stress-induced emissions. As shown above, exposure to stress conditions can turn a constitutively non-emitter or moderate emitter species into a strong emitter of mono- and sesquiterpenes. This might mean that the overall capacity of vegetation to emit volatiles has been strongly underestimated. Of course, the stress-elicited emissions occur only when there is a stress and emissions return to background levels when the stress is relieved (e.g., Fig. 2), but it is relevant to consider that these relatively short-termed emission peaks might not only importantly contribute to the total emission amount, but alter the timing of peak atmospheric concentrations of volatiles with major consequences for large-scale physiological and atmospheric processes.

3.2.1 Role of Volatiles in Altering Atmospheric Reactivity

Due to the large emissions, biogenic volatiles play major roles in biosphere-atmosphere processes. The chemical reactivity of non-oxygenated non-saturated compounds, in particular, non-oxygenated isoprenoids, is much larger than the reactivity of oxygenated volatiles, and thus, the role of different compound classes, oxygenated vs. non-oxygenated non-saturated compounds, in large-scale processes is different. Highly reactive compounds play a major role in ozone formation and quenching reactions in the troposphere. In particular, in a human-polluted air enriched with NO_x (sum of NO and NO_2), plant-generated volatiles contribute to photochemical ozone production and, in fact, control the rate of ozone formation in the atmosphere (Chameides et al. 1992; Fehsenfeld et al. 1992; Fuentes et al. 2000). In contrast, in non-polluted air with low NO_x, reactive hydrocarbons contribute to a reduction of ozone concentrations (Lerdau and Slobodkin 2002; Atkinson and Arey 2003; Loreto and Fares 2007).

The influence of elevated ozone concentration on photosynthetic productivity and volatile emissions can be different over the short- and the long-term. Increases in ozone concentration can strongly reduce plant photosynthetic productivity, but constitutive volatile emission itself can initially increase under moderately elevated atmospheric ozone concentration (Calfapietra et al. 2013 for a review). This might have global consequences as a reduction in photosynthetic CO_2 fixation can speed up elevation of atmospheric CO_2 concentration, and this in turn can increase the rate of temperature increase (Sitch et al. 2007). Due to the positive effect of temperature on volatile emissions, this is further expected to enhance volatile release and ozone formation and reduce carbon gain even more (Lerdau 2007; Sitch et al. 2007). Furthermore, the concentration of ozone, significantly driven by the concentration of reactive volatiles, is itself an important greenhouse gas that can contribute to global warming (Shindell et al. 2006), amplifying these patterns.

A severe ozone stress results in a reduction of constitutive volatile emissions as well (Calfapietra et al. 2013 for a review), but it also leads to elicitation of induced emissions in both constitutive emitters and non-emitters (Beauchamp et al. 2005; Hartikainen et al. 2009). Among the induced volatiles, mono- and sesquiterpenes are typically more reactive in ozone formation reactions than isoprene (Calogirou et al. 1996) and, thus, could temporarily even speed up ozone formation, especially because these emissions are induced in all species in vegetation. So far, the under-standing of ozone-dependent modifications in constitutive and induced emissions is only rudimentary, limiting quantitative assessment of species physiological responses in ozone formation potential of vegetation. Nevertheless, the available evidence suggests that the use of static emission factors estimated from screening studies that have considered only constitutive emitters and omission of physiologi-cal modifications in volatile emissions driven by ozone can lead to major uncertain-ties in predicting tropospheric ozone formation and quenching.

3.2.2 Volatiles in Altering Solar Radiation Scattering and Penetration and Ambient Temperature

Upon oxidation, the volatility of isoprenoids dramatically decreases, implying that they partition much more strongly to the liquid and solid phases than to the gas phase, creating secondary organic aerosols (SOA)(Kulmala et al. 2004a; Chen and Hopke 2009; Mentel et al. 2009; Kirkby et al. 2016). The presence of SOA decreases atmospheric clearness, thereby potentially reducing solar radiation penetration, but also increasing light scattering and, thus, the diffuse to total solar radiation ratio (Fig. 5, Malm et al. 1994; Farquhar and Roderick 2003; Misson et al. 2005). Because diffuse radiation penetrates deeper into the plant canopies and results in a more uniform distribution of solar radiation (Cescatti and Niinemets 2004), increases in diffuse to total solar radiation ratio enhance vegetation productivity (Gu et al. 2002, 2003; Mercado et al. 2009). Due to strong light effects on constitutive de novo vola-tile emissions, greater canopy light interception is expected to directly enhance these emissions (Fig. 5). In fact, the emissions are expected to increase even more

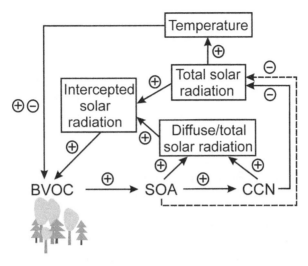

Fig. 5 Regional and global feedback relationships between constitutive biogenic volatile organic compound (BVOC) emissions and climatic drivers. The rates of BVOC emission are controlled by environmental drivers, whereas environmental drivers, in turn, are modified by BVOC emission generating feedback loops. There are two main types of BVOC emitters, the storage emitters and the de novo emitters (Grote et al. 2013; Copolovici and Niinemets 2016). In the storage emitters, BVOC is released from large storage pools, and the emissions depend only on temperature, increasing exponentially with increasing temperature (Niinemets et al. 2010b; Grote et al. 2013). In the case of de novo emitters, the emissions rely on immediately synthesized BVOCs and depend both on light intensity and temperature. These emissions increase asymptotically with increasing light intensity and increase exponentially with temperature up to an optimum temperature and decline thereafter (Niinemets et al. 2010b; Grote et al. 2013). For simplicity, the figure shows only the environmental controls on BVOC emission for de novo emitters. In the case of light, light interception by plant canopies depends on total solar radiation flux and the distribution of solar radiation between diffuse and direct components. Diffuse light drives canopy photosynthesis more efficiently because of its penetration to deeper canopy layers and resulting greater uniformity of radiation field and total light interception (Gu et al. 2003). Atmospheric volatiles enhance concentration of secondary organic aerosols (SOA) that increase the radiation scattering, but SOA also moderately reduce total radiation penetration through atmosphere (Spracklen et al. 2008; Chen and Hopke 2009; Kulmala et al. 2013). Increasing concentrations of SOA, in turn, enhance the concentration of cloud condensation nuclei (CCN) thereby contributing to enhanced cloudiness and thus, increasing radiation scattering too, but CNN more strongly reduce total solar radiation flux penetrating through the atmosphere to the vegetation (Roderick et al. 2001; Huff Hartz et al. 2005; Spracklen et al. 2008; Kulmala et al. 2013). The other key environmental driver, temperature, depends on total solar radiation flux, but can also directly depend on CCN as the result of reduced thermal radiation losses to the space, but this effect is more relevant at night and is therefore not shown in the figure. In addition to the feedbacks shown, more complex feedbacks can operate through modifications in atmospheric CO_2 and water vapor concentrations (e.g., Kulmala et al. 2013) as well as through induction of BVOC emissions upon abiotic and biotic stresses, severity of which depends on environmental drivers (section "Constitutive and Stress-Induced Volatile Emissions")

than the rate of photosynthesis, because de novo volatile emissions are more light-limited than photosynthetic carbon fixation, typically saturating at higher light intensities than photosynthesis (Niinemets et al. 1999, 2015). Although the connection between photosynthesis and storage emissions is less straightforward, enhanced carbon availability can also enhance the emissions in the storage emitters by increasing the size of the storage pools (Blanch et al. 2007, 2009, 2011).

Once formed in the atmosphere, the size of SOA particles increases in time due to condensation of atmospheric organics on particle surface (Kulmala et al. 2004a; Kirkby et al. 2016). These bigger particles can also serve as cloud condensation nuclei (CCN), especially if their hygroscopicity also increases as the result of further compound oxidations on particle surface or as more hydrophilic compounds condense onto the particle surface (Engelhart et al. 2008; Kulmala et al. 2013). Greater CCN concentrations imply a greater condensation sink and potentially higher cloudiness that can dramatically increase the diffuse to total solar radiation ratio but also strongly reduce the total solar radiation penetration (Spracklen et al. 2008; Still et al. 2009; Kulmala et al. 2013). Thus, although radiation penetration into deeper canopy layers is increased by enhanced cloudiness, the reduction in total radiation intensity reaching to the top of the vegetation is the dominating factor, ultimately reducing the vegetation productivity and the rate of volatile emission (Fig. 5). Furthermore, cooling due to increased cloudiness also directly reduces both de novo and storage volatile emissions (Fig. 5).

As the result of volatile effects on SOA and CCN concentrations, multiple feedback loops operate between solar radiation, temperature, volatile emission, and productivity at regional and global scales, and the overall effect of volatiles on climate depends on the relative significance of these loops (Fig. 5, Kulmala et al. 2013). In particular, both rising SOA concentrations and greater cloudiness can reduce the surface temperature, and this can directly reduce volatile formation due to the physiological controls on the emission rates, and this, in turn, is expected to inhibit further SOA and CCN formation (Fig. 5, Kulmala et al. 2013). On the other hand, enhanced SOA concentrations increase the fraction of diffuse radiation, thereby increasing the rate of volatile release and further enhancing SOA and CCN formation (Fig. 5, Kulmala et al. 2013). In contrast, dramatic reductions in total solar radiation by enhanced cloudiness are expected to lead to decreased volatile formation, thereby feedback-inhibiting SOA and CCN formation. Testing the quantitative significance of these feedback loops requires a combination of long-term data with regional- and global-scale modeling. The first such modeling exercise based on 15 years of measurements of vegetation carbon fixation fluxes and 6 years of measurements of emission fluxes of volatile organic compounds in a boreal forest ecosystem suggests that such feedback loops do indeed operate in nature (Kulmala et al. 2013). Due to both physiological and acclimation responses of volatile emissions to changes in environmental drivers, quantitative prediction of the feedback responses is complex, and clearly, more such case studies are needed to scale up to other regions and the globe and to quantitatively evaluate the role of volatile emissions on solar radiation penetration and surface temperature.

3.2.3 Volatiles as Modifiers of the Lifetime of Greenhouse Gases

As demonstrated above, reactive plant volatiles in polluted atmospheres can exacerbate the plant abiotic stress due to enhancing the key atmospheric pollutant, ozone, concentrations, while reactive volatiles in non-polluted atmospheres can reduce ozone concentrations. Thus reactive plant hydrocarbons can alter the rate of global warming by changes in vegetation CO_2 fixation capacity and, thus, by long-term modifications in atmospheric CO_2 concentration. Furthermore, ozone itself is a strong greenhouse gas (section "Constitutive Emissions of Specialized Volatiles", Shindell et al. 2006). In addition, alteration in solar radiation intensity, diffuse to total solar radiation ratio, and ambient temperature due to modifications in SOA and CCN concentrations directly affect the vegetation carbon fixation capacity and, thus, can strongly alter atmospheric CO_2 concentration as well, creating further major feedback loops (Kulmala et al. 2004b, 2013).

Apart from the highly reactive compounds, less reactive oxygenated hydrophilic compounds such as aldehydes, alcohols, and ketones, including, for instance, lipoxygenase pathway volatiles, but also saturated oxygenated volatiles, can also partition to particle phase and participate in SOA and CCN formation (Mentel et al. 2009). In addition, several of these volatiles can react with atmospheric OH radicals, reducing atmospheric OH radical concentration (Fall 2003; Sinha et al. 2010; Nölscher et al. 2012). Given that the reaction with OH radicals is the primary process reducing the atmospheric concentrations of the key greenhouse gas methane, a reduction of OH concentration due to biogenic volatiles increases the methane lifetime in the atmosphere (Jacob et al. 2005; Ashworth et al. 2013; Voulgarakis et al. 2013), thereby significantly contributing to global warming. In particular, plant-generated emissions of methanol, the oxygenated volatile with the greatest global source strength (Stavrakou et al. 2011; Guenther et al. 2012), can potentially contribute to the greatest degree to the increases in methane lifetime.

3.3 Trace-Gas-Driven Ecosystem Services

The vegetation capacity to emit volatiles has not generally been considered as an ecosystem service, and overall, the atmosphere is often not considered as part of ecosystem services (Cooter et al. 2013). In fact, due to the contribution of volatiles to ozone formation in NO_x-polluted atmospheres, volatile emission has even been considered an ecosystem "disservice" (Russo et al. 2016). However, from a biological perspective, plants can provide several key ecosystem services due to their trace gas emission. Although these ecosystem services are little-recognized in the community, they can have profound impacts on ecosystem performance. Among these biological services are:

- Direct enhancement of plant abiotic stress resistance by directly quenching reactive oxidative species generated in plant membranes upon abiotic stress (Vickers et al. 2009) or more specifically, improving heat stress tolerance by increasing membrane stability (Velikova et al. 2012; Sun et al. 2013)

- Increases in ecosystem resilience through communication among plants and with other organisms (e.g., plants and herbivore enemies), thereby ameliorating the biotic and abiotic stress impacts
- Enhancement of ecosystem capacity to uphold diversity, in particular, to maintain the integrity of multitrophic interactions

Although often "hidden," these services are crucial for stability and performance of ecosystems, and impairing some of these services can result in drastic deterioration of other ecosystem services. For instance, the carbon gain of chemically less diverse ecosystems such as monospecific tree plantations or crop fields can be much more vulnerable to deleterious pest attacks than that of more diverse ecosystems (Lerdau and Slobodkin 2002; Altieri and Nicholls 2004; Tooker and Frank 2012). This is highly relevant from the perspective of the "traditionally" perceived ecosystem services as pulpwood production of tree plantations or yield of crop fields, both of which are directly dependent on vegetation carbon gain.

From the perspective of atmospheric chemistry and large-scale biosphere-atmosphere processes, the balance between ecosystem service and "disservice" of volatile emissions importantly depends on the relative extent to which different atmospheric processes are affected by volatiles. The balance between different processes strongly varies regionally and depends on human effects on atmosphere. While in urban NO_x-polluted atmospheres, volatiles emitted by vegetation contribute to elevated ozone concentration and photochemical smog, in remote non-polluted atmospheres with low NO_x levels, plant-generated volatiles are expected to reduce atmospheric oxidant concentrations, including ozone concentrations (Lerdau and Slobodkin 2002). Analogously, reduction in atmospheric clearness and alteration of ambient air particle concentration by volatile contribution to SOA formation could be considered a disservice in urban habitats (Cooter et al. 2013), although it might also contribute to cooling of urban environments (Arneth et al. 2009).

From a global perspective, plant volatiles can provide three key ecosystem services:

- Improvement of ecosystem capacity to fix carbon by altering diffuse/direct radiation due to SOA formation
- Cooling of environment through production of SOA and CCN
- Reduction of the rate of global climate change through improved carbon gain and reduced transmission of solar radiation

The ultimate significance of these services depends on the relative magnitude of different environmental changes and volatile emissions as connected through multiple feedback loops (Fig. 5). It is, furthermore, highly likely that globally changing environmental drivers and CO_2 concentration alter the quantitative significance of the feedback loops due to the modifications in plant stress status, carbon fixation, and trace gas release as discussed in the next section.

3.4   Plant Stress, Volatile Emissions, and Trace-Gas-Dependent Ecosystem Services in Changing Climates

Climate change is expected to result in more severe heat stress worldwide (Kirtman et al. 2013; Field et al. 2014). In addition, climate change alters the distribution of precipitation with some areas predicted to become drier and some areas wetter (Kirtman et al. 2013; Field et al. 2014). Thus, the overall abiotic stress level is expected to increase in the future, but prediction of how enhanced stress level modifies volatile emission and ecosystem services is complicated. Increases in temperature can initially result in enhanced emissions of constitutive de novo emissions until the physiological optimum is exceeded (Loreto et al. 1998; Staudt and Bertin 1998; Rasulov et al. 2015). Beyond the physiological optimum, the rate of constitutive emissions decreases, but the stress volatile emissions are induced (Staudt and Bertin 1998; Kleist et al. 2012). These emissions under heat stress have been detected at the ecosystem scale (Karl et al. 2008), demonstrating that the stress emissions do contribute to large-scale atmospheric processes. Given that the stress emissions are induced in all plant species, more frequent heat waves in future climates can strongly enhance the overall release of volatiles. However, the quantitative information on the kinetics of elicitation of emissions under heat stress and interspecific variability of the capacity for heat-dependent release of stress volatiles is currently very limited, hindering scaling up from case studies to whole ecosystems, regions, and globe.

Global changes in environmental drivers have to be tempered in light of simultaneous modifications of ambient CO_2 concentrations. Elevation of atmospheric CO_2 concentration itself can improve plant carbon gain in drier climates due to reduction of the diffusion limitations on the CO_2 pathway from the ambient air through stomata and mesophyll to the chloroplasts where photosynthesis takes place (Niinemets et al. 2011; Flexas et al. 2016). In addition, elevated CO_2 can protect leaves from the heat stress, possibly by increasing leaf sugar concentrations that enhance the heat stability of cell and chloroplast membranes (Darbah et al. 2010; Sun et al. 2013). In the case of constitutive isoprene emissions, however, several studies have demonstrated that the capacity for isoprene emission decreases with a long-term increase in ambient CO_2 concentration similarly to the response of isoprene emissions to rapid changes in CO_2 concentration (Niinemets et al. 2010a; Possell and Hewitt 2011). Such an acclimatory response would mean that the elevation of CO_2 concentration impairs the heat stress protection by isoprene. However, in other studies, plants grown under elevated CO_2 had greater isoprene emission potential and improved heat stress resistance (Peñuelas and Staudt 2010; Sun et al. 2012, 2013). These controversial responses are evident when comparing model projections of emissions, which diverge greatly between models under future climate change (Keenan et al. 2009). This implies that prediction of constitutive emissions in future atmospheres is subject to large uncertainties and calls for more work on acclimation responses of constitutive emissions.

Apart from abiotic stresses, global environmental change is predicted to result in increasingly more frequent and severe pest and pathogen attacks (DeLucia et al. 2008; Chakraborty 2013). Such a greater biotic stress pressure can result from a reduction of constitutive defenses of vegetation due to reduced photosynthetic carbon availability under more severe abiotic stresses, shorter life cycles of current pests, and pathogens in hotter climates as well as invasion of new pests and pathogens facilitated by global trade and travel (Fig. 6, Vanhanen 2008; Huang et al. 2010; Gutierrez and Ponti 2014). Although biotic stress itself typically elicits volatile emissions for a relatively short period of time (Fig. 2), as the result of greater biotic stress pressure, the frequency of multiple sequential and simultaneous biotic attacks is likely to increase (Fig. 6). Thus, in the future climates, the emissions from biotic stresses are expected to continue longer and contribute to a larger extent to the overall emission of plant-produced volatiles.

The available evidence collectively suggests that global change enhances emissions induced by both abiotic and biotic stresses and might reduce constitutive emissions. Given that induced emissions occur in all plant species, vegetation trace

Fig. 6 Illustration of single and multiple biotic infections of *A. incana* leaves in the field. Mass infestations by the alder leaf beetle (*Agelastica alni*, **a**) and alder rust (*Melampsoridium hiratsukanum*, **b**, **c**) are frequently observed in European alder stands, and one might often also encounter combined infestations by both *A. alni* and *M. hiratsukanum* (**d**). In particular, the eastern Asian rust *M. hiratsukanum* that was first observed to lead to mass alder infestation in the mid-1990s in northern Europe (Põldmaa 1997; Hantula et al. 2009) has spread over much of the Europe by now, and infestations involving all alder trees in a given stand are common. Typically, the signs of infestations, orange urediniospores on the lower leaf surface, are observed in late summer and ultimately result in premature leaf senescence, extensive necrosis, and the leaf drying out with characteristic inward rolling of leaf margins (**c**)

gas emissions are expected to increase. However, due to stochastic nature of stress events, the emission kinetics of induced emissions is inherently much less predictable than the kinetics of constitutive emissions (Arneth and Niinemets 2010). Such stochasticity poses a challenge for model approaches trying to evaluate the influence of climate change on the feedback loops between constitutive and stress-induced volatile emissions, SOA, CCN, plant photosynthetic production, diffuse/direct radiation ratio, and warming and elevated CO_2 (Fig. 5). Future experimental work should fill the gaps in quantitative understanding of how stress-elicited volatile emissions scale with the severity of different abiotic and biotic stresses, how stress and altered atmospheric CO_2 modify constitutive emissions, and what is the biological variability in these responses. Armed with this knowledge, the community can start targeting the key research questions on the extent to which plant trace gas release can reduce the effects of climate change on vegetation and the extent to which it can reduce the rate of climate change.

4 Concluding Perspectives

Several widespread plant species are strong constitutive emitters of volatile isoprenoids such as isoprene and monoterpenes, and all plants can be induced to release volatiles upon abiotic and biotic stresses. These emissions play a variety of biological and biogeochemical roles, overall improving directly or/and indirectly the stress resistance of vegetation and altering the ambient environment at local, regional, and global scales. From a local perspective, release of volatiles can be considered both ecosystem service or ecosystem disservice depending on the human impact on atmospheric composition. In atmospheres polluted with nitrogen mono-oxides, plant volatiles contribute to ozone formation in the atmosphere, and thus, volatile release adversely affects the environment. In clean atmospheres, however, plant volatiles reduce atmospheric ozone levels and thus contribute to atmospheric cleansing. Furthermore, by enhancing vegetation stress resistance, the volatiles contribute to the maintenance of ecosystem integrity and, thus, contribute to all the well-perceived ecosystem services such as the capacity of forest ecosystems to provide wood and agroecosystems to provide agricultural products. In addition to controlling atmospheric ozone levels, plant volatiles provide a number of other key regulating services of local to global importance. In particular, plant volatiles alter atmospheric clearness due to the effects of volatiles on atmospheric particle concentrations and cloudiness. Modifications in atmospheric clearness in turn alter the ratio of diffuse to total solar radiation and atmospheric temperature with ultimate impacts on global vegetation productivity, rate of change in atmospheric CO_2 concentration, and rate of global change. In future hotter more stressful environments, stress-induced volatile emissions can be particularly relevant in driving the global feedbacks between volatile production, modifications in atmospheric oxidative status, clearness, and global change. Given that the role of volatiles in global coupled vegetation-climate models is still largely unaccounted, I argue that the plant/

atmosphere interface should be a high priority research target in future climate change studies.

Acknowledgments I thank Prof. Josep Peñuelas (Global Ecology Unit CREAF-CSIC-UAB, Barcelona, Catalonia) and Dr. Trevor F. Keenan (Earth and Environmental Sciences, Lawrence Berkeley National Lab, USA) for insightful comments on the MS. My work on plant volatiles has been supported by the Estonian Ministry of Science and Education (institutional grant IUT-8-3) and the European Commission through the European Research Council (advanced grant 322603, SIP-VOL+) and the European Regional Development Fund (Centre of Excellence EcolChange, TK 131).

References

Achotegui-Castells A, Danti R, Llusià J, Della Rocca G, Barberini S, Peñuelas J (2015) Strong induction of minor terpenes in Italian cypress, *Cupressus sempervirens*, in response to infection by the fungus *Seiridium cardinale*. J Chem Ecol 41:224–243

Altieri M, Nicholls C (2004) Biodiversity and pest management in agroecosystems, 2nd edn. Food Products Press, New York

Andreou A, Feussner I (2009) Lipoxygenases – structure and reaction mechanism. Phytochemistry 70:1504–1510

Arneth A, Niinemets Ü (2010) Induced BVOCs: how to bug our models? Trends Plant Sci 15:118–125

Arneth A, Monson RK, Schurgers G, Niinemets Ü, Palmer PI (2008) Why are estimates of global isoprene emissions so similar (and why is this not so for monoterpenes)? Atmos Chem Phys 8:4605–4620

Arneth A, Unger N, Kulmala M, Andreae MO (2009) Clean the air, heat the planet? Science 326:672–673

Ashworth K, Boissard C, Folberth G, Lathière J, Schurgers G (2013) Global modeling of volatile organic compound emissions. In: Niinemets Ü, Monson RK (eds) Biology, controls and models of tree volatile organic compound emissions, Tree physiology, vol 5. Springer, Berlin, pp 451–487

Atkinson R, Arey J (2003) Gas-phase tropospheric chemistry of biogenic volatile organic compounds: a review. Atmos Environ 37:197–219

Beauchamp J, Wisthaler A, Hansel A, Kleist E, Miebach M, Niinemets Ü, Schurr U, Wildt J (2005) Ozone induced emissions of biogenic VOC from tobacco: relations between ozone uptake and emission of LOX products. Plant Cell Environ 28:1334–1343

Blanch J-S, Peñuelas J, Llusià J (2007) Sensitivity of terpene emissions to drought and fertilization in terpene-storing *Pinus halepensis* and non-storing *Quercus ilex*. Physiol Plant 131:211–225

Blanch J-S, Peñuelas J, Sardans J, Llusià J (2009) Drought, warming and soil fertilization effects on leaf volatile terpene concentrations in *Pinus halepensis* and *Quercus ilex*. Acta Physiol Plant 31:207–218

Blanch J-S, Llusià J, Niinemets Ü, Noe SM, Peñuelas J (2011) Instantaneous and historical temperature effects on α-pinene emissions in *Pinus halepensis* and *Quercus ilex*. J Environ Biol 32:1–6

Blande JD, Tiiva P, Oksanen E, Holopainen JK (2007) Emission of herbivore-induced volatile terpenoids from two hybrid aspen (*Populus tremula* x *tremuloides*) clones under ambient and elevated ozone concentrations in the field. Glob Chang Biol 13:2538–2550

Blande JD, Holopainen JK, Niinemets Ü (2014) Plant volatiles in polluted atmospheres: stress responses and signal degradation. Plant Cell Environ 37:1892–1904

Bracho Nunez A, Knothe N, Liberato MAR, Schebeske G, Ciccioli P, Piedade MTF, Kesselmeier J (2009) Flooding effects on plant physiology and VOC emissions from Amazonian tree species from two different flooding environments: Varzea and Igapo. Geophys Res Abstr 11:EGU2009–EGU1497

Calfapietra C, Pallozzi E, Lusini I, Velikova V (2013) Modification of BVOC emissions by changes in atmospheric [CO_2] and air pollution. In: Niinemets Ü, Monson RK (eds) Biology, controls and models of tree volatile organic compound emissions, Tree physiology, vol 5. Springer, Berlin, pp 253–284

Calogirou A, Larsen BR, Brussol C, Duane M, Kotzias D (1996) Decomposition of terpenes by ozone during sampling on Tenax. Anal Chem 68:1499–1506

Cescatti A, Niinemets Ü (2004) Sunlight capture. Leaf to landscape. In: Smith WK, Vogelmann TC, Chritchley C (eds) Photosynthetic adaptation. Chloroplast to landscape, Ecological studies, vol 178. Springer, Berlin, pp 42–85

Chakraborty S (2013) Migrate or evolve: options for plant pathogens under climate change. Glob Chang Biol 19:1985–2000

Chameides WL, Fehsenfeld F, Rodgers MO, Cardelino C, Martinez J, Parrish D, Lonneman W, Lawson DR, Rasmussen RA, Zimmerman P, Greenberg J, Middleton P, Wang T (1992) Ozone precursor relationships in the ambient atmosphere. J Geophys Res 97:6037–6055

Chen X, Hopke PK (2009) A chamber study of secondary organic aerosol formation by linalool ozonolysis. Atmos Environ 43:3935–3940

Chen F, Tholl D, Bohlmann J, Pichersky E (2011) The family of terpene synthases in plants: a mid-size family of genes for specialized metabolism that is highly diversified throughout the kingdom. Plant J 66:212–229

Christianson DW (2008) Unearthing the roots of the terpenome. Curr Opin Chem Biol 12:141–150

Cooter EJ, Rea A, Bruins R, Schwede D, Dennis R (2013) The role of the atmosphere in the provision of ecosystem services. Sci Total Environ 448:197–208

Copolovici LO, Niinemets Ü (2005) Temperature dependencies of Henry's law constants and octanol/water partition coefficients for key plant volatile monoterpenoids. Chemosphere 61:1390–1400

Copolovici L, Niinemets Ü (2016) Environmental impacts on plant volatile emission. In: Blande J, Glinwood R (eds) Deciphering chemical language of plant communication, Signaling and communication in plants. Springer International Publishing, Berlin, pp 35–59

Copolovici LO, Filella I, Llusià J, Niinemets Ü, Peñuelas J (2005) The capacity for thermal protection of photosynthetic electron transport varies for different monoterpenes in *Quercus ilex*. Plant Physiol 139:485–496

Copolovici L, Kännaste A, Remmel T, Vislap V, Niinemets Ü (2011) Volatile emissions from *Alnus glutinosa* induced by herbivory are quantitatively related to the extent of damage. J Chem Ecol 37:18–28

Copolovici L, Kännaste A, Pazouki L, Niinemets Ü (2012) Emissions of green leaf volatiles and terpenoids from *Solanum lycopersicum* are quantitatively related to the severity of cold and heat shock treatments. J Plant Physiol 169:664–672

Copolovici L, Kännaste A, Remmel T, Niinemets Ü (2014a) Volatile organic compound emissions from *Alnus glutinosa* under interacting drought and herbivory stresses. Environ Exp Bot 100:55–63

Copolovici L, Väärtnõu F, Portillo Estrada M, Niinemets Ü (2014b) Oak powdery mildew (*Erysiphe alphitoides*)-induced volatile emissions scale with the degree of infection in *Quercus robur*. Tree Physiol 34:1399–1410

D'Alessandro M, Held M, Triponez Y, Turlings TCJ (2006) The role of indole and other shikimic acid derived maize volatiles in the attraction of two parasitic wasps. J Chem Ecol 32:2733–2748

Darbah JNT, Sharkey TD, Calfapietra C, Karnosky DF (2010) Differential response of aspen and birch trees to heat stress under elevated carbon dioxide. Environ Pollut 158:1008–1014

DeLucia EH, Casteel CL, Nabity PD, O'Neill BF (2008) Insects take a bigger bite out of plants in a warmer, higher carbon dioxide world. Proc Natl Acad Sci U S A 105:1781–1782

Dicke M, Baldwin IT (2010) The evolutionary context for herbivore-induced plant volatiles: beyond the 'cry for help'. Trends Plant Sci 15:167–175

Dicke M, van Loon JJA, Soler R (2009) Chemical complexity of volatiles from plants induced by multiple attack. Nat Chem Biol 5:317–324

Dindorf T, Kuhn U, Ganzeveld L, Schebeske G, Ciccioli P, Holzke C, Köble R, Seufert G, Kesselmeier J (2006) Significant light and temperature dependent monoterpene emissions from European beech (*Fagus sylvatica* L.) and their potential impact on the European volatile organic compound budget. J Geophys Res Atmos 111:D16305

Dong L, Jongedijk E, Bouwmeester H, Van Der Krol A (2016) Monoterpene biosynthesis potential of plant subcellular compartments. New Phytol 209:679–690

Ehn M, Thornton JA, Kleist E, Sipilä M, Junninen H, Pullinen I, Springer M, Rubach F, Tillmann R, Lee B, Lopez-Hilfiker F, Andres S, Acir I-H, Rissanen M, Jokinen T, Schobesberger S, Kangasluoma J, Kontkanen J, Nieminen T, Kurtén T, Nielsen LB, Jørgensen S, Kjaergaard HG, Canagaratna M, Dal Maso M, Berndt T, Petäjä T, Wahner A, Kerminen V-M, Kulmala M, Worsnop DR, Wildt J, Mentel TF (2014) A large source of low-volatility secondary organic aerosol. Nature 506:476–479

Engelhart GJ, Asa-Awuku A, Nenes A, Pandis SN (2008) CCN activity and droplet growth kinetics of fresh and aged monoterpene secondary organic aerosol. Atmos Chem Phys 8:3937–3949

Falara V, Akhtar TA, Nguyen TTH, Spyropoulou EA, Bleeker PM, Schauvinhold I, Matsuba Y, Bonini ME, Schilmiller AL, Last RL, Schuurink RC, Pichersky E (2011) The tomato terpene synthase gene family. Plant Physiol 157:770–789

Fall R (2003) Abundant oxygenates in the atmosphere: a biochemical perspective. Chem Rev 103:4941–4952

Fall R, Benson AA (1996) Leaf methanol – the simplest natural product from plants. Trends Plant Sci 1:296–301

Farquhar GD, Roderick ML (2003) Pinatubo, diffuse light, and the carbon cycle. Science 299:1997–1998

Farré-Armengol G, Filella I, Llusià J, Primante C, Peñuelas J (2015) Enhanced emissions of floral volatiles by *Diplotaxis erucoides* (L.) in response to folivory and florivory by *Pieris brassicae* (L.) Biochem Syst Ecol 63:51–58

Fehsenfeld F, Calvert J, Fall R, Goldan P, Guenther AB, Hewitt CN, Lamb B, Liu S, Trainer M, Westberg H, Zimmerman P (1992) Emissions of volatile organic compounds from vegetation and the implications for atmospheric chemistry. Glob Biogeochem Cycles 6:389–430

Field CB, Barros VR, Dokken DJ, Mach KJ, Mastrandrea MD, Bilir TE, Chatterjee M, Ebi KL, Estrada YO, Genova RC, Girma B (eds) (2014) Climate change 2014: impacts, adaptation, and vulnerability. Part A: global and sectoral aspects. Contribution of Working Group II to the fifth assessment report of the Intergovernmental Panel on Climate Change. Cambridge University Press, Cambridge, UK/New York, USA

Fineschi S, Loreto F, Staudt M, Peñuelas J (2013) Diversification of volatile isoprenoid emissions from trees: evolutionary and ecological perspectives. In: Niinemets Ü, Monson RK (eds) Biology, controls and models of tree volatile organic compound emissions, Tree physiology, vol 5. Springer, Berlin, pp 1–20

Flexas J, Díaz-Espejo A, Conesa MA, Coopman R, Douthe C, Gago J, Gallé A, Galmés J, Medrano H, Ribas-Carbo M, Tomàs M, Niinemets Ü (2016) Mesophyll conductance to CO_2 and Rubisco as targets for improving intrinsic water use efficiency in C_3 plants. Plant Cell Environ 39:965–982

Fuentes JD, Lerdau M, Atkinson R, Baldocchi D, Bottenheim JW, Ciccioli P, Lamb B, Geron C, Gu L, Guenther A, Sharkey TD, Stockwell W (2000) Biogenic hydrocarbons in the atmospheric boundary layer: a review. Bull Am Meteorol Soc 81:1537–1575

Gray DW, Goldstein AH, Lerdau M (2006) Thermal history regulates methylbutenol basal emission rate in *Pinus ponderosa*. Plant Cell Environ 29:1298–1308

Gray DW, Breneman SR, Topper LA, Sharkey TD (2011) Biochemical characterization and homology modeling of methylbutenol synthase and implications for understanding hemiterpene synthase evolution in plants. J Biol Chem 286:20582–20590

Grote R, Monson RK, Niinemets Ü (2013) Leaf-level models of constitutive and stress-driven volatile organic compound emissions. In: Niinemets Ü, Monson RK (eds) Biology, controls and models of tree volatile organic compound emissions, Tree physiology, vol 5. Springer, Berlin, pp 315–355

Gu L, Baldocchi D, Verma SB, Black TA, Vesala T, Falge EM, Dowty PR (2002) Advantages of diffuse radiation for terrestrial ecosystem productivity. J Geophys Res 107. https://doi.org/10.1029/2001JD001242

Gu L, Baldocchi DD, Wofsy SC, Munger JW, Michalsky JJ, Urbanski SP, Boden TA (2003) Response of a deciduous forest to the Mount Pinatubo eruption: enhanced photosynthesis. Science 299:2035–2038

Guenther A (2013a) Biological and chemical diversity of biogenic volatile organic emissions into the atmosphere. ISRN Atmos Sci 2013:786290

Guenther A (2013b) Upscaling biogenic volatile compound emissions from leaves to landscapes. In: Niinemets Ü, Monson RK (eds) Biology, controls and models of tree volatile organic compound emissions, Tree physiology, vol 5. Springer, Berlin, pp 391–414

Guenther AB, Jiang X, Heald CL, Sakulyanontvittaya T, Duhl T, Emmons LK, Wang X (2012) The model of emissions of gases and aerosols from nature version 2.1 (MEGAN2.1): an extended and updated framework for modeling biogenic emissions. Geosci Model Dev 5:1471–1492

Gutierrez AP, Ponti L (2014) Analysis of invasive insects: links to climate change. In: Ziska LH, Dukes JS (eds) Invasive species and global climate change. CABI Publishing, Wallingford, pp 45–61

Hantula J, Kurkela T, Hendry S, Yamaguchi T (2009) Morphological measurements and ITS sequences show that the new alder rust in Europe is conspecific with *Melampsoridium hiratsukanum* in eastern Asia. Mycologia 101:622–631

Harley P, Greenberg J, Niinemets Ü, Guenther A (2007) Environmental controls over methanol emission from leaves. Biogeosciences 4:1083–1099

Hartikainen K, Nerg A-M, Kivimäenpää M, Kontunen-Sopplea S, Mäenpää M, Oksanen E, Rousi M, Holopainen T (2009) Emissions of volatile organic compounds and leaf structural characteristics of European aspen (*Populus tremula*) grown under elevated ozone and temperature. Tree Physiol 29:1163–1173

Helmig D, Revermann T, Pollmann J, Kaltschmidt O, Hernandez AJ, Bocquet F, David D (2003) Calibration system and analytical considerations for quantitative sesquiterpene measurements in air. J Chromatogr A 1002:193–211

Himanen SJ, Blande JD, Klemola T, Pulkkinen J, Heijari J, Holopainen JK (2010) Birch (*Betula* spp.) leaves adsorb and re-release volatiles specific to neighbouring plants – a mechanism for associational herbivore resistance? New Phytol 186:722–732

Holopainen JK, Nerg A-M, Blande JD (2013) Multitrophic signalling in polluted atmospheres. In: Niinemets Ü, Monson RK (eds) Biology, controls and models of tree volatile organic compound emissions, Tree physiology, vol 5. Springer, Berlin, pp 285–314

Huang S-H, Cheng C-H, Wu W-J (2010) Possible impacts of climate change on rice insect pests and management tactics in Taiwan. Crop Environ Bioinform 7:269–279

Huff Hartz KE, Rosenørn T, Ferchak SR, Raymond TM, Bilde M, Donahue NM, Pandis SN (2005) Cloud condensation nuclei activation of monoterpene and sesquiterpene secondary organic aerosol. J Geophys Res Atmos 110:D14208, https://doi.org/10.11029/12004JD005754

Hüve K, Christ MM, Kleist E, Uerlings R, Niinemets Ü, Walter A, Wildt J (2007) Simultaneous growth and emission measurements demonstrate an interactive control of methanol release by leaf expansion and stomata. J Exp Bot 58:1783–1793

Jacob DJ, Field BD, Li Q, Blake DR, de Gouw J, Warneke C, Hansel A, Wisthaler A, Singh HB, Guenther A (2005) Global budget of methanol: constraints from atmospheric observations. J Geophys Res Atmos 110:D08303, https://doi.org/10.01029/02004JD005172

Jiang Y, Ye J, Li S, Niinemets Ü (2016a) Regulation of floral terpenoid emission and biosynthesis in sweet basil (*Ocimum basilicum*). J Plant Growth Regul 35:921–935

Jiang Y, Ye J, Veromann L-L, Niinemets Ü (2016b) Scaling of photosynthesis and constitutive and induced volatile emissions with severity of leaf infection by rust fungus (*Melampsora laricipopulina*) in *Populus balsamifera* var. *suaveolens*. Tree Physiol 38:856–872

Karl T, Guenther A, Turnipseed A, Patton EG, Jardine K (2008) Chemical sensing of plant stress at the ecosystem scale. Biogeosciences 5:1287–1294

Karlik JF, Winer AM (2001) Measured isoprene emission rates of plants in California landscapes: comparison to estimates from taxonomic relationships. Atmos Environ 35:1123–1131

Kask K, Kännaste A, Talts E, Copolovici L, Niinemets Ü (2016) How specialized volatiles respond to chronic and short-term physiological and shock heat stress in *Brassica nigra*. Plant Cell Environ 39:2027–2042

Keenan T, Niinemets Ü, Sabate S, Gracia C, Peñuelas J (2009) Process based inventory of isoprenoid emissions from European forests: model comparisons, current knowledge and uncertainties. Atmos Chem Phys 9:4053–4076

Kirkby J, Duplissy J, Sengupta K, Frege C, Gordon H, Williamson C, Heinritzi M, Simon M, Yan C, Almeida J, Tröstl J, Nieminen T, Ortega IK, Wagner R, Adamov A, Amorim A, Bernhammer A-K, Bianchi F, Breitenlechner M, Brilke S, Chen X, Craven J, Dias A, Ehrhart S, Flagan RC, Franchin A, Fuchs C, Guida R, Hakala J, Hoyle CR, Jokinen T, Junninen H, Kangasluoma J, Kim J, Krapf M, Kürten A, Laaksonen A, Lehtipalo K, Makhmutov V, Mathot S, Molteni U, Onnela A, Peräkylä O, Piel F, Petäjä T, Praplan AP, Pringle K, Rap A, Richards NAD, Riipinen I, Rissanen MP, Rondo L, Sarnela N, Schobesberger S, Scott CE, Seinfeld JH, Sipilä M, Steiner G, Stozhkov Y, Stratmann F, Tomé A, Virtanen A, Vogel AL, Wagner AC, Wagner PE, Weingartner E, Wimmer D, Winkler PM, Ye P, Zhang X, Hansel A, Dommen J, Donahue NM, Worsnop DR, Baltensperger U, Kulmala M, Carslaw KS, Curtius J (2016) Ion-induced nucleation of pure biogenic particles. Nature 533:521–526

Kirtman B, Power SB, Adedoyin JA, Boer GJ, Bojariu R, Camilloni I, Doblas-Reyes FJ, Fiore AM, Kimoto M, Meehl GA, Prather M, Sarr A, Schär C, Sutton R, van Oldenborgh GJ, Vecchi G, Wang HJ (2013) Near-term climate change: projections and predictability. In: Stocker TF et al (eds) Climate change 2013: the physical science basis. Contribution of Working Group I to the fifth assessment report of the Intergovernmental Panel on Climate Change. Cambridge University Press, Cambridge, UK/New York, USA, pp 953–1028

Kleist E, Mentel TF, Andres S, Bohne A, Folkers A, Kiendler-Scharr A, Rudich Y, Springer M, Tillmann R, Wildt J (2012) Irreversible impacts of heat on the emissions of monoterpenes, sesquiterpenes, phenolic BVOC and green leaf volatiles from several tree species. Biogeosciences 9:5111–5123

Kosina J, Dewulf J, Viden I, Pokorska O, Van Langenhove H (2013) Dynamic capillary diffusion system for monoterpene and sesquiterpene calibration: quantitative measurement and determination of physical properties. Int J Environ Anal Chem 93:637–649

Kreuzwieser J, Rennenberg H (2013) Flooding-driven emissions from trees. In: Niinemets Ü, Monson RK (eds) Biology, controls and models of tree volatile organic compound emissions, Tree physiology, vol 5. Springer, Berlin, pp 237–252

Kreuzwieser J, Kühnemann F, Martis A, Rennenberg H, Urban W (2000) Diurnal pattern of acetaldehyde emission by flooded poplar trees. Physiol Plant 108:79–86

Kreuzwieser J, Harren FJM, Laarhoven LJJ, Boamfa I, te Lintel HS, Scheerer U, Huglin C, Rennenberg H (2001) Acetaldehyde emission by the leaves of trees – correlation with physiological and environmental parameters. Physiol Plant 113:41–49

Kulmala M, Laakso L, Lehtinen KEJ, Riipinen I, Dal Maso M, Anttila T, Kerminen V-M, Hõrrak U, Vana M, Tammet H (2004a) Initial steps of aerosol growth. Atmos Chem Phys 4:2553–2560

Kulmala M, Suni T, Lehtinen KEJ, Dal Maso M, Boy M, Reissell A, Rannik Ü, Aalto P, Keronen P, Hakola H, Bäck J, Hoffmann T, Vesala T, Hari P (2004b) A new feedback mechanism linking forests, aerosols, and climate. Atmos Chem Phys 4:557–562

Kulmala M, Nieminen T, Chellapermal R, Makkonen R, Bäck J, Kerminen V-M (2013) Climate feedbacks linking the increasing atmospheric CO_2 concentration, BVOC emissions, aerosols and clouds in forest ecosystems. In: Niinemets Ü, Monson RK (eds) Biology, controls and models of tree volatile organic compound emissions, Tree physiology, vol 5. Springer, Berlin, pp 489–508

Lerdau M (2007) A positive feedback with negative consequences. Science 316:212–213

Lerdau M, Slobodkin K (2002) Trace gas emissions and species-dependent ecosystem services. Trends Ecol Evol 17:309–312

Li Z, Sharkey TD (2013) Molecular and pathway controls on biogenic volatile organic compound emissions. In: Niinemets Ü, Monson RK (eds) Biology, controls and models of tree volatile organic compound emissions, Tree physiology, vol 5. Springer, Berlin, pp 119–151

Liavonchanka A, Feussner N (2006) Lipoxygenases: occurrence, functions and catalysis. J Plant Physiol 163:348–357

Llusià J, Peñuelas J, Sardans J, Owen SM, Niinemets Ü (2010) Measurement of volatile terpene emissions in 70 dominant vascular plant species in Hawaii: aliens emit more than natives. Glob Ecol Biogeogr 19:863–874

Llusià J, Sardans J, Niinemets Ü, Owen SM, Peñuelas J (2014) A screening study of leaf terpene emissions of 43 rainforest species in Danum Valley Conservation Area (Borneo) and their relationships with chemical and morphological leaf traits. Plant Biosyst 148:307–317

Loreto F, Fares S (2007) Is ozone flux inside leaves only a damage indicator? Clues from volatile isoprenoid studies. Plant Physiol 143:1096–1100

Loreto F, Förster A, Dürr M, Csiky O, Seufert G (1998) On the monoterpene emission under heat stress and on the increased thermotolerance of leaves of *Quercus ilex* L. fumigated with selected monoterpenes. Plant Cell Environ 21:101–107

Loreto F, Nascetti P, Graverini A, Mannozzi M (2000) Emission and content of monoterpenes in intact and wounded needles of the Mediterranean pine, *Pinus pinea*. Funct Ecol 14:589–595

Loreto F, Pinelli P, Manes F, Kollist H (2004) Impact of ozone on monoterpene emissions and evidence for an isoprene-like antioxidant action of monoterpenes emitted by leaves. Tree Physiol 24:361–367

Malm WC, Gebhart KA, Molenar J, Cahill T, Eldred R, Huffman D (1994) Examining the relationship between atmospheric aerosols and light extinction at Mount Rainier and North Cascades National Parks. Atmos Environ 28:347–360

Mentel TF, Wildt J, Kiendler-Scharr A, Kleist E, Tillmann R, Dal Maso M, Fisseha R, Hohaus T, Spahn H, Uerlings R, Wegener R, Griffiths PT, Dinar E, Rudich Y, Wahner A (2009) Photochemical production of aerosols from real plant emissions. Atmos Chem Phys 9:4387–4406

Mercado LM, Bellouin N, Sitch S, Boucher O, Huntingford C, Wild M, Cox PM (2009) Impact of changes in diffuse radiation on the global land carbon sink. Nature 458:1014–1017

Micheli F (2001) Pectin methylesterases: cell wall enzymes with important roles in plant physiology. Trends Plant Sci 6:414–419

Misson L, Lunden M, McKay M, Goldstein AH (2005) Atmospheric aerosol light scattering and surface wetness influence the diurnal pattern of net ecosystem exchange in a semi-arid ponderosa pine plantation. Agric For Meteorol 129:69–83

Monson RK, Grote R, Niinemets Ü, Schnitzler J-P (2012) Tansley review. Modeling the isoprene emission rate from leaves. New Phytol 195:541–559

Nemecek-Marshall M, MacDonald RC, Franzen JJ, Wojciechowski CL, Fall R (1995) Methanol emission from leaves. Enzymatic detection of gas-phase methanol and relation of methanol fluxes to stomatal conductance and leaf development. Plant Physiol 108:1359–1368

Niinemets Ü, Tenhunen JD, Harley PC, Steinbrecher R (1999) A model of isoprene emission based on energetic requirements for isoprene synthesis and leaf photosynthetic properties for *Liquidambar* and *Quercus*. Plant Cell Environ 22:1319–1336

Niinemets Ü, Hauff K, Bertin N, Tenhunen JD, Steinbrecher R, Seufert G (2002a) Monoterpene emissions in relation to foliar photosynthetic and structural variables in Mediterranean evergreen *Quercus* species. New Phytol 153:243–256

Niinemets Ü, Reichstein M, Staudt M, Seufert G, Tenhunen JD (2002b) Stomatal constraints may affect emission of oxygenated monoterpenoids from the foliage of *Pinus pinea*. Plant Physiol 130:1371–1385

Niinemets Ü, Loreto F, Reichstein M (2004) Physiological and physicochemical controls on foliar volatile organic compound emissions. Trends Plant Sci 9:180–186

Niinemets Ü, Arneth A, Kuhn U, Monson RK, Peñuelas J, Staudt M (2010a) The emission factor of volatile isoprenoids: stress, acclimation, and developmental responses. Biogeosciences 7:2203–2223

Niinemets Ü, Monson RK, Arneth A, Ciccioli P, Kesselmeier J, Kuhn U, Noe SM, Peñuelas J, Staudt M (2010b) The leaf-level emission factor of volatile isoprenoids: caveats, model algorithms, response shapes and scaling. Biogeosciences 7:1809–1832

Niinemets Ü, Flexas J, Peñuelas J (2011) Evergreens favored by higher responsiveness to increased CO_2. Trends Ecol Evol 26:136–142

Niinemets Ü, Kännaste A, Copolovici L (2013) Quantitative patterns between plant volatile emissions induced by biotic stresses and the degree of damage. Front Plant Sci Front Plant-Microbe Interact 4:262

Niinemets Ü, Sun Z, Talts E (2015) Controls of the quantum yield and saturation light of isoprene emission in different-aged aspen leaves. Plant Cell Environ 38:2707–2720

Noe SM, Copolovici L, Niinemets Ü, Vaino E (2008) Foliar limonene uptake scales positively with leaf lipid content: "non-emitting" species absorb and release monoterpenes. Plant Biol 10:129–137

Nölscher AC, Williams J, Sinha V, Custer T, Song W, Johnson AM, Axinte R, Bozem H, Fischer H, Pouvesle N, Phillips G, Crowley JN, Rantala P, Rinne J, Kulmala M, Gonzales D, Valverde-Canossa J, Vogel A, Hoffmann T, Ouwersloot HG, Vilà-Guerau de Arellano J, Lelieveld J (2012) Summertime total OH reactivity measurements from boreal forest during HUMPPA-COPEC 2010. Atmos Chem Phys 12:8257–8270

Pazouki L, Niinemets Ü (2016) Multi-substrate terpenoid synthases: their occurrence and physiological significance. Front Plant Sci 7:1019

Pazouki L, Kanagendran A, Li S, Kännaste A, Rajabi Memari H, Bichele R, Niinemets Ü (2016) Mono- and sesquiterpene release from tomato (*Solanum lycopersicum*) leaves upon mild and severe heat stress and through recovery: from gene expression to emission responses. Environ Exp Bot 132:1–15

Peng J, van Loon JJ, Zheng S, Dicke M (2011) Herbivore-induced volatiles of cabbage (*Brassica oleracea*) prime defence responses in neighbouring intact plants. Plant Biol 13:276–284

Peñuelas J, Llusià J (2004) Plant VOC emissions: making use of the unavoidable. Trends Ecol Evol 19:402–404

Peñuelas J, Staudt M (2010) BVOCs and global change. Trends Plant Sci 15:133–144

Põldmaa K (1997) Explosion of *Melampsoridium* sp. on *Alnus incana*. Folia Cryptogamica Estonica 31:48–51

Portillo-Estrada M, Kazantsev T, Talts E, Tosens T, Niinemets Ü (2015) Emission timetable and quantitative patterns of wound-induced volatiles across different damage treatments in aspen (*Populus tremula*). J Chem Ecol 41:1105–1117

Possell M, Hewitt CN (2011) Isoprene emissions from plants are mediated by atmospheric CO_2 concentrations. Glob Chang Biol 17:1595–1610

Possell M, Loreto F (2013) The role of volatile organic compounds in plant resistance to abiotic stresses: responses and mechanisms. In: Niinemets Ü, Monson RK (eds) Biology, controls and models of tree volatile organic compound emissions, Tree physiology, vol 5. Springer, Berlin, pp 209–235

Rajabi Memari H, Pazouki L, Niinemets Ü (2013) The biochemistry and molecular biology of volatile messengers in trees. In: Niinemets Ü, Monson RK (eds) Biology, controls and models of tree volatile organic compound emissions, Tree physiology, vol 5. Springer, Berlin, pp 47–93

Rasulov B, Hüve K, Bichele I, Laisk A, Niinemets Ü (2010) Temperature response of isoprene emission in vivo reflects a combined effect of substrate limitations and isoprene synthase activity: a kinetic analysis. Plant Physiol 154:1558–1570

Rasulov B, Bichele I, Hüve K, Vislap V, Niinemets Ü (2015) Acclimation of isoprene emission and photosynthesis to growth temperature in hybrid aspen: resolving structural and physiological controls. Plant Cell Environ 38:751–766

Roderick ML, Farquhar GD, Berry SL, Noble IR (2001) On the direct effect of clouds and atmospheric particles on the productivity and structure of vegetation. Oecologia 129:21–30

Rosenkranz M, Schnitzler J-P (2013) Genetic engineering of BVOC emissions from trees. In: Niinemets Ü, Monson RK (eds) Biology, controls and models of tree volatile organic compound emissions, Tree physiology, vol 5. Springer, Berlin, pp 95–118

Rottenberger S, Kleiss B, Kuhn U, Wolf A, Piedade MTF, Junk W, Kesselmeier J (2008) The effect of flooding on the exchange of the volatile C_2-compounds ethanol, acetaldehyde and acetic acid between leaves of Amazonian floodplain tree species and the atmosphere. Biogeosciences 5:1085–1100

Russo A, Escobedo FJ, Zerbe S (2016) Quantifying the local-scale ecosystem services provided by urban treed streetscapes in Bolzano, Italy. AIMS Environ Sci 3:58–76

Sharkey TD, Chen XY, Yeh S (2001) Isoprene increases thermotolerance of fosmidomycin-fed leaves. Plant Physiol 125:2001–2006

Shindell D, Faluvegi G, Lacis A, Hansen J, Ruedy R, Aguilar E (2006) Role of tropospheric ozone increases in 20th-century climate change. J Geophys Res Atmos 111:D08302, 083. https://doi.org/10.01029/02005JD006348

Simon V, Dumergues L, Ponche J-L, Torres L (2006) The biogenic volatile organic compounds emission inventory in France. Application to plant ecosystems in the Berre-Marseilles area (France). Sci Total Environ 372:164–182

Sinha V, Williams J, Lelieveld J, Ruuskanen TM, Kajos MK, Patokoski J, Hellen H, Hakola H, Mogensen D, Boy M, Rinne J, Kulmala M (2010) OH reactivity measurements within a boreal forest: evidence for unknown reactive emissions. Environ Sci Technol 44:6614–6620

Sitch S, Cox PM, Collins WJ, Huntingford C (2007) Indirect radiative forcing of climate change through ozone effects on the land-carbon sink. Nature 448:791–794

Spracklen DV, Bonn B, Carslaw KS (2008) Boreal forests, aerosols and the impacts on clouds and climate. Philos Trans R Soc Lond A 366:4613–4626

Staudt M, Bertin N (1998) Light and temperature dependence of the emission of cyclic and acyclic monoterpenes from holm oak (*Quercus ilex* L.) leaves. Plant Cell Environ 21:385–395

Staudt M, Lhoutellier L (2011) Monoterpene and sesquiterpene emissions from *Quercus coccifera* exhibit interacting responses to light and temperature. Biogeosciences 8:2757–2771

Staudt M, Mir C, Joffre R, Rambal S, Bonin A, Landais D, Lumaret R (2004) Isoprenoid emissions of *Quercus* spp. (*Q. suber* and *Q. ilex*) in mixed stands contrasting in interspecific genetic introgression. New Phytol 163:573–584

Stavrakou T, Guenther A, Razavi A, Clarisse L, Clerbaux C, Coheur P-F, Hurtmans D, Karagulian F, De Maziére M, Vigouroux C, Amelynck C, Schoon N, Laffineur Q, Heinesch B, Aubinet M, Rinsland C, Müller J-F (2011) First space-based derivation of the global atmospheric methanol emission fluxes. Atmos Chem Phys 11:4873–4898

Still CJ, Riley WJ, Biraud SC, Noone DC, Buenning NH, Randerson JT, Torn MS, Welker JM, White JWC, Vachon R, Farquhar GD, Berry JA (2009) Influence of clouds and diffuse radiation on ecosystem-atmosphere CO_2 and $CO^{18}O$ exchanges. J Geophys Res Biogeosci 114:G01018

Stocker TF, Qin D, Plattner G-K, Tignor M, Allen SK, Boschung J, Nauels A, Xia Y, Bex V, Midgley PM (eds) (2014) Climate change 2014: the physical science basis. Contribution of Working Group I to the fifth assessment report of the Intergovernmental Panel on Climate Change. Cambridge University Press, Cambridge, UK/New York, USA

Sun Z, Niinemets Ü, Hüve K, Noe SM, Rasulov B, Copolovici L, Vislap V (2012) Enhanced isoprene emission capacity and altered light responsiveness in aspen grown under elevated atmospheric CO_2 concentration. Glob Chang Biol 18:3423–3440

Sun Z, Hüve K, Vislap V, Niinemets Ü (2013) Elevated [CO_2] magnifies isoprene emissions under heat and improves thermal resistance in hybrid aspen. J Exp Bot 64:5509–5523

Tholl D, Lee S (2011) Terpene specialized metabolism in *Arabidopsis thaliana*. Arabidopsis Book 9:e0143

Tholl D, Sohrabi R, Huh J-H, Lee S (2011) The biochemistry of homoterpenes – common constituents of floral and herbivore-induced plant volatile bouquets. Phytochemistry 72:1635–1646

Tooker JF, Frank SD (2012) Genotypically diverse cultivar mixtures for insect pest management and increased crop yields. J Appl Ecol 49:974–985

Vanhanen H (2008) Invasive insects in Europe – the role of climate change and global trade. Dissertationes Forestales 57. Faculty of Forest Sciences, University of Joensuu, 33 pages

Velikova V, Sharkey TD, Loreto F (2012) Stabilization of thylakoid membranes in isoprene-emitting plants reduces formation of reactive oxygen species. Plant Signal Behav 7:139–141

Vickers CE, Gershenzon J, Lerdau MT, Loreto F (2009) A unified mechanism of action for volatile isoprenoids in plant abiotic stress. Nat Chem Biol 5:283–291

von Dahl C, Hävecker M, Schlögl R, Baldwin IT (2006) Caterpillar-elicited methanol emission: a new signal in plant-herbivore interaction? Plant J 46:948–960

Voulgarakis A, Naik V, Lamarque J-F, Shindell DT, Young PJ, Prather MJ, Wild O, Field RD, Bergmann D, Cameron-Smith P, Cionni I, Collins WJ, Dalsøren SB, Doherty RM, Eyring V, Faluvegi G, Folberth GA, Horowitz LW, Josse B, MacKenzie IA, Nagashima T, Plummer DA, Righi M, Rumbold ST, Stevenson DS, Strode SA, Sudo K, Szopa S, Zeng G (2013) Analysis of present day and future OH and methane lifetime in the ACCMIP simulations. Atmos Chem Phys 13:2563–2587

Widegren JA, Bruno TJ (2010) Vapor pressure measurements on low-volatility terpenoid compounds by the concatenated gas saturation method. Environ Sci Technol 44:388–393

Wilkinson MJ, Monson RK, Trahan N, Lee S, Brown E, Jackson RB, Polley HW, Fay PA, Fall R (2009) Leaf isoprene emission rate as a function of atmospheric CO_2 concentration. Glob Chang Biol 15:1189–1200

Winters AJ, Adams MA, Bleby TM, Rennenberg H, Steigner D, Steinbrecher R, Kreuzwieser J (2009) Emissions of isoprene, monoterpene and short-chained carbonyl compounds from *Eucalyptus* spp. in southern Australia. Atmos Environ 43:3035–3043

Zhang P-J, Zheng S-J, van Loona JJA, Boland W, David A, Mumm R, Dicke M (2009) Whiteflies interfere with indirect plant defense against spider mites in lima bean. Proc Natl Acad Sci U S A 106:21202–21207

Towards Functional Green Infrastructure in the Baltic Sea Region: Knowledge Production and Learning Across Borders

Marine Elbakidze, Per Angelstam, Lucas Dawson, Alena Shushkova, Vladimir Naumov, Zigmārs Rendenieks, Liga Liepa, Laura Trasūne, Uladzimir Ustsin, Natalia Yurhenson, Siarhei Uhlianets, Michael Manton, Austra Irbe, Maxim Yermokhin, Aleksandra Grebenzshikova, Anton Zhivotov, and Marharyta Nestsiarenka

1 Introduction

Biodiversity, in terms of species, their habitats and ecological processes, forms the natural capital that ultimately provides multiple ecosystem services for human well-being. However, in spite of explicit and implicit policy visions on biodiversity conservation in Europe, the state of biodiversity is continuing to deteriorate (EC 2015). The loss of biodiversity has been more rapid in the past 50 years than at any time in human history (EC 2015). Up to 25% of Europe's animal species are facing extinction, 65% of habitats of importance in the European Union (EU) are in an unfavourable conservation status, and ecological processes have been modified (EC 2015).

To tackle the degradation, fragmentation and finally loss of representative ecosystems, there is a need to protect, manage and restore habitats for wildlife, ecosystem

M. Elbakidze (✉) · P. Angelstam · V. Naumov
School for Forest Management, Swedish University of Agricultural Sciences,
Skinnskatteberg, Sweden
e-mail: marine.elbakidze@slu.se

L. Dawson
Department of Physical Geography and Quaternary Geology, Stockholm University,
Stockholm, Sweden

A. Shushkova · U. Ustsin · N. Yurhenson
SSPA "The Scientific and Practical Center of the National Academy of Sciences of Belarus
for Biological Resources", Minsk, Belarus

Z. Rendenieks · L. Trasūne
Faculty of Geography and Earth Sciences, University of Latvia, Riga, Latvia

L. Liepa
Faculty of Forestry, Latvia University of Agriculture, Jelgava, Latvia

S. Uhlianets · M. Yermokhin · A. Zhivotov
State Scientific Institution "The Institute of Experimental Botany named after V.F. Kuprevich
of the National Academy of Sciences of Belarus", Minsk, Belarus

© Springer International Publishing AG, part of Springer Nature 2018
A. H. Perera et al. (eds.), *Ecosystem Services from Forest Landscapes*,
https://doi.org/10.1007/978-3-319-74515-2_4

services and human wellbeing. Green infrastructure (GI), an overarching policy concept that highlights the importance of natural capital for biodiversity and human wellbeing, is identified as one of the key policy priorities for the EU (European Commission 2013). GI is expected to make a significant contribution to providing ecological, economic, cultural and social benefits to human society through natural solutions. Likewise, scholars envision GI as a promising land management approach that is able to reconcile various interests of different stakeholder groups in obtaining multiple ecosystem services from landscapes, whilst simultaneously maintaining biodiversity (Ewers et al. 2009; Lafortezza et al. 2009, 2013).

Conceptually, GI evolved more than a century ago in the United Kingdom and the USA from approaches linking urban parks and other green space to benefit people and from biodiversity conservation measures to counteract habitat degradation, fragmentation and loss (Allen 2014; Benedict and McMahon 2002; Lafortezza et al. 2013). GI's long and diverse conceptual development trajectory has led to multiple definitions and interpretations (Benedict and McMahon 2002; Weber et al. 2006). In this chapter we use the definition of GI provided by the EU (2013) as a "strategically planned network of high quality natural and seminatural areas with other environmental features, which is designed and managed to deliver a wide range of ecosystem services and protect biodiversity in both rural and urban settings". Ecosystem services are the benefits that people obtain, directly or indirectly, from ecosystems (Costanza et al. 1997; Daily et al. 1997; MA 2005; Lele et al. 2013). These include provisioning, regulating, cultural and supporting services (MA 2005; TEEB 2010). Following Lele et al. (2013) and Huntsinger and Oviedo (2014), we also acknowledge that some ecosystem services are rather social-ecological services because the role of humans in the past and at present has had a considerable effect on composition, structure and functions of contemporary ecosystems. As Diaz et al. (2011) specified, social actors in many sectors and at multiple levels steer ecosystems by manipulating land covers, functional diversity and other ecosystem properties in order to obtain a particular ecosystem service. For example, the production and flow of multiple ecosystem services in both forest landscapes (Naumov et al. 2016) and cultural woodlands (Garrido et al. 2017a, b) are the direct result of interactions of land managers and ecosystems.

M. Manton
Faculty of Forest Science and Ecology, Aleksandras Stulginskis University, Kaunas, Lithuania

School for Forest Management, Swedish University of Agricultural Sciences, Skinnskatteberg, Sweden

State Environmental Institution National Park "Braslavskie Ozera", Braslav, Belarus

A. Irbe
Zemgale Planning Region Administration, Jelgava, Latvia

A. Grebenzshikova
Pskovlesproekt Company, Pskov, Russian Federation

M. Nestsiarenka
State Environmental Institution National Park "Braslavskie Ozera", Braslav, Belarus

GI is supposed to fulfil two main functions; one is related to biodiversity conservation, and thus the maintenance of natural capital, and the other to human wellbeing. GI is spatially explicit and should be closely linked to spatial planning at multiple spatial scales and levels of governance in rural and urban landscapes. Given that many ecosystem services depend on maintaining GI networks across national borders, functional GI requires cross-border cooperation and integration among both states and sectors in a region.

The Baltic Sea Region is a good example of current challenges in maintaining functional GI networks at multiple levels. A key factor for achieving ecological sustainability in the Baltic Sea Region is the cross-border coordination and integration of the efforts of relevant countries and among regional and local authorities. Handling eutrophication of the Baltic Sea, supplying resources towards a green economy and maintaining functional habitat networks for aquatic and terrestrial species demanding large areas are some examples. The EU Strategy for the Baltic Sea Region (European Commission 2009) highlights the need for reconciling economic, environmental and social objectives through the leading principles of sustainable development (Baker 2006), and its outputs and consequences (Rauschmayer 2009). This strategy aims at functional coordination and more efficient use of financial resources and existing cooperation schemes between Sweden, Finland, Estonia, Latvia, Lithuania, Poland, Germany and Denmark. The European Commission (2009) notes that the involvement of stakeholders needs to be strengthened at parliamentary, regional government and civil society levels. EU InterReg and other funding mechanisms for neighbourhood collaboration have adopted a broader geographical scope than the EU, and non-EU countries participate also actively in work with the Baltic Sea Region Cooperation. These include Belarus, Iceland, Norway and the Russian Federation. However, a major challenge is the diversity of governance systems, historical, social-economic and cultural legacies in the Baltic Sea Region. Therefore, pan-European, EU and national policies are likely to be perceived and implemented differently. An urgent need is thus to overcome multiple barriers among different stakeholders within the Baltic Sea Region in order to enhance knowledge production and learning towards functional GI networks at multiple levels. Forests are the main natural terrestrial ecosystems in the Baltic Sea Region. For this reason it is crucial to understand the extent to which management and governance of forest landscapes satisfies different dimensions of sustainable development by maintaining functional GI (e.g. Angelstam et al. 2011; Elbakidze et al. 2015).

The aim of this chapter is threefold. First, we present our methodology to integrative research, including both knowledge production and learning towards functional GI. Second, we summarise the results of applying this methodology through a transnational partnership among academic and nonacademic stakeholders in Belarus, Latvia, Russian Federation and Sweden. Third, we discuss the potential for cross-border learning and knowledge production to sustain a wide range of ecosystem services.

2   Integrative Research in Theory and Practise

2.1   The Baltic Sea Region as a Laboratory for Learning and Knowledge Production

The Baltic Sea Region represents a steep West-East gradient in Europe's land ownership and political culture. To capture this gradient, we selected four Pilot Areas representing four Baltic Sea Region countries, namely, Sweden, Latvia, Belarus and the Russian Federation (Fig. 1 and Photos 1, 2, and 3). The main criterion for the selection of each Pilot Area was the presence of a cluster of partners that expressed an interest and willingness to participate in integrative research.

Fig. 1 Location of Pilot Areas in the Baltic Sea Region: 1 Bergslagen region in Sweden. 2 Zemgale planning region in Latvia. 3 Braslav municipality in Belarus. 4 Strugi-Krasnye municipality in the Russian Federation

Photo 1 Lakes and forests as providers of multiple ecosystem services for human wellbeing are important components of green infrastructure in Sweden (Photo – Marine Elbakidze)

Photo 2 Kemeri National Park, Latvia. Protection and restoration of wetlands in Latvia is an important land management strategy towards the development of functional green infrastructure (Photo – Marine Elbakidze)

Photo 3 Cultural woodlands and lakes are the priority landscapes for urban and rural residents in Russia and Belarus (Photo – Marine Elbakidze)

2.2 Seven Steps for Knowledge Production and Learning

Two important bridging factors that can support the maintenance of representative and functional GI networks are knowledge production and learning (Angelstam et al. 2013a, 2017). The first is about producing evidence-based knowledge about ecological tipping points and "safe operating spaces for humanity" that define ecological sustainability (Rockström et al. 2009) and identifying measures required for managing, restoring and recreating habitats for wild species and humans, as well as ecosystem functions (e.g. Dawson et al. 2017). The second implies landscape stewardship to coordinate governance by integrating stakeholders from public, private and civil sectors at multiple levels (Elands and Wiersum 2001; Angelstam and Elbakidze 2017). The place-based integration of these two bridging factors has been termed a landscape approach (Axelsson et al. 2011, 2013; Sabogal et al. 2015). We followed a seven-step approach to integrate academic and nonacademic stakeholders in each Pilot Area as well as among them (Fig. 2) (Angelstam and Elbakidze 2017).

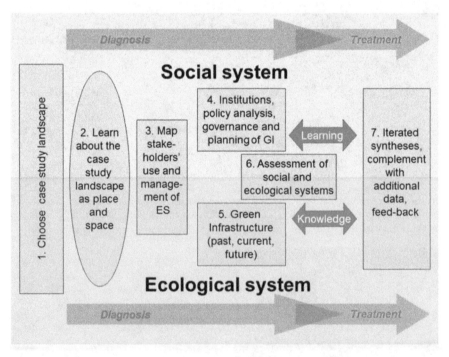

Fig. 2 Overview of integrative research, or knowledge production and learning, towards functional green infrastructure reported in this chapter. The place-based diagnoses of social and ecological systems form the base for treatment by learning through quantitative and qualitative data collection, iterated synthesis and feedback (See more in Angelstam and Elbakidze 2017)

3   Pilot Areas

3.1   *Bergslagen Region as the Pilot Area in Sweden*

The Bergslagen region (hereafter - Bergslagen) is an informal historic region in south central Sweden (Fig. 3 and Photo 1). It has a very long history of land use focusing on intensive sustained-yield forestry (Angelstam et al. 2013a, b, 2015) and ranges from the Mälardalen Valley's agricultural landscapes and urban centres to forested, remote rural upland areas and highlands in Dalarna County.

There are four main types of land ownership in Bergslagen: non-industrial private (57%), large forest companies (28%), the state forest company Sveaskog (10%) and public bodies such as municipalities, the National Property Board and forest commons (5%). Whilst natural resources such as wood, hydro-energy and minerals continue to serve as the basis for commodity production in rural landscapes, many municipalities are economically vulnerable (Tillväxtverket 2011), and the public and service sectors dominate the regional economy. Urbanisation has increased the disconnection between people and nature (Laird et al. 2010), and intensification and modernisation of natural resource use have resulted in depopulation of rural areas

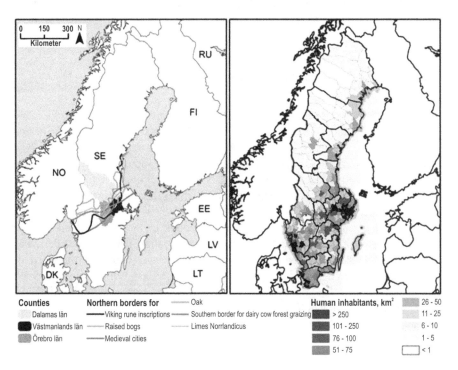

Fig. 3   The three counties, Örebro, Västmanland and Dalarna in Bergslagen, were chosen as a Pilot area in Sweden (left). The left map shows the steep biogeographic and cultural transition zone, and the right map demonstrates a steep urban-rural gradient within the study area

(Bryden and Hart 2004). At the same time, non-material values such as sense of place and inspiration are seen as increasingly important for the wellbeing of citizens.

3.2 Zemgale Planning Region as the Pilot Area in Latvia

Zemgale Planning Region (ZPR) is one of five highest-level planning units in Latvia. It covers the area of 073211 hectares, including 20 administrative districts, which corresponds to 17% of the country. The area consists mostly of flat lowlands with low hills in western and northeastern parts. Land covers include fertile agricultural land, forest, bogs and urban areas (Fig. 4 and Photo 2). The largest urban centres are Jelgava, Bauska, Jēkabpils and Dobele.

The total forested area within the ZPR is 446,774 ha, or 41% of the total area, and approximately 91% of the forest is used for wood production. There are more than 200 protected areas. Forest and agricultural lands differ greatly in ownership structure. Almost all agricultural lands were privatised after the restoration of state independence in 1991. However, ownership of Latvian forestland is divided between the State and private owners. In the ZPR 51% of forests are state-owned; the rest is owned by private owners and municipalities.

3.3 Braslav Municipality as the Pilot Area in Belarus

Braslav municipality is located in the northwestern part of the Vitebsk Region and covers an area of 22,000 ha (Fig. 5 and Photo 3). Most of the municipality is located within the Braslav ridge in the south, which is a part of the Polotsk lowland. The main part of the territory is occupied by a hilly moraine landscape. Agricultural land covers about 43% of the territory, forest and lakes about 35% and 10%, respectively. The district is crossed by important highways, which link the transport systems of Belarus and Latvia.

All forested and some agricultural land belongs to the state. Braslav Lakes National Park is the largest land user in the municipality. The National Park was established in 1995 and now occupies 64,500 ha (approximately one-third of Braslav municipality). The administration of the park manages the nature-protected area and all forest land of Braslav municipality and some agricultural lands. Agricultural production is the mainstay of the local economy, followed by forestry and tourism.

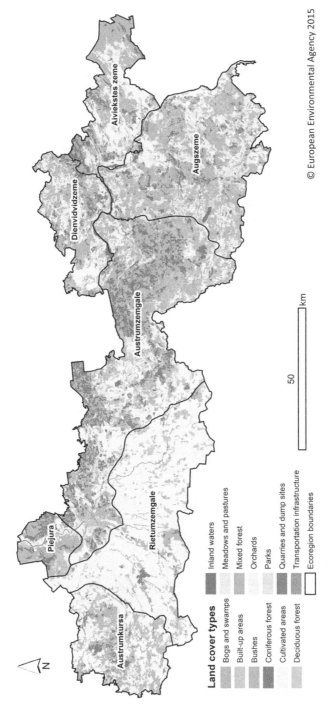

Fig. 4 Land cover types in Zemgale Planning Region (simplified CORINE Land Cover 2012 classification)

Fig. 5 Location of Bralsav municipality, the Belarusian Pilot Area

3.4 Strugi-Krasnye Municipality as the Pilot Area in the Russian Federation

Strugi-Krasnye municipality is located in Russia's westernmost Pskov Oblast within the Luga upland. The municipality covers an area of 316,400 ha (Fig. 6). Forests belong to the southern taiga subzone. The state is the only forest owner. Intensive forest logging and a lack of silvicultural activities during the twentieth century led to alteration of forest age and tree species structure. Young- and middle-aged stands now dominate, and there is a prevalence of deciduous trees (birch and aspen) over coniferous (pine and spruce) in the forest regeneration. Wood export and local wood processing are the main economic activities. The municipality is making efforts to support tourism development and to maintain rural areas and cultural heritage.

This Pilot Area hosted the Pskov Model Forest (2000–2008) that aimed at developing new regional forestry norms for intensification of forest management to sustain the wood resource base, primarily for international forest companies using the Nordic intensive sustained yield approach. Due to the Model Forest's activities, partnership among local and regional stakeholders has emerged that still exist in different constellations (Elbakidze et al. 2010).

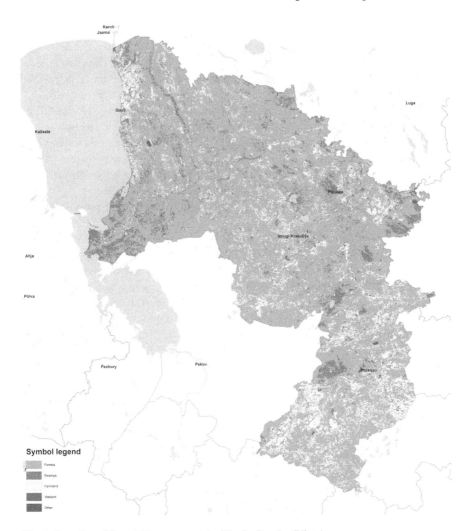

Fig. 6 Location of Strugi-Krasnye municipality, the Russian Pilot Area

4 Application of the Seven-Step Approach in Pilot Areas

After selecting Pilot Areas and learning about their landscape histories (steps 1–2, Fig. 1), researchers and practitioners jointly (i) diagnosed the ability of the ecological system to provide functional habitats for species and multiple ecosystem services, and (ii) assessed the capability of the social system to steer GI's functionality towards ecological sustainability and human wellbeing (steps 3–6).

In each Pilot Area, we used spatial data for modelling capacity of forest landscapes (Fig. 2) to deliver provisioning and supporting ecosystem services (Naumov et al. in press). We also conducted 400 structured interviews with rural and urban residents in each Pilot Area to identify the range of ecosystem services that were important for human wellbeing and to map land covers that delivered the demanded ecosystem services (Elbakidze et al. 2017). Additionally, nonacademic actors identified champions, or successful environmental managers, in each Pilot Area who had pioneered projects in land management/governance important for functional GI (Dawson et al. 2017). Researchers conducted interviews and group-modelling workshops with champions, identifying lessons to be learned as well as useful governance, planning and management strategies that may be transferable to future GI projects at national, regional and local levels. Finally, national experiences in land management and governance that were considered as important for functional GI in each Pilot Area were analysed.

In step 7, researchers and practitioners jointly developed and proposed treatments in terms of production of socially robust knowledge about what functional GI requires and how to carry out governance, planning and management of GI. To enhance learning, in step 7 the outcomes of steps 3–6 were presented and discussed during 5-day travelling workshops (Box 1) and 3-day round-table discussions (Box 2) in each Pilot Area (Photo 4). In order to structure the discussions and produce common knowledge, we used conceptual group-modelling workshops (Box 3) with the main topics: (i) how to improve knowledge regarding GI at different levels of governance and (ii) how to balance ecological and economic interests in land management/governance towards functional GI in the Baltic Sea Region.

Finally, dissemination of evidence-based knowledge about functional GI, analyses of data about ecological and social systems in Pilot Areas and facilitation of collaboration were performed to increase the knowledge of stakeholders and to enhance public awareness about GI. Popular style publications, jointly produced by researchers and practitioners, are placed on the website www.euroscapes.org as Euroscapes News, Communications and Reports, which are freely available in English (Box 4). All activities were organised in such a way that the partners have time to communicate and discuss experiences, issues and knowledge that were identified as important.

Box 1 Travelling Workshop

A travelling workshop, or atelier, is an applied integrated research tool pioneered by the Gund Institute for Ecological Economics (Farley and Costanza 2010). The main elements of travelling workshop are (1) stakeholder participation, including researchers and practitioners, (2) problem-based discussion and learning and (3) practical communication of results. The travelling workshops brought together 36 practitioners and 20 researchers from the four countries (Belarus, Latvia, Russia and Sweden). In each Pilot Area, the travelling workshop was used to see and collaboratively learn and discuss issues about GI. The outcomes are (a) joint understanding of ecological sustainability and GI as defined in international and national policies; (b) applied new and practical knowledge on planning processes for functional GI locally, regionally and in a transnational context; and (c) an improved set of baseline information for practitioners, including digital maps and an assessment of GI for the Pilot Areas (Photo 4).

Photo 4 Travelling workshops are tools for integration of researchers and practitioners for learning and knowledge production. The left picture, the travelling workshop in the Russian Pilot Area; the right picture, travelling workshop in the Latvian Pilot Area

Box 2 Round-Table Discussion
This is a way to learn about stakeholders' opinions, values and interests and to exchange knowledge related to biodiversity conservation and human well-being in each Pilot Area. Round-tables consisted of 20–25 partners that were invited to (1) discuss methods and approaches to spatial planning for functional GI in each Pilot Area and in the relevant country in general, (2) learn from "champions" in natural resource management and governance regarding landscape interventions that are beneficial for functional GI and (3) discuss how to improve spatial planning for GI in the Baltic Sea Region.

Box 3 Conceptual Group Modelling
This is a participatory tool whereby a group of stakeholders analyse a complex reality and together develop a joint systems-based understanding of problems, suitable for decision-making (e.g. Vennix 1999). The participants were introduced to system analysis and causal loop diagram (CLD) notation. With the researchers as facilitators, the participants assessed, discussed and developed CLDs during the workshop as a means of structuring dialogue and visualising mental models.

> **Box 4 Dissemination**
> *Euroscapes News* as a brief documentation of recent activities. Target: jour-
> nalists in all kinds of media. *Euroscapes Communications* as a popular docu-
> mentation of the partnership's outcomes and activities, based on knowledge
> generated within the project. Target: stakeholders at all levels. *Euroscapes
> Reports* as a comprehensive in-depth documentation of the project's results.
> Target: planners, natural resource managers and researchers.

5 Green Infrastructure in the Baltic Sea Region: Examples in Different Contexts

5.1 Functionality of Protected Areas as GI for Biodiversity: Examples from Sweden and Latvia

Current forest and environmental policies in Sweden imply that forests should be
managed so that all naturally occurring species are maintained in viable populations. This requires maintenance of functional networks of representative natural
forest and cultural woodland habitats. Angelstam et al. (2011) reviewed the policy
implementation process regarding protected areas for biodiversity conservation in
Sweden 1991–2010 and analysed how ecological knowledge was used to formulate
interim short-term and strategic long-term biodiversity conservation goals and the
development of a hierarchical spatial planning approach. Following policy statements to maintain viable populations of all naturally occurring forest species, ecologically and biologically based strategic quantitative long-term forest protection
targets were formulated using a quantitative gap analysis for the country's main
ecoregions. The discrepancy between the long-term policy goal for protected areas,
based on the gap analysis, and what was actually protected in 1997, was very large,
resulting in the need for additional area protection, including existing non-protected
forests with high conservation value. Subsequently, a short-term interim target was
formulated in terms of area for forest protection 1998–2010, as well as a long-term
landscape restoration target.

In order to demonstrate the need to assess the functionality of forest habitat networks, Angelstam et al. (2011) used the Bergslagen region as a case study to estimate the functionality of old Scots pine, Norway spruce and deciduous forest
habitats, as well as cultural woodland, in different forest regions, and to assess the
extent to which operational biodiversity conservation planning processes took place
among forest owner categories and responsible government agencies. It was concluded that Swedish policy pronouncements capture the contemporary knowledge
about biodiversity and conservation planning well. However, the existing areas of
protected and set-aside forests were presently too low and with too poor connectivity to satisfy current forest and environmental policy ambitions. Forest owners and
planners did not plan for forest biodiversity conservation spatially across ownership

borders in order to maintain a functional GI. It can also be noted that despite the fact that there are many different landowners in Sweden, there are few forest and conservation planners. The existing areas of high conservation value forests in Sweden are presently too small and too fragmented in relation to the current forest and environmental policy ambitions. Bridging this gap requires continued protection, management and restoration to create representative and functional habitat networks. This calls for both improved evidence-based knowledge about states and trends of forest and woodland biodiversity and the establishment of neutral platforms for collaboration and partnership development to improve integration among different actors.

In Latvia, GI networks are formed by nature protection areas, which are under special state-level protection, in order to safeguard and maintain biodiversity and ecosystem services important for human wellbeing. Rendenieks and Nikodemus (2015) analysed the functionality of protected areas as GI by comparing the coverage of focal species' habitats with formally and voluntarily protected areas. Most protected areas have to fulfil multiple functions (conservation of biodiversity, tourism, recreation, education, etc.) (Borgström et al. 2013, Vanwambeke et al. 2012). The analysis shows that formally protected areas had limited functionality as GI in successful provision of multiple ecosystem services. The complicated design of protected areas in Latvia impeded planning processes and the functionality of some types of protected areas. In some cases, overlapping protected areas were managed by different organisations, for example, by both the Nature Conservation Agency and Latvian State Forest Agency.

Trasune and Nikodemus (2015) show that many protected areas in Latvia either did not have management plans or these were otherwise outdated – information regarding nature protected areas, the planned activities and individual regulations of protection and use included in the spatial planning documents were insufficient or inaccurate.

In the case of voluntarily protected areas, the protection of forest stands relies on the incentives from the state forest management company. Several studies suggest clustering old and ecologically valuable stands by carefully planned harvesting actions (Kurttila et al. 2002) in order to create large tracts of seminatural stands (Öhman 2000). In this context, eco-forests in state forests are an important incentive, but the protection of forest stands inside these areas can be very limited. The actual functionality of eco-forest areas and the significance of woodland key habitats at the regional level require additional studies.

Additionally, a lack of protected areas in private forests creates problems in planning of GI at regional and national levels, because public and private forests are spatially interspersed. This problem is not new, and in some countries forest stands with particular importance within the functional network can be set aside through schemes providing compensations to private forest owners. The woodland key habitat initiative is one such scheme. Forests form the basis for GI in both urban and rural areas in Latvia; therefore, the importance of forest management and forest planning is particularly high. Spatial modelling and ecological planning at a landscape scale are not utilised in Latvian forest management planning. This sector primarily utilises traditional approaches to management planning, which in most cases do not consider spatial objectives.

5.2 Development of the National Ecological Network in Belarus

The national ecological network concept first appeared in Belarus in 2002 when the core areas of European importance, and existing and potential ecological corridors between them, were mapped as a result of a project aimed at developing the Pan-European ecological network for Central and Eastern Europe. The national ecological network concept was legitimised into national legislation in 2010, and in 2012 the government initiated a project to identify and prepare main network components. This project was finalised in 2015 and is currently being implemented.

The scheme of the national ecological network consists of core areas, ecological corridors, and buffer zones (Fig. 7). The core areas are formed by nature protection areas, established according to the nature conservation legislation and/or natural and seminatural areas that are formally protected according to the Forest Code, Water Code or other legal documents. The ecological corridors include some formally protected areas that are not included in core areas, providing links among

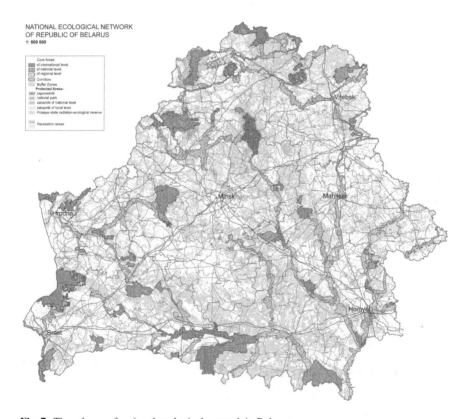

Fig. 7 The scheme of national ecological network in Belarus

them. Different types of formally protected areas that are not included in the core areas or ecological corridors form the buffer zones.

In spite of favourable legal conditions for the development of the national ecological network, there are several challenges in the implementation phase. The main challenge is to establish additional protected areas at the local level or to change land management regimes that is in conflict with nature conservation targets. For example, most planned forest corridors had been excluded from the ecological network because the State Forest Agency did not approve establishment of these elements on their land. Additionally, intensification of forestry and agriculture, fragmentation of natural habitats due to urbanisation and grey infrastructure development are key obstacles in the development of the national ecological network.

5.3   Green Infrastructures for Human Wellbeing in the Baltic Sea Region: Which Natural and Seminatural Areas Do People Need for Their Wellbeing?

A comprehensive analysis of which ecosystem services are important for people in both rural and urban areas in a specific socio-ecological context is needed in order to identify human-related functions of GI that delivers a wide range of ecosystem services for human wellbeing (EU 2013, Forest Research 2010).

We conducted surveys in each Pilot Area to allow (1) identification of natural and seminatural areas as potential GI hubs for human wellbeing, (2) an improved understanding of the multiple ecosystem services people obtain from these hubs and (3) estimation of the regional availability of GI hubs. Comparison of results from all Pilot Areas shows that there are many similarities among the preferences of urban and rural people regarding natural and seminatural areas that are perceived as important for human wellbeing. The majority of local residents identified lake, wooded grasslands, old-growth forest, mature pine forest and rural farmsteads as important for their personal wellbeing. The selected land covers represent both aquatic and terrestrial ecosystems that are located within a gradient from old-growth forests without any active management to managed pine at age of final felling and seminatural areas under traditional land use (Figs. 8 and 9).

Each of these natural and seminatural areas was associated with multiple ecosystem services in each Pilot Area. Whilst cultural services were the most frequently mentioned in the Swedish and Latvian Pilot Areas, provisioning services were the most acknowledged in Belarusian and Russian Pilot Areas. Recreation/tourism and inspiration associated with old-growth forests were the most frequently mentioned ecosystem services by the Swedish respondents, whilst Belarusian and Russian respondents also strongly associated with wild food and habitats for species. Other ecosystem services most frequently mentioned by respondents were subsistence food provided by wood pastures, sense of place by rural farmstead and timber/wild food by mature pine forest.

Fig. 8 Priority local landscapes important for human wellbeing in Sweden

Fig. 9 Priority local landscapes important for human wellbeing in Belarus

5.4 Balancing Wood Production and Biodiversity Conservation: Examples from Sweden and Latvia

Traditionally, forestry aims to maximise a limited set of provisioning services such as timber, pulpwood and biomass. At the same time, current policies at both international (European Commission 2013) and national levels also aim at restoring and maintaining supporting and cultural services. However, past trajectories of forest use and management affect the opportunities for the sustainable provision of multiple ecosystem services at present and in the future. The Baltic Sea Region is a good example of a region with different forest and woodland histories. Sweden, for example, has a long history of forest use and has successfully promoted intensification through maximum sustained yield forestry (Sverdrup and Stjernquist 2013). Recently, intensification efforts have expanded to include increasing the supply of forest biomass for bioenergy. In Latvia, the intensification of wood production has emerged during the last two/three decades. In terms of biodiversity conservation, Sweden has not been able to maintain the historical range of variability of boreal forests. Consequently, efforts towards protection, management (Angelstam 1998; Similiä and Junninen 2012; Stanturf 2015) and restoration (Miller and Hobbs 2007) have been made to restore biodiversity in terms of species, and their habitats (Brumelis et al. 2011). In Latvia, there is a shift towards monocultural Scots pine forests with an age distribution dominated by young- and middle-aged stands. Thus, there are almost no mature and overmature Norway spruce stands remaining (Tērauds et al. 2011).

A balance between wood production and biodiversity conservation is one of the targets of forest management and governance towards functional GI. Naumov et al. (in press) analysed and compared indicators for wood production and biodiversity applied for selected forest management units in all four Pilot Areas. The study shows that there was a conflict between intensified wood production and the protection of forest biodiversity. In particular, intensified wood production in Sweden and Latvia has caused the reduction, and even extinction, of both old growth and deciduous forest.

Regarding wood production, Sweden's Bergslagen region had the largest areas for sustainable wood production, followed by Braslav in Belarus and Zemgale in Latvia. NW Russia's Pskov region had the lowest values, except for the biomass indicator. Regarding biodiversity conservation, the overall rank was opposite. Mixed and deciduous stands were maintained in Russia, Belarus and Latvia. Thus, the regional forest history provides different challenges in terms of satisfying both wood production and biodiversity conservation objectives in a forest management unit. This gives an opportunity for exchanging experiences among different regional contexts about how to achieve both objectives.

5.5   Champions in Land Management and Governance: Examples from the Baltic Sea Region

Analysis of good practices in land management/governance important for GI development within the Baltic Sea Region shows that the main focus of the Pilot Areas is on terrestrial and aquatic landscape restorations at different levels (Table 1). Given the long history of landscape degradation through habitat degradation, fragmentation and loss in the Baltic Sea Region, efforts to maintain functional GI are highly dependent on the success of innovative landscape restoration projects. There is thus an urgent need to better understand how successful landscape restoration takes place in countries with different governance, socio-economic and cultural contexts at a variety of scales (Dawson et al. 2017).

Identified champions from each Pilot Area gave in-depth interviews regarding their respective success stories and its specific environmental management and gov-

Table 1 Examples of successful environmental management and governance in each Pilot Area that have been innovated by champions from different societal sectors (civil, private or public) who operated at local and national levels

	Civil	Private	Public
National	Wetland restoration (Belarus)	Forest landscape restoration (Latvia)	Wetland restoration (Belarus) Forest landscape restoration (Sweden) Wetland restoration (Latvia)
Local		Wetland restoration (Latvia) Value-added production from NWFPs (Belarus)	Wetland restoration (Sweden) River restoration (Sweden)

ernance context. Champions guided project partners from all four Pilot Areas on excursions to their implementation areas in order to show what had been achieved. Questions and discussions during these trips helped to clarify many issues related to the history of a certain landscape restoration project, the main stakeholders involved as well as the opportunities and challenges related to the specific context. Finally, group workshops were held where champions participated together with project partners to examine a broad series of issues regarding the causal structure and system dynamics of landscape restoration projects.

In general, landscape restoration projects were indicative of an adaptive approach to environmental management and governance, adopting an experimental approach and seeking to integrate project-derived learning via a series of both short- and long-term processes. However, the degree to which learning was formalised into shareable knowledge differed among projects. Champions emerged as such as a result of a complex set of legitimising processes arising from various aspects of the project work, although they also brought a set of competences and a suitable personal background prior to project start. Whilst champions were chiefly responsible for initiating and developing project visions, many landscape restoration projects came about as a response to various unforeseeable crises, either in landscapes themselves, in governance systems or in both. Champions had the requisite skills and background to be able to perceive a window of opportunity as well as a suitably strategic position to exploit it. Champions consciously employed a variety of pedagogical tools and approaches to engage stakeholders and managed risk and expectations by breaking projects up into small, concrete, practical steps. Another key success factor was the degraded initial state of the landscape itself, which reduced the exposure of projects to both risk and criticism, and led to positive responses from stakeholders who perceived even minor changes in the landscape as improvements. A common barrier for many of the restoration projects was the lack of suitable long-term financing, which often led to progress being made in an ad hoc, piecemeal fashion (Dawson et al. 2017).

5.6   Strategic Spatial Planning as a Collaborative Learning Process: An Example from Sweden

Europe is characterised by a high diversity of planning systems originating from political, cultural and institutional differences within and among countries (Albrechts 2006; Albrechts et al. 2010). This diversity in planning systems presents an opportunity for learning towards effective GI policy implementation. Contemporary European Union (EU) and pan-European policies (e.g. Council of Europe 2006; UNECE 2008; European Council 2011) stress the importance of spatial planning for the long-term sustainability of regions and for the integration of territorial economic and social requirements with ecological and cultural functions.

Sweden is often described as an example of a European country where strategic spatial planning includes developed participatory mechanisms grounded on its long democratic traditions. Researchers, or academic stakeholders, analysed the extent to which the comprehensive planning process was characterised as a collaborative learning process through analysis of the main attributes of public-led strategic spatial planning in Bergslagen, the Pilot Area in Sweden (Elbakidze et al. 2015). The main questions were: Is municipal spatial planning a collaborative learning process among actors and stakeholders or is it a technical project? What are the main drivers for collaborative learning in spatial planning? This participatory research identified that the attributes of strategic spatial planning needed for collaborative learning were either absent or undeveloped. All municipalities experienced challenges in coordinating complex issues regarding long-term planning to steer territorial development and to solve conflicts among competing interests. Stakeholder participation was identified as a basic condition for social learning in planning. Together, academic and nonacademic stakeholders identified the causal structure behind stakeholder participation in municipal planning processes, including main drivers and feedback loops. The joint conclusion was that there is a need for arenas allowing and promoting stakeholder activity, participation and inclusion that combines both bottom-up and top-down approaches, where evidence-based collaborative learning can occur.

The results of the investigation were presented at a conceptual group-modelling workshop to nonacademic stakeholders, or practitioners, dealing with strategic spatial planning that represented seven municipalities in Bergslagen. With the researchers as facilitators, the participants assessed, discussed and developed the CLD during the workshop (Fig. 10 and Box 3). A draft of the resulting CLD was distributed by e-mail to workshop participants for final validation and review. The results were subsequently presented as a popular style publication and sent to all interested stakeholders in Bergslagen.

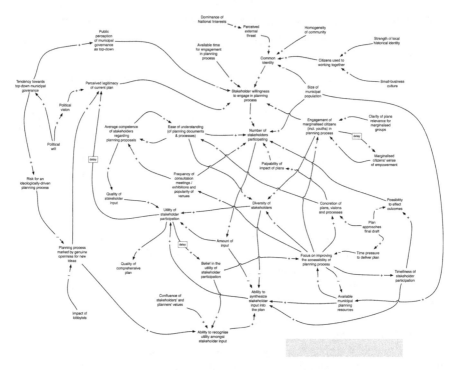

Fig. 10 Causal loop diagram (CLD) made by Bergslagen's stakeholders identifying the underlying system structure driving stakeholder participation in municipal spatial planning. CLDs indicate the relationships between various system variables by the use of arrows pointing from one variable to each variable that it directly influences. These relationships can either be reinforcing, represented by a positive (+) sign, or balancing, represented by a negative (−) sign (Elbakidze et al. 2015)

6 Challenges and Opportunities for Development of Functional Green Infrastructure in the Baltic Sea Region

6.1 Challenges

The overview of attempts to develop functional GI in the Pilot Areas representing countries with different landscape histories as well as political, social and cultural contexts in the Baltic Sea Region illustrates that in spite of differences in the contexts, there are similar sets of challenges in the development and maintenance of GI functions for both biodiversity and human wellbeing.

The first set of challenges concerns maintaining sufficient amounts of representative terrestrial, freshwater and coastal ecosystems with functional connectivity. Examples from Sweden and Latvia show that protection of forest habitats on productive sites are underrepresented between both protected areas and in the production landscape as a result of historical and current forestry use. Protected areas are often located as isolated islands in an intensively managed production landscape, resulting in poor functionality of habitat networks. Also, the proportion of forest land reserved for species conservation remains low in relation to evidence-based policy targets.

Regarding the ecological dimension of GI, systematic analyses of the functionality of different types of ecosystems in landscapes need to be made according to the given level of ambition (Lazdinis and Angelstam 2004). This includes (1) estimation of regional gaps in the amount and representativity of ecosystems; (2) analyses of the functionality of the habitat networks, for example, in terms of hosting viable populations; and (3) understanding of how operational protection, management and restoration measures can be combined in practice at different spatial scales.

Protected area networks are often presented as the backbone of GI for biodiversity conservation. Sufficient representation of ecosystems with different disturbance regimes in protected area networks is crucial for the conservation of species, habitats and processes (Angelstam and Kuuluvainen 2004; Shorohova et al. 2011). However, the functionality of networks, the management of protected areas and the qualities of the surrounding matrix are usually not analysed.

Regarding the Baltic Sea Region, one might firstly consider the difference between ecosystems and habitats among ecoregions and countries. Secondly, for the conservation of species in a concrete forest biome, the functionality of the protected area network of a particular ecosystem type needs to be assessed. Spatial modelling of the size, quality and juxtaposition of protected areas can be used for such network functionality assessments (Andersson et al. 2013). Several studies show that the functionality of small set-asides is often insufficient in relation to contemporary ecological sustainability policies (Elbakidze et al. 2011). Conservation management towards landscape restoration can thus contribute to filling the gap between present amounts of habitat and what is needed to satisfy policy goals (Hanski 2000; Mansourian et al. 2006). Finally, the land-use in the matrix composition surrounding protected areas matters. To understand the role of protected areas for ecological sustainability, other set-asides at different spatial scales also need to be mapped and their duration and management regimes understood. Key challenges are to measure, aggregate, and assess these efforts in a landscape or an eco-region so that it is possible to communicate the consequences of conservation efforts at different spatial scales to various stakeholders (Angelstam and Bergman 2004).

The second set of challenges concerns maintaining land management regimes that support natural and seminatural areas for human wellbeing. Wooded grasslands and villages are among the main types of landscape that provide multiple ecosystem services for human wellbeing in all Pilot Areas. However, there are challenges to maintain such areas in each country. The importance of wooded grasslands throughout Europe is recognised (Bergmeier et al. 2010; Eichhorn et al. 2006). However, these landscapes continue to be subjected to changing processes leading to degradation and fragmentation (Garrido et al. 2017a, b). Urban sprawl, land abandonment, and the intensification of agriculture and forestry all represent great uncertainty for the long-term conservation of valuable wooded grassland in the Baltic Sea Region.

For example, regional public officials estimate that two-thirds of former oak-wooded grasslands with high nature values in Sweden need to be restored to sustain biodiversity and ecosystem services important for people. However, restoration of wooded grasslands is of limited effect unless they are maintained in the long term by traditional land management. Therefore, the role of farmers and management practices to maintain wooded grasslands is fundamental. However, such management practices are currently of marginal profitability in all Pilot Ares, thus endangering the overall land-use system and the provision of ecosystem services important for people (Garrido et al. 2017a, b).

Villages and rural areas in general, face significant challenges linked to isolation and the deterioration of cultural and social assets primarily due to net rural-to-urban out-migration in all Pilot Areas. At the same time, rural landscapes are perceived as attractive and healthy living environments. This implies that the sustainability of rural areas is subject to both opportunities and threats and requires integrated spatial planning and design of GI prior to the development of land-use plans.

The third set of challenges relates to the development of stakeholder collaboration models that allow people from different sectors working together within and across all spatial scales and level of governance to achieve their diverse goals whilst maintaining a sustainable use of ecosystem services across the Baltic Sea Region. In all Pilot Areas there are, unfortunately, clear barriers between evidence-based knowledge and biodiversity and ecosystem service policy targets on the one hand, and the generally sector-based planning and management of ecosystems on the other. A key-bridging factor is to coordinate and integrate stakeholders' efforts among sectors at multiple levels. However, this remains challenging, regardless of governance structure and is not made easier by rhetoric stressing continued pressure on valuable natural systems, or a piecemeal approach to management practices and set-asides. There is a need for arenas allowing and promoting stakeholder activity, participation and inclusion, i.e. where people representing diverse societal sectors at different levels can interact in both bottom-up and top-down processes and where collaborative learning can occur. Model Forests, Biosphere Reserves and Community Led Local Development (Leader) are examples of good approaches to partnership development used in our Pilot Areas (e.g. Besseau et al. 2008; Axelsson et al. 2013).

Understanding and managing regions and landscapes to improve ecosystem function and the delivery of ecosystem services are fundamental to navigating sustainable development. This requires both compass and gyroscope (Lee 1993). Compass is about knowing the states and trends of biodiversity in regions, using evidence-based knowledge about ecological targets, tipping points and measures to manage, restore and recreate habitats for species. Gyroscope is about developing close interaction between researchers, stakeholders and policy makers including, e.g. managers and users in the field, businesses, policy actors, local administrations and citizens. The adequacy of communication strategies and processes are a key component in developing such ongoing interaction. An oft-overlooked challenge to successful landscape governance and management, the development of diversified communication strategies and tools, enables meaningful dialogue between multiple stakeholders with a variety of backgrounds and competences. This is of particular importance given the social-ecological complexity of the Baltic Sea Region, including governance legacies.

Our study shows the increasing importance of integrated spatial planning in the sustainable use and governance of natural capital towards functional GI for human wellbeing. Public sector-led spatial planning is an important tool for the holistic integration of economic, environmental, cultural and social policy agendas and for rescaling issues from international and national levels to regional and local levels (Albrechts et al. 2010).

Several key research questions remain: Which critical ecological and socio-economic features are most important for guiding the establishment, management and evaluation of GI? How can the net impacts of global and socio-economic pressures and responses on GI functionality be transparently assessed and presented at various platforms for stakeholders and policy makers? How does the level of social learning between actors and sectors vary among pan-European regions with different landscape histories and political/governance cultures?

6.2   Opportunities

There are at least five main sets of opportunities for maintaining, improving or restoring multiple functions of GI in the Baltic Sea Region. The first set relates to favourable international environmental policies towards establishment of functional GI in the Baltic Sea Region. For example, contemporary European Union (EU) and Pan-European policies (e.g. Council of Europe 2006; UNECE 2008; European Council 2011) stress the importance of spatial planning for the long-term sustainability of regions. To support long-term, large-scale and balanced territorial development, spatial planning is expected to integrate territorial economic and social requirements with ecological and cultural functions.

The second set of opportunities relates to the abundance of existing applied/practical knowledge in biodiversity conservation upon which the further development and integration of strategic planning for functional GI across the Baltic Sea Region could be based. For example, Sweden has a rich experience in strategic conservation planning for biodiversity. At its core is an evidence-based regional gap analysis that defines the amount of functional habitat network needed to maintain long-term biodiversity. The emergence of systematic GI planning in Sweden provides a useful, evidence-based hierarchical approach in terms of (1) strategic planning based on regional gap analyses and evidence-based ecological knowledge regarding the necessary types and amounts of habitat required to reach policy ambitions, (2) tactical spatial planning towards functional GI and (3) operational planning and work in terms of protection, management and restoration of viable populations of species, habitats and ecosystem processes. Another example is forest zoning in the Russian Federation to satisfy biodiversity conservation and social functions. This forest management approach was originally developed already in the nineteenth century, and in 1943 all forests in the USSR were divided into three groups (Koldanov 1992; Algvere 1966). Forests with mainly protective and social functions belonged to the first (I) group; the second (II) group included forests with multiple functions and had certain limitations for exploitation. Forests available for exploitation belonged to the third (III) group (Teplyakov et al. 1998). Many studies show that spatial zonation is more efficient than integrated stand-level approaches at delivering the broad portfolios of benefits from forests requested by sustainable forest management policy (e.g. Mönkkönen et al. 2014). The Russian national FSC forest certification standard also prescribes zoning around villages to support rural livelihoods by securing access to, and maintenance, non-wood forest products (Stryamets et al. 2015). Zoning regulations thus help to achieve both biodiversity conservation and human wellbeing. By spatially segregating forest landscapes, multipurpose forest management can be achieved at the landscape level, including wood production. Other countries have also developed zoning approaches. For instance, Sweden's state forest company Sveaskog uses its Ekopark concept (Angelstam and Bergman 2004) to manage forest for biodiversity conservation and wood production. Another approach developed in Swedish forest management planning is the operational division of forest land into four classes representing a gradient from wood production to biodiversity conservation. Such approaches, applied throughout a landscape, may benefit sustained-yield wood production, biodiversity conservation and social values such as outdoor recreation.

The third set of opportunities is connected with existing landscape approach initiatives in the Baltic Sea Region that have gained rich experience in management and governance of landscapes based on sustainable development principles. Axelsson et al. (2011) presented a practical operationalisation of the landscape approach using five core normative attributes: (1) a sufficiently large area that matches management requirements and challenges to deliver desired natural benefits, (2) multilevel and multi-sector stakeholder collaboration that promotes sustainable development as a social process, (3) commitment to and understanding of sustainability as an aim among stakeholders, (4) integrative knowledge production and (5) sharing of experience, results and information, to develop local tacit and general explicit knowledge. Biosphere Reserves (BR) and Model Forest (MF) are two examples of landscape approach initiatives (Elbakidze et al. 2010, 2013). Each BR is intended to fulfil three core functions: (1) a conservation function to conserve genetic resources, species, ecosystems, habitats and landscapes; (2) a development function to foster sustainable economic and human development and (3) a logistic support function, to support research, monitoring, education, training and establishment of demonstration sites and to promote environmental awareness related to local, national and global issues of conservation and sustainable development (UNESCO 1995). The MF concept was developed in Canada in the early 1990s. An MF can be understood as a process designed to establish a partnership and a forum for collaboration to solve a wide spectrum of issues related to the implementation of sustainability policies. The key functions of MFs are to test new ideas and develop innovations related to sustainable development, as agreed to by MF partners, with the goal of developing the adaptive capacity of the local social-ecological system to deal with uncertainty and change. These initiatives, BRs and MFs, as "learning sites for sustainable development" could be used as "pools of applied knowledge" and "a source of inspiration" that empower stakeholders to take part in knowledge production and learning for GI development.

A fourth set of opportunities for functional GI in the Baltic Sea Region concerns the potential of landscape restoration projects. The typically low present use value of, or perceived lack of plausible alternative uses for, currently degraded landscapes in itself may provide an opportunity for GI. Perceptions of risk – both environmental risks and failed project risks – associated with most types of LR project interventions are often low, unless they have direct implications for public safety. In this way, currently degraded landscapes may avoid the strict value protection focus associated with nature conservation regulations and other institutional frames, allowing greater management flexibility. This minimises project delivery delays and leads to quicker GI outcomes. Similarly, landscape restoration projects are often developed as a response to a crisis, either in ecological or societal systems, or both. In this manner, the projects are able to present themselves as timely low-risk solutions to current high-risk problems. In addition, LR projects appear to provide an emergent platform for innovative approaches to environmental governance and management. Investments in LR thus represent a low-risk win-win opportunity for societal experimentation with participatory and holistic governance approaches at a variety of scales, whilst simultaneously improving the connectivity and functionality of GI (Dawson et al. 2017).

The fifth set of opportunities is related to multi- and transdisciplinary research projects that have been practised in the Baltic Sea Region. Such projects provide important conditions for collaborative learning concerning functional GI. Transdisciplinary research projects have excellent potential as trust-building platforms, as the goals of such projects are often simple and unifying – the co-production of new knowledge – and shared explicitly among all participants. Another important element is the adoption of a genuinely multidirectional flow of knowledge and competence between nations and regions. In this respect, the historically institutionalised view of unidirectional, aid-oriented knowledge flows from West to East requires review.

7 Conclusion

GI is a policy term that captures the need for functional ecosystems that deliver ecosystem services (European Commission 2013). To tackle the increasing loss and fragmentation of natural and seminatural areas important for GI in the Baltic Sea Region, there is a need to protect, manage and restore functional habitat networks for biodiversity, ecosystem services and human wellbeing. Supporting implementation of GI policy requires informed collaborative and evidence-based spatial planning across sectors and levels of governance in forest, rural and urban landscapes.

We argue for a two-step approach. The first step, diagnosis, examines how institutions and actors steer GI functionality by spatial planning, outputs related to planning processes and planning tools, as well as consequences on the ground for ecological sustainability and human wellbeing. The second step, treatment, aims at the production of socially robust, evidence-based knowledge about the requirements of functional GI in terms of thresholds and tipping points in ecosystems (Rockström et al. 2009) and how to govern, plan and manage GI. Stakeholders in the Baltic Sea Region have much to gain from increased multilateral, learning-based collaborations regarding all aspects of sustainable forest landscapes. Such collaborations have the potential to serve as laboratories for cross-border governance and management, which may also be pertinent in other domains.

Acknowledgements This study was made with funding from the Swedish Institute [grant number 10976/2013] to Marine Elbakidze and from the Swedish Research Council Formas [grant number 2011-1737] to Per Angelstam.

References

Albrechts L (2006) Shifts in strategic spatial planning? Some evidence from Europe and Australia. Environ Plan A 38:1149–1170

Albrechts L, Healey P, Kunzmann K (2010) Strategic spatial planning and regional governance in Europe. J Am Plan Assoc 69(2):113–129

Algvere KV (1966) Forest economy in the U.S.S.R. An analysis of Soviet Competitive potentialities. Studia Forestalia Suecica 39. Royal College of Forestry, Stockholm, Sweden

Allen WL (2014) A green infrastructure framework for vacant and underutilized urban land. J Conserv Plan 10:43–51

Andersson K, Angelstam P, Elbakidze M et al (2013) Green infrastructures and intensive forestry: need and opportunity for spatial planning in a Swedish rural–urban gradient. Scand J For Res 28(2):143–165

Angelstam P (1998) Maintaining and restoring biodiversity in European boreal forests by developing natural disturbance regimes. J Veg Sci 9(4):593–602

Angelstam P, Barnes G, Elbakidze M, Marais C, Marsh A, Polonsky S, Richardson DM, Rivers N, Shackleton RT, Stafford W (2017) Collaborative learning to unlock investments for functional ecological infrastructure: Bridging barriers in social-ecological systems in South Africa. Ecosyst Serv 27:291–304

Angelstam P, Elbakidze M (2017) Forest landscape stewardship for functional green infrastructures in Europe's West and East: diagnosing and treating social-ecological systems. In: Pleininger T, Bieling C (eds) The science and practice of landscape stewardship. Oxford University Press

Angelstam P, Dönz-Breuss M, Roberge JM (eds) (2004) Targets and tools for the maintenance of forest biodiversity. Ecol Bull 51:1–510

Angelstam P, Andersson K, Axelsson R et al (2011) Protecting forest areas for biodiversity in Sweden 1991–2010: policy implementation process and outcomes on the ground. Silva Fenn 45(5):111–1133

Angelstam P, Elbakidze M, Axelsson R et al (2013a) Knowledge production and learning for sustainable landscapes: seven steps using social-ecological systems as laboratories. Ambio 42(2):116–128

Angelstam P, Andersson K, Isacson M et al (2013b) Learning about the history of landscape use for the future: consequences for ecological and social systems in Swedish Bergslagen. Ambio 42(2):150–131

Angelstam P, Andersson K, Axelsson R et al (2015) Barriers and bridges for Sustainable Forest Management: the role of landscape history in Swedish Bergslagen. In: Kirby KJ, Watkins D (eds) Europe's changing woods and forests: from wildwood to cultural landscapes. CABI, Wallingford, pp 290–305

Axelsson R, Angelstam P, Elbakidze M et al (2011) Sustainable development and sustainability: landscape approach as a practical interpretation of principles and implementation concepts. J Land Ecol 4(3):5–30

Axelsson R, Angelstam P, Myhrman L (2013) Evaluation of multi-level social learning for sustainable landscapes: perspective of a development initiative in Bergslagen, Sweden. Ambio 42(2):241–253

Baker S (2006) Sustainable development, Routledge, Taylor & Francis Group, London and NY

Benedict MA, McMahon ET (2002) Green infrastructure: smart conservation for the 21st century. Renew Res J 20(3):12–17

Bergmeier E, Petermann J, Schröder E (2010) Geobotanical survey of wood-pasture habitats in Europe: diversity, threats and conservation. Biodivers Conserv 19(11):2995–3014

Besseau P, Bonnell B, Muni K (2008) Ustoychivoe razvitie partnerskih otnosheniy dlya ustoichivogo upravleniya lesnymi landshaftami: opyt modelnyh lesov. Ustoychivoe lesopolzovanie 18(2):2–8

Borgström S, Lindborg R, Elmqvist T (2013) Nature conservation for what? Analyses of urban and rural nature reserves in southern Sweden 1909–2006. Landsc Urban Plan 117:66–80

Brumelis G, Jonsson BG, Kouki J et al (2011) Forest naturalness in northern Europe: perspectives on processes, structures and species diversity. Silva Fennica 45(5):807–821

Bryden J, Hart J (2004) A new approach to rural development in Europe: Germany, Greece, Scotland and Sweden. The Edwin Mellen Press, Ceredigion

Costanza R, d'Agre R, De Groot RS, Farber S et al (1997) The value of the world's ecosystem services and natural capital. Nature 385:253–260

Council of Europe (2006) European conference of ministers responsible for regional spatial/planning (CEMAT), Lisbon

Daily GC, Alexander S, Ehrlich PR et al (1997) Ecosystem services: benefits supplied to human societies by natural ecosystems. Issue Ecol 2:1–18

Dawson L, Elbakidze M, Angelstam P, Gordon J (2017) Governance and management of landscape restoration at multiple scales: learning from successful environmental managers in Sweden. J Environ Manag 197:24–40

Diaz S, Quetier F, Caceres D et al (2011) Linking functional diversity and social actor strategies in a framework for interdisciplinary analysis of nature's benefits to society. Proc Natl Acad Sci 108(3):895–902

EC (2015) Mid-term review of the EU biodiversity strategy to 2020. Retrieved from http://ec.europa.eu/environment/nature/biodiversity/comm2006/pdf/mid_term_review_sum22693mary.pdf

Eichhorn MP, Paris P, Herzog F, Incoll LD, Liagre F, Mantzanas K, . . . & Dupraz C (2006) Silvoarable systems in Europe—Past, present and futureprospects. Agrofor Syst 67(1):29–50

Elands B, Wiersum K (2001) Forestry and rural development in Europe: an exploration of socio-political discourses. Forest Policy Econ 3:5–16

Elbakidze M, Angelstam P, Sandström C, Axelsson R (2010) Multi-stakeholder collaboration in Russian and Swedish model Forest initiatives: adaptive governance towards sustainable forest management? Ecol Soc 15(2):14

Elbakidze M, Angelstam P, Andersson K, Nordberg M, Pautov Y (2011) How does forest certification contribute to boreal biodiversity conservation? Standards and outcomes in Sweden and NW Russia. For Ecol Manag 262(11):1983–1995

Elbakidze M, Angelstam P, Sandström C et al. (2013) Biosphere Reserves for conservation and development in Ukraine? Legal recognition and establishment of the Roztochya initiative. Env Cons 40(2):157–166

Elbakidze M, Dawson L, Andersson K et al (2015) Is spatial planning a collaborative learning process? A case study from a rural–urban gradient in Sweden. Land Use Policy 48:270–285

European Commission (2009) European Union Strategy for the Baltic Sea Region. Retrived from
http://ec.europa.eu/regional_policy/sources/docoffic/official/communic/baltic/com_baltic_en.pdf
European Commission (2013) Green Infrastructure (GI) — enhancing Europe's natural capital.
Communication from the commission to the European Parliament, the Council, the European
Economic and Social Committee and the Committee of the regions. European Commission:
Environment, Brussels
European Council (2011) Territorial agenda of the European Union 2020. Towards an Inclusive,
Smart and Sustainable Europe of Diverse Regions. Agreed at the Informal Ministerial Meeting
of Ministers responsible for Spatial Planning and Territorial Development on 19th May 2011.
Gödöllő, Hungary
Ewers R, Kapos V, Coomes D et al (2009) Mapping community change in modified landscapes.
Biol Conserv 142:2872–2880
Farley J, Constanza R (2010) Payments for ecosystem services: from local to global. Ecol Econ
69(11):2060–2068
Forest Research (2010) Benefits of green infrastructure report by forest research. [20.10.2015]
http://ec.europe.eu/environment/nature/info/pubs/docs/greeninfrastructure.pdf/retrived
Garrido P, Elbakidze M, Angelstam P (2017a) Stakeholders' perceptions on ecosystem services
in Östergötland's (Sweden) threatened oak wood-pasture landscapes. Landsc Urban Plan
157:96–104
Garrido P, Elbakidze M, Angelstam P et al (2017b) Stakeholder perspectives of wood pasture ecosystem services: a case study from Iberian dehesas. Land Use Policy 60:324–333
Hanski I (2000) Extinction debt and species credit in boreal forests: modelling the consequences of
different approaches to biodiversity conservation. Ann Zool Fennici 37:271–280
Huntsinger L, Oviedo JL (2014) Ecosystem services are social-ecological services in a traditional
pastoral system: the case of California's Mediterranean rangelands. Ecol Soc 19(1):8
Koldanov VY (1992) Ocherki Istorii Sovetskogo Lesnogo Chozyastva. Ekologiya, Moscow, Russia
Kurttila M, Uuttera J, Mykrä S et al (2002) Decreasing the fragmentation of old forests in land-
scapes involving multiple ownership in Finland: economic, social and ecological consequences.
For Ecol Manage 166(1):69–84
Lafortezza R, Carru G, Sanesi G, Davies C (2009) Benefits and well-being perceived by people
visiting green spaces in periods of heat stress. Urban For Urban Green 8:97–108
Lafortezza R, Davies C, Sanesi G, Konijnendijk C (2013) Green infrastructure as a tool to support
spatial planning in European urban regions. iForest - Biogeosci For 6(3):102–108
Laird SA, McLain R, Wynberg RP (eds) (2010) Wild product governance: finding policies that
work for non-timber forest products. Earthscale, London
Lee KN (1993) Compass and gyroscope. Island Press, Washington DC
Lele S, Springate-Baginski O, Lakerveld R et al (2013) Ecosystem services: origins, contributions,
pitfalls, and alternatives. Conserv Soc 11(4):343–358
MA (2005) Ecosystems and human wellbeing: synthesis. Island Press, Washington DC
Mansourian S, Vallauri D, Dudley N (eds) (2006) Forest restoration in landscapes, beyond planting
trees. Springer, New York
Miller JR, Hobbs RJ (2007) Habitat restoration—do we know what We're doing? Restor Ecol
15(3):382–390
Mönkkönen M, Juutinen A, Mazziotta A, Miettinen K, Podkopaev D, Reunanen P, Salminen H,
Tikkanen O-P (2014) Spatially dynamic forest management to sustain biodiversity and eco-
nomic returns. J Environ Manag 134:80–89
Naumov V, Angelstam P, Elbakidze M (2016) Barriers and bridges for intensified wood production
in Russia: insights from the environmental history of a regional logging frontier. Forest Policy
Econ 66:1–10
Naumov V, Angelstam P, Manton M et al. (in press). Balancing wood production and biodiversity
conservation in boreal forest management units: regional European landscape history matters.
Env Conser
Öhman K (2000) Creating continuous areas of old forest in long-term forest planning. Can J For
Res 30(11):1817–1823

Rauschmayer F, Berghöfer A, Omann I, Zikos D (2009) Examining processes or/and outcomes? Evaluation concepts in European governance of natural resources. Environ Policy Gov 19(3):159–173

Rendenieks Z, Nikodemus O (2015) Protected areas as green infrastructures in Latvia? Zemgale planning region as an example. Euroscapes Report on www.euroscapes.org

Rockstrom J, Steffen W, Noone K et al (2009) A safe operating space for humanity. Nature 461(7263):472–475

Sabogal C, Besacier C, McGuire D (2015) Forest and landscape restoration: concepts, approaches and challenges for implementation. Unasylva 66(3):3–10

Shorohova E, Kneeshaw D, Kuuluvainen T, Gauthier S (2011) Variability and dynamics of old-growth forests in the circumboreal zone: implications for conservation, restoration and management. Silva Fennica 45(5)

Similiä M, Junninen K (2012) Ecological restoration and management in boreal forests: best practices from Finland. Metsähallitus, Natural Heritage Services, Vantaa

Stanturf JA (2015) Restoration of boreal and temperate forests, 2nd edn. CRC Press, London

Stryamets N, Elbakidze M, Ceuterick M, Angelstam P, Axelsson R (2015) From economic survival to recreation: contemporary uses of wild food and medicine in rural Sweden, Ukraine and NW Russia. J Ethnobiol Ethnomed 11(1)

Sverdrup H, Stjernquist I (2013) Developing Principles and Models for Sustainable Forestry in Sweden. Springer Science and Business Media. ISBN 978-94-015-9888-0

TEEB (2010) The Economics of Ecosystems and Biodiversity: Mainstreaming the Economics of Nature: A Synthesis of the Approach, Conclusions and Recommendations of TEEB. The Economics of Ecosystem and Biodiversity. Retrieved from http://www.teebweb.org/publication/mainstreaming-the-economics-of-nature-a-synthesis-of-the-approach-conclusions-and-recommendations-of-teeb/

Teplyakov V, Kuzmichev E, Baumgartner D, Everett R (1998) A history of Russian forestry and its leaders. Washington State University, Pullman

Tērauds A, Brūmelis G, Nikodemus O (2011) Seventy-year changes in tree species composition and tree ages in state-owned forests in Latvia. Scand J For Res 26(5):446–456

Tillväxtverket (2011) Genuint sårbara kommuner. Företagandet, arbetsmarknaden och beroendet av enskilda större företag. [Genuinely vulnerable municipalities. Business, labor market and the dependency of single large companies]. Rapport 0112, Stockholm, Tillväxtverket, p 72 (in Swedish)

Trasune L, Nikodemus O (2015) Planning of green infrastructure through nature protection plans for specially protected areas: Zemgale planning region as an example. Euroscapes Report on www.euroscapes.org

UNECE (2008) Spatial planning – key instrument for development and effective governance, with special reference to countries in transition. Economic Commission for Europe, CE/HBP/146, Geneva

UNESCO (1995) The Seville Strategy and the Statutory Framework of the World Network of Biosphere Reserves. UNESCO, Paris

Vanwambeke S, Meyfroidt P, Nikodemus O (2012) From USSR to EU: 20 years of rural landscape changes in Vidzeme, Latvia. Landsc Urban Plan 105(3):241–249

Vennix J (1999) Group model-building: tackling messy problems. Syst Dyn Rev 15(1):379–401

Weber T, Sloan A, Wolf J (2006) Maryland's green infrastructure assessment: development of a comprehensive approach to land conservation. Landsc Urban Plan 77:94–110

Sustainable Planning for Peri-urban Landscapes

Daniele La Rosa, Davide Geneletti, Marcin Spyra, Christian Albert, and Christine Fürst

1 Peri-urban Landscapes

In the past decades, different typologies of peripheral landscapes have emerged as a result of dynamic processes of urban development and relative change in natural, seminatural, and agricultural areas. Historically, the concept of periphery has expressed a distance (or separation) with respect to a core, in terms of geographic, economic, political, and social factors (Bourne 2000). The addition of new urban agglomerations far from existing poles, the "peripheralization" of areas that had no peripheral characters previously following changes in economic and social conditions (e.g., migration), and the development of infrastructure are the most relevant of these processes. Peripheries have been characterized by particular features such as remoteness, isolation, and harsh natural conditions, but, on the other hand, they could sometimes offer favorable conditions to attract new urban developments.

Among the different types of peripheries, peri-urban contexts are located somewhere in between the urban core and the rural landscape (Meeus and Gulinck 2008) and represent an "uneasy phenomenon" (Allen 2003) to be defined, both geographically and conceptually. Attempts in establishing a comprehensive set of criteria for the definition of peripheral landscapes which make it possible to capture their

D. La Rosa (✉)
Department Civil Engineering and Architecture, University of Catania, Catania, Italy
e-mail: dlarosa@darc.unict.it

D. Geneletti
Department of Civil, Environmental and Mechanical Engineering, University of Trento, Trento, Italy

M. Spyra · C. Fürst
Institute for Geosciences and Geography, Dept. Sustainable Landscape Development, Martin Luther University Halle, Halle, Germany

C. Albert
Institute of Environmental Planning, Leibniz Universität Hannover, Hannover, Germany

© Springer International Publishing AG, part of Springer Nature 2018   89
A. H. Perera et al. (eds.), *Ecosystem Services from Forest Landscapes*,
https://doi.org/10.1007/978-3-319-74515-2_5

different features (Piorr et al. 2011) are incomplete or not robust enough to ensure transferability to other geographical contexts besides the ones where they have been elaborated. The changing, dynamic, complex, and heterogeneous nature of peripheral landscapes remains irreducible to single interpretations and approaches for their definitions and therefore for their planning and management.

Many attempts have been made to identify and classify peri-urban areas using parameters such as urban centrality, hierarchy, urban–rural relationships, and the degree of urbanity and remoteness (OECD 2002; Dijkstra and Poelman 2008; EUROSTAT 2010). All these research showed the limit of using administrative units such as NUTS levels as geographical units of the analysis, therefore not considering that the spatial extent of peri-urban areas cannot be reducible to administrative boundaries.

For European countries and in the context of the PLUREL project, Zasada et al. (2013) delineated a method to identify degrees of peri-urbanity by using population density of particular classes of Corine land use/land cover and logistic regression models. These authors showed that peri-urban areas occur in the United Kingdom, the Netherlands, Belgium, Northern Italy, Western and Southern Germany along the Rhine valley, and in Southern Poland, mainly within bigger urban conurbations or metropolitan areas.

This chapter outlines the main characteristics and peculiarities of peri-urban landscapes and introduces examples of planning approaches and topics that can be found in current research about sustainable planning. We will refer to peri-urban landscapes as those areas that are partly located outside the more compact part of a city and can spread to the surrounding rural area following low-density patterns of development and covering larger areas than peri-urban neighbors of single municipalities. They are characterized by low density and a mixture of diverse land uses, including non-urban and seminatural uses (Gallent et al. 2006). Peri-urban landscapes represent a diffuse and blurred territory where urban and rural development processes meet, mix, and interact at the edge of the cities.

Being at the edge of cities' limits, peri-urban areas are planned through diverse instruments or schemes: they can be planned by municipal land-use plans regulating the use of the land within administrative border of the municipality (master plans, land-use plans, zoning regulations) or be under the spatial jurisdiction of metropolitan, regional, or landscape plans implemented by regional authorities. This means that peri-urban areas can be planned under diverse planning levels, therefore requiring an appropriate coordination (see section "The need of a metropolitan planning and governance for the peri-urban").

One of the most common features of peri-urbanization processes deals with the progressive colonization of the agricultural and forest landscapes through different land-use changes (Geneletti et al. 2017). Peri-urban areas are progressively acknowledged as areas with peculiar features. Some authors highlight that new functions, not properly urban or fully rural, emerge in these spaces (Korthals Altes and van Rij 2013).

Peri-urban landscapes cannot be understood only in terms a progressive intensification of urban functions in the rural or seminatural environment but rather as a space of "interaction between urban and rural elements" (Rauws and de Roo 2011, in Loupa Ramos et al. 2013).

The following subsections of this chapter introduce and describe different types of peri-urban landscapes, particularly focusing on some categories of forest, agricultural, and other ecosystems that can be found in peri-urban contexts, often highly mixed and intertwined with other human uses of the land. These categories represent general families of landscapes that can be found in peripheral contexts of Europe, where a varied range of mixed land uses and land covers can be observed in areas where the influence of humans is dominant. Despite the urban environment these categories belong to, they refer to particular urban ecosystems characterized by low or null presence of built-up areas. Some particular types of ecosystems, such as private green spaces or domestic gardens (DTLR 2002), will not be included in the previous categories because they are mostly part of urban patches and private owned.

As it will be described, these landscapes are able to provide important functions and relative ecosystem services, such as biodiversity in urban areas, production of O_2, reduction of air pollutants and noise, regulation of microclimates, reduction of heat island effect, and supply of recreational value, and play a fundamental role in health, well-being, and social safety (La Rosa and Privitera 2013; Vejre et al. 2010).

1.1 Peri-urban Forests

In this chapter, urban and peri-urban forests are considered as the most natural ecosystems in an urban–rural context, whose composition, structural diversity, and overall character rely greatly on the demands for (non-monetary) goods and services. An accepted definition of urban forestry is the one based on Miller (1997), who describes urban forestry as "the art, science, and technology of managing trees and forest resources in and around urban community ecosystems for the physiological, sociological, economic, and aesthetic benefits trees provide society." Therefore, urban forestry is strictly related to the positive impacts of trees on human well-being (Fig. 1).

According to Forrest et al. (1999), a range of possible definitions of urban forests have been used in different European countries, demonstrating how the concept and term are open to different interpretations and planning approaches. These definitions highlight once again one of the most important features of urban forest, that is, their ability to connect the human need for the natural environment in urban areas with life support systems of a persistent forest ecosystem. This connection substantially contributes to the well-being of urban societies.

A comprehensive review of definitions of an urban forest is provided in Konijnendijk (2003): in this work the author focuses on the difficulties in finding a shared definition of what is meant by "urban" or "forest." The term "forest," for instance, may be related to its more traditional definition, while in urban areas terms such as "other wooded land" and "trees," used by FAO for its forest resource assessments (FAO 2002), can be particularly more appropriate to describe urban parks, gardens, and street trees. By including small woods, parks, and gardens with

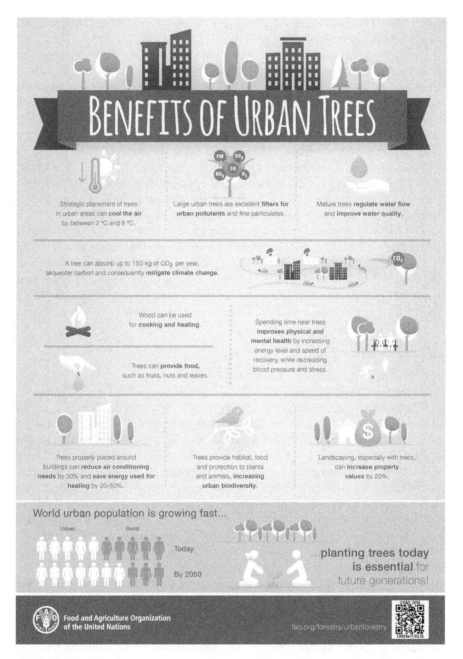

Fig. 1 Benefits of trees in urban contexts (Source: FAO, available at http://www.fao.org/documents/card/en/c/427898a5-e452-4dbb-87ed-4b25286de3b4)

an area size or canopy cover that are below thresholds for "forest," the traditional forest concept has been broadened considerably.

This definition can be also extended to peri-urban contexts, located between the urban core and rural or (semi)natural surroundings, where the size of forests could be larger but proximity and accessibility by urban residents of the city centers are lower. Peri-urban forests form a kind of mixed system, with higher societal influence on management objectives compared to other sides, but still acting as connective element to rural sites with their demands on classical forest ecosystem services.

Furthermore, as it will be discussed later on, peri-urban forests can suffer from high pressure of urban development or request for further farmland. However, the difference between urban and peri-urban could be very smooth and difficult to be defined, as boundaries of cities are extremely difficult to identify, especially in large or sprawled urban areas or metropolitan regions.

Peri-urban forests are particularly under pressure as they are continuously used for recreation and (non-) monetary provisioning services (mushrooms, berries, hunting, drinking water), while they can supply many regulating ecosystem services (e.g., providing cool, clean, and fresh air to the urban environments, protection against flooding).

As the actor groups in peri-urban forests are much more complex compared to pure urban or rural forests, societal processes can be considered as the key drivers in how intense and with which key objectives peri-urban forest planning and management are conducted. Being part of urban systems, some actors might expect well-designed road infrastructure for hiking, cycling, horse riding, or country skiing and relative good accessibility and the availability of parking space for these activities. This might require, for example, more investments into the nice design of forest edges with more mixed or deciduous tree species and more structural diversity. Indirectly, the increased usage of these forests for recreational activities leads also to more needs for protective measures, for example, against further urban development (see section "Peri-urbanization and sprawl processes") or forest fires (fire strips). On the other hand, in their more rural context, peri-urban forests are expected to provide also jobs and traditional forest products such as timber (lumber, fuel wood, industrial wood) for creating income and contribute to sustainable rural development. In addition, expectations to conserve a high biodiversity levels are enhanced through the more intense perception of biodiversity from urban contexts.

Urban forest structure is a determinant of ecosystem function, which has been documented as a mean of mitigating environmental quality problems associated with the urban-built environment (Nowak et al. 2006). The structure and subsequent function of the urban forest will therefore determine the provision of ecosystem services and goods (De Groot et al. 2010). Thus, by modifying the structure of the urban forests, as well as their size and composition, planners may be able to modify certain ecosystem functions in order to maximize human well-being in cities. From a planning perspective, peri-urban forests should connect the more rural landscape parts with the rest of the urban green infrastructure to ensure that all relevant cultural and regulating services are sustained (see section "Trace-Gas Driven Ecosystem Services"). On the other hand, disservices from the movement of some species such as foxes, wild boar, or other animals might become an issue for peri-urban residents.

1.2   Farmlands and Peri-urban Agriculture

Agricultural areas, in use or abandoned, are one of the most typical landscapes of peri-urban contexts and can be the result of fragmentation processes due to urbanization pressures (Fig. 2). Agriculture in metropolitan areas contrasts sharply with its non-urban counterpart. As observed by Heimlich (1989), the longer areas are affected by urban pressures, the greater the adaptation they reflect in some farm characteristics. Since these areas are part of wider metropolitan contexts, their

a

b

Fig. 2 Example of cultivated vineyards (left 2**a**) and abandoned agricultural terraces (right 2**b**) in Italy located in peri-urban contexts in Sicily (southern Italy)

services assume higher importance for the number of people that can benefit from them (Swinton et al. 2007). In fact, agriculture both provides and receives services that extend beyond the provision of food, fiber, and fuel, so that only in their absence do they become most apparent (Fig. 3). Among the managed ecosystems, farmlands offer special potential because of their variety of generated ecosystem services. This potential arises from both their broad spatial extent and human management objectives focused on biotic productivity (Swinton et al. 2007). At the same time, agriculture offers an important potential to diminish its dependence on external agrochemical inputs by reliance on enhanced management of supporting ecosystem services (Fig. 3).

New Forms of Urban Agriculture (NFUA) are typical in peri-urban contexts and are characterized by high level of multifunctionality and general post-productive attitude (Zasada 2011). Urban agriculture is defined as "the growing, processing, and distribution of food and non-food plant and tree crops in farmlands that are mainly located on the fringe of an urban area" (Zezza and Tasciotti 2010). A growing evidence from empirical and experimental research also suggests that incorporating NFUA into the urban environments may improve the sustainability level of cities, taking advantage of the multiple benefits and services that can be generated.

Urban agriculture is particularly present in developing countries and often produces perishable products such as fruits and vegetables. This type of agriculture

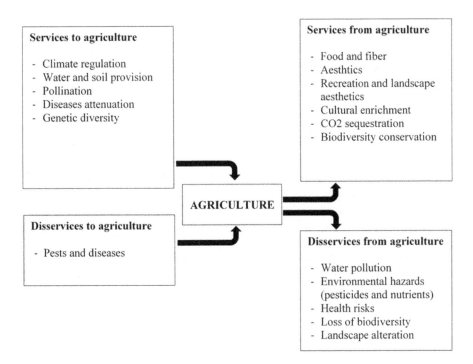

Fig. 3 Main ecosystem services and disservices of agriculture (Modified from Swinton et al. (2007))

meets a local and growing urban demand for food, but it also generates an intensifying conflict "between the maintenance of local agricultural production and the rapid and often uncontrolled consumption of land by growing urban activities and infrastructures" (Aubry et al. 2012).

In China, peri-urban agriculture has also been characterized by the specialization and diversification of traditional agriculture. In the Beijing areas, such NFUA have been mostly initiated by local residents and include agro-tourism, enterprise-based food processing, high-tech agro-enterprises/agro-parks, and farmer collective activities (Yang et al. 2016). In Europe and North America, NFUA are emerging in response to low-density urbanization patterns and aim at producing "local" food as a way to enhance food security by shortening food supply chains (Benis and Ferrao 2017).

A highly differentiated set of NFUA can be found in peri-urban landscapes (La Rosa et al. 2014). Urban farms represent a partnership of mutual commitment between farms and communities of users/supporters which provide a direct and short link between the production of agricultural goods and their consumption. Community-supported agriculture consists of agricultural practices that are directly economically supported by users and communities that take advantage of food produced in the supported farms. They can provide environmental benefits due to an environmentally friendly production process as well as reduced "food miles" thanks to the proximity of production and consumption. Allotment gardens are more oriented to generate social values, including active participation in the management of gardens by particular social groups such as children and retired or unemployed people. Finally, agricultural parks are larger agroforestry systems where food production (mainly by private farms) is promoted and safeguarded along with more general rural and seminatural landscapes. They are public-managed areas that support existing wildlife management and protection and promote the fruition and access of the park, therefore providing important cultural and aesthetics services.

Peri-urban agriculture differs from urban agriculture often practiced by urban residents as part-time activity on available open spaces. Peri-urban agriculture is characterized by small- or medium-sized farms in urban fringe areas, where these farms have to deal with both market globalization and urban urbanization processes (Clark and Munroe 2013). In between globalization and urbanization, peri-urban agriculture is therefore struggling to re-create networks of food provision that are alternative to the global agri-food system that is consumed in cities (Paül and McKenzie 2013).

Farmlands within or near towns are no longer considered simply as reserves of land for future urbanization and are becoming a challenging issue in urban planning that aims at conserving and enhancing productive function and ecosystem services provided by urban and peri-urban farmlands. There are two key concepts that must be kept in consideration by planners when dealing with urban and peri-urban agriculture: the sustainability of the production, both at the farm (internal) and territorial (external) level, and the multifunctionality of the activities achieved by agriculture (Aubry et al. 2012).

1.3   Other Types of Peri-urban Landscapes

Peri-urban landscapes include other forms of seminatural ecosystems, which, in Europe, are mainly made up of shrub and grass vegetation, typical green elements with a limited height of less than 5–6 meters. In Mediterranean areas, shrublands are ecosystems with a long history of grazing by domestic animals, and their biome can reach its maximum extent. Much of these formations are considered a subclimax developed on degraded and eroded soils and maintained in part by fire and goats or sheep. In arid and semiarid areas, such as Mediterranean landscapes, low amounts of rain do not allow for a continuous vegetation cover, resulting in a typically patchy landscape. In addition to their role in plant interactions, shrubs strongly modify plant dispersal patterns by processes such as trapping of water-, wind-, and bird-dispersed seeds (Aguiar and Sala 1999). Thus, they are a key element for community structure and dynamics in semiarid ecosystems, and they act as "hot spots" of diversity in these areas (Pugnare and Lázaro 2000).

In peri-urban contexts, grasslands are habitats that can be present in parks, brownfields, and other derelict land, disused quarries, and along roads or transportation buffers. The type of grassland varies with the geographic features, acidity of soils, and moisture level (dry or damp grassland). They also support a range of grasses and wildflowers, such as grasses, forbs, shrubs, trees, vertebrate animals, and invertebrate animals. Remnant seminatural grasslands, in particular those serving as habitat fragments, are essential to the maintenance of diverse terrestrial arthropod communities in human-dominated landscapes. These temperate biomes are extremely important, as they include diverse and productive terrestrial ecosystems that are among the most threatened in the world, suffering from pressures by urbanization and agricultural processes. In many urban contexts, these areas are often restricted to linear remnants along roads and railways. These linear patches are, however, at a great risk of edge effects that alter vegetation composition by promoting exotic species invasion (Forman 1995). Understanding and mitigating these impacts are of increasing importance for biodiversity conservation in peri-urban areas.

Another important category of peri-urban landscapes are urban lawns, typical and frequent urban biotopes in cities, especially found in urban parks, private gardens, playing fields, golf courses, public places (squares, plazas, etc.), schoolyards, and along streets, roads, and tramways. The presence of lawns is also widespread in private gardens and front- and backyards, especially in suburban areas, where detached houses represent the typical form of urban settlement.

All these types of peri-urban landscapes are particularly sensitive to human activities and continuously under pressure from them. Characterizing the different drivers of changes in peri-urban landscapes can provide relevant information to set up policies aimed at their protection and management.

2 Drivers of Changes in Peri-urban Landscapes

All types of landscapes described in the previous section may provide a complete array of ecosystem services, as also described in chapters "Ecosystem Services From Forest Landscapes: An Overview" and "Ecosystem Services From Forest Landscapes: Where We Are and Where We Go".

However, peri-urban landscapes have a unique characteristic that makes them highly different from other landscapes: their proximity to or partial inclusion in urban systems makes them particularly vulnerable to pressures by urban development or related activities. Peri-urban landscapes have gone through a series of socioeconomic transitions that have deeply modified their territorial assets and spatial land uses. Particularly, agricultural and seminatural areas have been deeply affected by low-density urban developments. Such developments have fragmented farmlands and seminatural areas, producing not-continuous, low-density, and highly mixed urban patterns.

2.1 Peri-urbanization and Sprawl Processes

Despite a decreasing population in many European countries, urban expansion due to spatial development pressure has been an impressive driver of very high consumption of land and agricultural resources. In the period between 1990 and 2000, at least 2.8% of Europe's land experienced a change in use "including significant increase in urban areas" (Commission of the European Communities 2006). The European Environment Agency (EEA) has described the process of urban sprawl "as the physical pattern of low-density expansion of large urban areas, under market conditions, mainly into the surrounding agricultural areas" (EEA 2006). It is an urban development process that "separates where people live from where they shop, work, recreate and educate—thus requiring cars to move between zones" (Sierra Club 1999).

Sprawl is the leading edge of urban growth, and it is usually related to limited planning control in land allocation. Urban development is usually patchy, scattered, and strung out, with a tendency for discontinuity. It leapfrogs over areas, leaving agricultural enclaves (Fig. 4). Sprawling cities, the opposite of compact cities, are full of empty spaces that indicate the inefficiencies in development and highlight the consequences of uncontrolled growth (EEA 2006). More recently, EEA has advocated for a reuse of developed land that is not used anymore to address the risks of further sprawl (EEA 2015).

Among all definitions that can be found in the literature, some recurrent terms highlight the main (negative) features of sprawl: "spreading," "scattered," "low density," "car dependent," "environmental externalities," and "social disparities." The externalities and impacts of sprawl on the environment and landscape have been the focus of several studies and include the loss of fragile environmental lands, increases in air pollution and energy consumption, decreases in the aesthetic appeal of the landscape, the loss or fragmentation of farmland and forests, a reduction in biodiversity, increases in water runoff and risks of flooding, and ecosystem fragmentation (Johnson 2001).

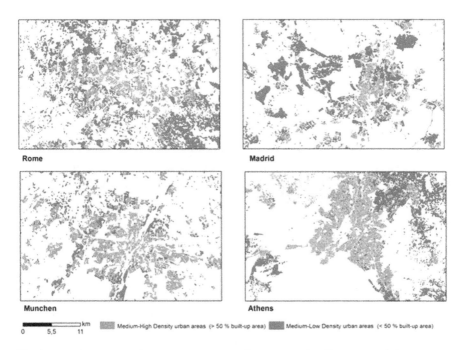

Fig. 4 Low-density peri-urban areas in the metropolitan areas of four European capitals according to 2012 Urban Atlas data (EEA 2015). It can be seen that low-density peri-urban patches are largely the prevailing categories of urban areas when considering the metropolitan contexts

Land sustains many ecosystem functions (e.g., production of food, habitat for species, recreation, water retention, and storage) that are directly linked with existing land uses. Impacts on natural areas are also exacerbated by the increased proximity and accessibility of urban activities to these areas, which in the past were farther from "urban influence." This proximity produces stress on ecosystems and species through noise and air pollution. Moreover, the fragmentation caused by transport infrastructures and other urban-related activities creates significant barrier effects that can degrade the ecological functions of natural habitats. From an ecological point of view, fragmentation can heavily modify corridor spaces for species or can isolate populations by reducing habitats to extend below the minimum area required for the survival of these species. The loss of agricultural and forest land also has major impacts on biodiversity, involving the risk of losing some valuable biotopes for many species, particularly birds.

According to the EEA (2006), in Europe urban development tends to "consume the best agricultural lands, displacing agricultural activity to both less productive areas (requiring higher inputs of water and fertilizers) and more remote upland locations (with increased risk of soil erosion). In addition, the quality of farmlands that are not urbanized but in the vicinity of sprawling cities has also been reduced." The loss of agricultural land is often directly connected to land consumption due to sprawl processes (Thompson and Stalker Prokopy 2009). There are several consequences to this: landscape fragmentation and simplification, loss of biodiversity, decreased agricultural land value, and increased externalities of urban sprawl (Johnson 2001). New urbanizations often occur in proximity to already urbanized areas or existing infrastructure because the price of agricultural land is lower if compared to residential zone land. Agricultural land usually becomes a highly attractive target for investors and urban developers (EEA, 2006). For these reasons, the hazard of loss of agricultural land may be potentially higher in areas close to already urbanized lands or roads. In contemporary metropolitan contexts, rural land and its agroecological features are exposed to dramatic pressures that are driven by the expansion of the urban influence on areas that once were considered as purely rural (Donadieu 1998). In this context, farmlands suffer from a wide range of pressures by urbanization processes. These pressures are physical, environmental, and socioeconomic (EEA 2006).

Urban developments in peri-urban contexts are not continuous and show low-density patterns so that outside the main city, the landscape is characterized by a strong degree of farmland fragmentation and mixes of urban and non-urban uses. The relationship between the agricultural landscape and the city has produced a particular contemporary peri-urban landscape, where residential low-density settlements are mixed with farmlands that have been partially modified and reduced by urbanizations. A low-density settlement has widely become the main landmark of new metropolitan areas.

More and more people in Europe are moving away from the center of metropolitan areas, apparently attracted by the imagined quality of life in these rural settings, to live in residential developments built on converted peri-urban farmlands. "The detached terrace-houses and semi-detached houses condense the new type of residential landscape in the metropolitan peripheries of the cities of southern Europe" (Munoz 2003), and settlements belonging to different municipalities, once far from another, are getting closer and closer and become parts of larger metropolitan areas.

In these new metropolitan areas, the concept of rural–urban fringe, as appeared in the geography and planning literature from the 1930s (Whitehand 1988), is today more and more smooth, and it may be difficult to distinguish what is urban from what is rural. A chaotic set of land uses is "a product of post-war planning legislation that has partly fossilized some patterns of use, but it is also a reflection of dynamic change as certain components of these areas have grown as part of complex and singular developments" (Gant et al. 2011). Moreover, in new metropolitan contexts, rural land and the relative ecosystems are exposed to dramatic pressures that are driven by the expansion of the urban influence on areas that once were considered as purely rural (Donadieu 1998).

What's left today of the seminatural and agricultural areas in peri-urban landscapes? A different mix in types and sizes of residual and non-urbanized areas deeply characterizes metropolitan landscapes in many European regions, such as farmlands (abandoned or still in use), small orchards, wood and shrub areas, local parks, regional parks, reserves and natural protected areas, and grasslands (Fig. 5).

Gallent and Shaw (2007) identified a number of anthropic land uses in the transition zone from urban to rural of the greenbelts in the United Kingdom: (i) service functions and commercial activities, (ii) noisy and unsociable uses pushed away

Fig. 5 Examples of non-urbanized areas of different types and sizes in peri-urban contexts: agricultural spaces and other non-urbanized areas are intertwined with low-density urban settlements in the metropolitan area of Rome

from people, (iii) transient uses such as markets, (iv) bulk retail, (v) light manufacturing, (vi) warehousing and distribution, (vii) public institutions, (viii) degraded farmland, (ix) fragmented residential development (often centered on road junctions), and (x) areas of unkempt rough or derelict land awaiting reuse. These land-use patterns are very similar to the ones that can be found in other European contexts, with various ranges of size and extent.

2.2 Climate Change

Climate change has been predicted to have many consequences for human health arising from the direct and indirect impacts of changes in temperature and precipitation (Patz et al. 2005). One of the primary public health concerns is an increase in the intensity and frequency of heat waves, which have been linked with heat stroke, hyperthermia, and increased mortality rates (Tan et al. 2004).

These consequences appear to be more dramatic in urban and peri-urban areas, which will be especially vulnerable to the negative aspects of climate change (such as more frequent and severe floods and heat waves), due to the higher concentration of people and human activities, although at a lesser extent than in dense urban areas. Climate change impacts on peri-urban landscapes include impacts on the peri-urban agriculture systems: for example, impacts of flooding, groundwater salinization, sea level rise, heat stress, drought, and changes in resources availability are likely to intensify with climate change and especially in Africa and Asia (Padgham et al. 2015). Therefore, the existence of peri-urban agriculture can be threatened by the convergence of urban development (as discussed in the previous section) and climate change pressures.

Thus, there is a pressing need to evaluate strategies that may adapt against further increases in temperature in peri-urban areas and the associated negative impacts on human health. The most common adaptation strategy is to "green" urban areas, essentially by increasing the abundance and cover of vegetation (Gill et al. 2008). As a complement to such adaptation measures, particularly in peri-urban contexts, there is a need to ensure that future land-use development does not worsen the current risk level (especially hydrological risk), either through influencing the hazards themselves or through affecting the future vulnerability and adaptive capacity of the urban system.

Spatial planning of peri-urban landscapes therefore has a critical role to play in mitigating the severity of hazards and in reducing the levels of exposure and vulnerability experienced by the urban system. Different scales of planning from macroscale land-use planning to microscale urban design are both important to this process, responding to the different scales over which risk and vulnerability are expressed (O'Brien et al. 2004).

This recognizes that although many aspects of adaptive behavior associated with vulnerability reduction strategies are the result of a decision-making process that operates at an individual level, the government and other policy makers can address this process through their activities. Given the length of time involved in the strategic planning process, and the long lifetime of urban infrastructure, it is critical that decision-making aimed at mitigation of or adaptation to climate change does not reinforce negative feedback in any part of the process (Lindley et al. 2006). The urgency for information to assist with "climate conscious" planning is evident and asks for detailed tools for the assessment of different urban features that are involved in climate change processes.

For peri-urban forests, an increased awareness of climate change leads in many countries to an increase in the harvesting of fuel wood through private actors, so that in trend, less woody debris are available for supporting biodiversity and for being incorporated in the organic matter cycles. Other indirect drivers connected with urban activities are larger emissions of pollutants, namely, NOx and particulate matter. These disturb matter cycles, might lower the competitiveness of some species, and thus shape the development of forest species communities. Forests close to large urban congestions often suffer from decline and are more vulnerable to climate change. Typical climax communities might now develop due to these disturbances.

2.3 Farmland Abandonment

Among the many available definitions, farmland abandonment can be defined as the cessation of land management which might lead to modifications in biodiversity and ecosystem services provision (Terres et al. 2015). There are several different reasons for it, and these reasons are often concurrent, hardly separable, and context specific. Drivers of abandonment depend on the result of their co-occurrence and interactions (Coppola 2004) and can be natural/geographical constraints (including changes in geo-climatic features), land degradation, socioeconomic factors, or political changes in national or regional assets.

Terres et al. (2015) classified the driving forces of abandonment into unsuitable environmental conditions, low farm stability and viability, and the regional context. They identified the most relevant drivers as low farm income, low farm dynamism/ adaptation capacity, aging farmer population, low farmer qualifications in farm management, small farm size, and enrollment in specific agricultural schemes. Drivers from the regional context were identified as the presence of weak land markets, previous farmland abandonment, and remoteness and low population density.

In peri-urban contexts, processes of farmland abandonment are also linked to sprawl processes (Thompson and Stalker Prokopy 2009). Urban development and agriculture compete for the same land, as farmlands closer or adjacent to urban areas are ideal places for urban expansion. Farmers' reasons for selling farmlands in this process are clear, as they can get substantial financial benefits by the sale of farmland for new housing or other urban developments, especially in times of a general crisis of agriculture. On the other hand, agricultural soils need to be conserved, since they are almost nonrenewable resources and soil sealing reduces or eliminate soils' capacity to perform their essential functions.

Farmland abandonment can also generate contrasting perceptions in people living in peri-urban areas (Benjamin et al. 2007). Abandoned farmlands can be seen as "useless spaces" with no proper status or even as elements not aesthetically pleasant or even unsafe. But they can also generate poetic connotation and feelings of freedom or be considered as important ecological spaces where natural field succession processes are taking place.

This contrasting perception by residents and neighbor farmers should be carefully considered when imagining new planning scenarios for abandoned farmlands. In fact, because of their proximity to city but also to existing farmlands or forests, abandoned farmlands in peri-urban areas represent an interesting opportunity for the sustainable spatial planning of metropolitan areas, as they can be considered as new components of new agricultural landscapes (see section "Planning New Forms of Agriculture in peri-urban contexts"). Proximity to the city can provide an advantage for diversification and innovation, offering new opportunities for farmers to sustain or even increase their income by reaching new short-distance markets (Benjamin et al. 2007).

3   Existing Sustainable Planning Approaches for Urban Peripheral Landscapes

Sustainable planning can be considered as a combination of knowledge, science, and creativity to design, evaluate, and implement a set of justified actions in the public domain, which encompass the different dimensions of sustainability such as environment, economy, and social sphere (Friedmann 1987; Berke and Conroy 2000). In this section, we present some examples of planning approaches, solutions, and topics proposed by current academic research and planning practice that might be suitable to be applied to define new planning scenarios aimed at conserving and/or enhancing the sustainability of peri-urban landscapes as defined in section "Peri-urban Landscapes".

3.1   Planning and Design of Peri-urban Greenery

One of the objectives of sustainable spatial planning is to promote equitable access to social and economic resources and therefore improve environmental health of people living in urban contexts (Berke and Conroy 2000). To this end, socially inclusive planning approaches to greenery in peripheral urban contexts should maximize its social benefits based on convergence of human interests (accessibility and qualities of goods and services, culturally appropriate development and fulfillment, self-reliance, etc.), considering equity and disparity within the current population and between present and future generations (van Herzele et al. 2005). This is particularly relevant in peripheries worldwide, where access to resources is often limited or disputed among different social groups. Since access to green spaces is important to human health and well-being, the reduction of the uneven distribution of green spaces within cities (especially those most populated) is one of the key objectives of sustainable planning (e.g., Dai 2011), as urban areas with lowest green land covers have been related to residents with lower socioeconomic status (Aquino and Gainza 2014).

However, within the large body of research on accessibility to greenery (e.g., Neuvonen et al. 2007; Schipperijn et al. 2010; Sugiyama et al. 2008; Swanwick 2009; La Rosa 2014), peri-urban areas have been less explored. Green spaces located outside the urban core such as seminatural areas, woodlands, fringe forests, country/ agricultural parks, and peri-urban open spaces are appreciated by users for their recreation and leisure activities even more than intensively maintained green areas (Žlender and Ward Thompson 2017), because they are able to provide a diverse kind of "nature" and satisfy different recreational needs (Rupprecht et al. 2015).

Žlender and Ward Thompson (2017) recently compared two cities (Ljubljana and Edinburgh) with relative different green space strategies for the peripheries (green wedges for Ljubljana, greenbelts for Edinburgh) and demonstrated how the specific strategy of each city affects people's access and their use of peri-urban greenery. While the strategy of green wedges for Ljubljana is used by people because they reach the city center from periphery, the greenbelts in Edinburgh are mostly used for recreational purposes much less than the green spaces within the city (Žlender and Ward Thompson 2017).

This research also highlighted the importance of preference for greenery of different social groups as important information for urban planners. Results from the same authors showed that residents of the most central parts of cities preferred seminatural green spaces and other linear greenery (e.g., green corridors) that can be easily accessed from home. Tu et al. (2016) explored the heterogeneity of people's preferences for green spaces by using a choice experiment in Nancy (France). Authors showed that the willingness to pay for having peri-urban forests in the vicinity of their home increases with the frequency of forest visits, although the respondents' preferences varied significantly with income differences and the possible ownership of private green (as a substitute for being close to parks).

Shkaruba et al. (2017) explored how green space planning has been affected by the interplays of socialist and post-socialist systems, in the context of rural–urban peripheries of two middle-sized cities in Belarus (Mahiliou̯) and Russia (Pskov). Authors discussed how planning options in the two cities are looking for a compromise between a compact city cherished by the socialist planning tradition (still supported by existing spatial regulations and frameworks) and an increasing tendency toward urban sprawl as the western way of modern development. These options have consequences for green spaces that remain somehow under high pressure by urban development: in fact, the most common outcomes of urban development include ecosystem fragmentation, major disturbance of ecosystems, and loss of forest and other valuable ecosystems, and these negative outcomes can be the results of planning choice or failures of planning implementation (Shkaruba et al. 2017).

Conedera et al. (2015) performed a quantitative survey in a peri-urban area of the Southern Alps in Switzerland about the importance of green and the frequency of the visits to green spaces. Results showed that maintaining a visual relation with the green area and vegetation is important to the perceived general quality of life for the peri-urban residents that live far from the city center and closer to the mountain slopes. These findings suggested that land planners and managers should consider the proximity of the place of residence and the background green of the mountain slopes, for example, by ensuring and conserving visibility of the greenery when designing urban development.

3.2 Ecosystem Services-Based Planning

The integration of ecosystem services into spatial planning has recently attracted interest of current research about sustainability issues (see chapter "Ecosystem Services From Forest Landscapes: An Overview"). Spatial planning processes lead to decisions that usually modify land uses and may affect the quantity, quality, and distribution of a wide set of ecosystem services that are benefited by humans. Hence, it is crucial to use information on ecosystem services to support planning processes (Geneletti 2013).

Many scholars believe that ecosystem services might be able to improve decisions on land use by adding the information on the services (with relative values) provided by ecosystems in an urban context and also highlighting trade-offs among different planning scenarios (Albert et al. 2014; Dorning et al. 2015). Several authors have suggested that the ES concept has a potential to facilitate land-use planning and landscape governance by facilitating knowledge exchange between involved stakeholders and connect them at different spatial scales or administrative levels (Opdam et al. 2015). Particularly, the spatial dimension of ES is a key issue for involving stakeholders in the planning process, since they are usually more interested in knowing where a decision is made rather than the reasons behind the decision itself (Fürst et al. 2014).

However, the integration of ES in real planning processes and the use of information coming from ES assessments are still not consolidated and/or not yet producing relevant results in terms of improved sustainability, especially for urban systems (Haase et al. 2014).

There are several reasons for this incomplete integration, such as differences in terminology, the emphasis on existing assessment methods and economic values, and the dominant scale of application (Opdam et al. 2015). Also, the lack of binding norms in national planning systems hampers or delays the integration. To this end, Woodruff and BenDor (2016) believe that the missing integration between ES and planning is also due to the inability of plan quality guidance to incorporate ecosystem services and to guide practitioners in how to include ES information to improve spatial plans.

Geneletti et al. (2017) showed that in peri-urban landscapes, ES-based planning approaches have been rarely applied and that the research on these contexts is still limited and under development. Some exceptions are present in researches that explore how to plan new spatial configuration of remnant peri-urban agricultural lands and other types of non-urbanized areas in new planning scenarios (Lee et al.

2015; La Rosa and Privitera 2013). The management and protection of services by agro-ecosystems is considered crucial in the context of urban growth of peri-urban landscapes and thus appropriate tools to inform and guide planning choices for highly complex landscapes such as those in peri-urban areas. Focusing on farmlands as part of the peri-urban green infrastructure (see section "Trace-Gas Driven Ecosystem Services"), Lee et al. (2015) proposed a set of metrics to assess ecosystem services with landscape composition and configuration metrics for each of the research sites. Results for the case study of a plain area in Taiwan showed that agroecosystem services are related with the spatial configuration of paddy rice fields and that it is possible to guide the agricultural land-use change to optimize spatial configuration and therefore to conserve the agroecosystem services—especially the regulation of potential flooding events.

La Rosa and Privitera (2013) developed a planning scenario of new land uses for existing open unmanaged spaces in peri-urban contexts of South Italy, by evaluating their suitability to new land uses that increase the overall provision of ecosystem services for the entire metropolitan area. The obtained results showed a new spatial configuration of land use that provide municipalities or other metropolitan public bodies in charge of spatial planning (provinces, metropolitan areas) different possibilities for the planning policies aimed at the conservation and increased provision of ecosystem services.

The high complexity of peri-urban contexts in terms of pressure on land and possible conflicts that the use of land can generate characterizes the work by Gret-Regamey et al. (2016) that developed a spatial decision support tool to support the allocation of new urban development zones for the city and hinterland of Thun in Switzerland. The tool evaluates different alternatives of new urban developments based on ecosystem services and locational factors, and it reveals that when ecosystem services are taken into account, the most suitable locations of developments are given by the more compact part of urban centers rather than those in the peri-urban areas. This means that the ecosystem services provided in peri-urban areas were considered important to be conserved by the stakeholders that used the tool.

3.3   Nature-Based Solutions and Green Infrastructure

In urban contexts, there is a growing interest in using and deploying natural ecosystems to provide solutions to several urban issues and improve the overall sustainability of urban environments (Cohen-Shacham et al. 2016). These nature-based solutions provide sustainable, cost-effective, multipurpose, and flexible alternatives for various planning objectives and can significantly enhance resilience of cities. They can be many types of "actions to protect, sustainably manage, and restore natural or modified ecosystems that address societal challenges effectively and adaptively, simultaneously providing human well-being and biodiversity benefits" (Cohen-Shacham et al. 2016). Furthermore, by reshaping the built environment, nature-based solutions can enhance the inclusivity, equitability, and livability of cities, regenerate deprived districts through urban regeneration programs, improve mental and physical health and quality of life for the citizens, reduce urban violence, and decrease social tensions through better social cohesion (particularly for some vulnerable social groups, such as children, elderly, and people of with low socioeconomic status).

Many definitions are available for green infrastructure (GI) (see Pulighe et al. 2016 for a comprehensive review): among the available definitions, Tzoulas et al. (2007) define GI as "all natural, semi-natural and artificial networks of multifunctional ecological systems within, around and between urban areas, at all spatial scales." This definition emphasizes the holistic ecosystem vision of urban environments (including the abiotic, biotic, and cultural functions) and claim for multiscale approaches able to take into account the scale-dependent relationships of ecological processes occurring in cities, with particular reference to the human health and well-being of citizens and residents. For these reasons, GI can be considered as a nature-based solution that has become the focus of increasing interest in sustainability science and planning.

In particular, for peri-urban areas, GI aims at the following actions:

(i) Environmental protection and integration of agriculture into urban context, providing specific new urban agricultural land-use types such as agricultural parks, community-supported agriculture, and allotment gardens. These land uses can provide various improvements, such as increasing local food production in the city, becoming areas for leisure, and supporting the integration of socially deprived population groups.

(ii) Development of suburban green areas in order to provide a more equal distribution of public parks and gardens.

(iii) Enhancement of current urban green spaces by improving quality, usability, and accessibility (La Greca et al. 2011; O'Brien et al. 2017).

According to these objectives, the planning of GI in peri-urban contexts should also include agriculture and farmlands. If green areas act as an infrastructure for the well-being of contemporary society, agricultural areas must be included in this infrastructure of spaces providing ES.

Planning GI requires different strategic objectives to be defined for peripheral landscapes, such as environmental protection, leisure, local green services, and urban agriculture. This might allow the identification of new metropolitan scenarios of land uses (La Rosa and Privitera 2013). In fact, planning strategies for peripheral landscapes should be related to the entire urban and peri-urban surroundings, and metropolitan areas appear to be the most appropriate scale for such scenarios.

3.3.1 Sustainable Urban Drainage Systems

Urbanization processes are responsible for altering natural flow patterns in terms of runoff volumes and peaks. Conventional storm water systems are pushed beyond their drainage capacity and may lead to more frequent and intense floods.

Urban planning can deeply affect the hydrologic response of catchments. Then, understanding potential effects of urban development on the water runoff drainage system represents a crucial issue in the planning process (Miguez et al. 2009), and the use of sustainable urban drainage systems (SUDS) can help minimizing these effects.

Specifically, SUDS are particular NBS that consist "of a range of technologies and techniques used to drain storm water/surface water in a manner that is more sustainable than conventional solutions" (Fletcher et al. 2014). They are based on the philosophy of mimicking the natural predevelopment site hydrology and follow the principles and goals of low-impact development (Ahaiblame et al. 2012). Conventional techniques collect and channel water out of the catchment as fast as possible through structural storm water conveyance systems (channels, pipes, pumps, regulators, and end-of-pipe solutions) at the outlet of a drainage area. On the contrary, SUDS aim at keeping water on-site as much as possible using landscape features and natural processes (Pappalardo et al. 2017).

Despite the relevance of peripheral contexts in current processes of urban development, limited attention has been given to the hydrological impacts of urbanization on previously rural areas. Existing research confirms the evident changes in hydrological regime in peri-urban areas and particularly underlines the complexity of catchments that present a mix of fast and slow hydrologic response as a result of combining artificial with natural flow pathways (Miller et al. 2014).

Two challenges are raised for the adoption of SUDS in peripheral urban landscapes (Barbedo et al. 2014): (i) to promote the preservation of existing (semi)natural ecosystems with related functions and services and (ii) to apply new technologies for the transformation of land and water resources. Peri-urban landscapes are subject to major socioeconomic pressures for further development and land transformations, posing a big challenge to the implementation of measures aimed at the regulating ecosystem services of water runoff.

Barbedo et al. (2014) use a model to test hypothetical changes in the land uses of a coastal city in Brazil. Authors tested how different scenarios of urban densification can respond to the needs of a growing population while safeguarding cultural landscapes of high environmental value. They demonstrated how water flow regulation services of runoff can be improved and that restoring natural functions of peri-urban floodplains may reduce events of urban flooding.

Pappalardo et al. (2016) modeled the effect of urban development for a peri-urban catchment in Italy, evaluating the potential impact of development on the urban storm water drainage systems (Fig. 6). Authors compared flow peak catchment releases under scenarios of pre- and post-urban development and derived a set of flow release restrictions to be included in the local land-use master plan in order to ensure hydraulic invariance in the two scenarios. Results from the modeling showed that release restrictions could be achieved by SUDS modeled for runoff events with low return periods (1–3 years) and that release restrictions should be defined among areas involved in the urban development proportionally to the extent and type of these developments.

Fig. 6 Modeling of the effects of new urban developments on runoff for a peri-urban basin in Sicily (Italy): in white the areas for the new urban developments for which the release restrictions are defined (Modified from Pappalardo et al. (2016))

3.4 Planning New Forms of Agriculture in Peri-urban Contexts

Spatial planners and decision-makers are required to consider New Forms of Urban Agriculture (NFUA), as defined in section "Farmlands and peri-urban agriculture" in peri-urban contexts, since in these areas, low-density urban development keeps growing and threatening agricultural lands (European Environmental Agency 2006). To this end, a better understanding of the different features of current peri-urban landscapes would allow identification of the land uses that are most suitable to fulfill the multifunctional aims of NFUA and take part of new planning scenarios (La Rosa et al. 2014).

Areas for urban agriculture can be planned and designed in different forms and to different scales to contribute to biodiversity conservation and provide a massive range of ecological benefits for urban residents (Deelstra and Girardet 2000). The integration of urban agriculture into densely populated areas might greatly extend opportunities for mixing food production with social, cultural, and recreational functions of urban green spaces (Taylor Lovell 2010).

To be a feasible alternative in cities and cohabit with other urban land uses, urban agriculture should include ecological and cultural functions in addition to the direct benefits of food production (Taylor Lovell 2010). A transition from traditional agriculture into a multifunctional one can produce several benefits for society (Zasada 2011), thanks to the localization of farms near or inside dense urban areas and the consequent easier transfer of services and goods from the agriculture activities to the urban environment.

Urban planning needs to include in planning scenarios for peri-urban landscapes a wide range of functions including urban agriculture and other typologies of green spaces for leisure, biodiversity protection, and recreation. This scenario has to be designed according to the specific features of geographical contexts (Hough 2004). However, the integration of urban agriculture in land-use planning has been seldom considered in top-down urban planning, and urban agriculture practices have often been implemented from the bottom-up and spontaneously (Taylor Lovell 2010).

As an example of planning of NFUA, a recent research by La Rosa et al. (2014) proposed a GIS-based multi-criteria model to check the suitability of land-use transitions of current open spaces (farmlands, abandoned farmlands, seminatural areas, mainly located in the peripheral areas of the city) to New Forms of Urban Agriculture, by delineating scenarios that aim to increase the provision of ecosystem services such as food production in urban contexts and access to green spaces. The model returned some scenarios for NFUA that integrate urban agriculture in peri-urban contexts of the city and provide useful information for urban planning policies aimed at reaching a multifunctional and sustainable land use for current urban open spaces and protecting existing productive farmland from urban development pressures (Fig. 7).

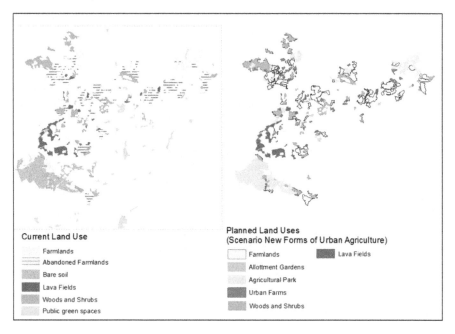

Fig. 7 Example of planned scenarios of New Forms of Urban Agriculture for the peri-urban context of Catania, Italy (Modified from La Rosa et al. (2014))

In an analysis of the peri-urban agriculture in the Beijing peri-urban area, Yang et al. (2016) highlighted the importance of multifunctionality and diversity in agricultural development literature. The authors also recognized the role of the local municipal government in promoting bottom-up local initiatives for the inclusion of these activities into land-use plans. However, both the built-up land and lands needed for peri-urban agriculture activities require collective land with ambiguous property rights, which hinders large-scale projects of peri-urban agriculture and discourages long-term investments (Yang et al. 2016). Provè et al. (2016) suggested that NFUA could hardly benefit from a governance strategy that only stimulates advocacy and institutional support. Adding more specific needs coming from the urban world (e.g., request for specific goods or creating local markets) and integrating other functions (e.g., leisure and tourism) can stimulate peri-urban agriculture toward its full potential. Furthermore, NFUA can be part of municipal programs and investments for public greenery and environmental conservation, but their planning cannot be reduced to the administrative boundaries of a single municipality as their extent go beyond these boundaries. To this end, synergies and coordination among different administrative levels should be pursued within larger metropolitan regions (see next section).

4 Planning and Challenges for peri-urban Landscapes

4.1 The Need of a Metropolitan Planning and Governance for the peri-urban

Current literature is increasingly debating the role of peri-urban areas as part of wider metropolitan contexts (Ros-Tonen et al. 2015; Salet and Savini 2015) that range from large urban agglomerations to smaller local metropolitan areas. This character reflects the manifold links and relations between peri-urban areas and core centers, either in terms of geographical assets, physical connections, and flows (of goods and people) or in terms of processes, including population and urbanization growth and other more specific processes, for example, eco-gentrification (Goodling et al. 2015).

This argument is in line with the current debate on the most effective administrative level at which to plan peri-urban systems (Kline et al. 2014). Understanding the continuous changes that occur in the functional and sociopolitical relations between the urban core and peri-urban areas and framing them in their institutional and administrative context are a prerequisite for effective planning (Salet et al. 2015).

The metropolitan condition of peripheries requires working on more complex relations than those between a specific peri-urban area and its reference core city or rural landscape represented by the traditional core–periphery model. First and most important, peripheries as part of metropolitan systems need to face cross-administrative boundary phenomena and to address the interlinked issues that are relevant at different scales (e.g., the relation between mobility and the urban form).

This is particularly relevant for the accessibility of several urban functions (health, commercial, retail, parks, and other recreational activities) that could be limited or not present in peri-urban areas. In some instances, functions can be shared among different municipalities belonging to the same metropolitan area. Contemporary metropolitan areas require forms and instruments of spatial governance that are able to integrate different planning levels and sectors but are often not presented in national planning instruments or schemes.

Although the dynamics of the transformation of peri-urban areas are certainly not independent from the dynamics of the more central parts in the city, peri-urban areas of developing metropolises exhibit specific characteristics that make their governance (as defined below) a distinctive challenge that deserves the attention of planners. These specific and interlinked characteristics result from a combination of rapid socio-ecological transformations, conflicting stakes and interests, environmental vulnerability, and a lack of an adequate political–administrative jurisdiction.

As an example of the relation between peri-urban areas and metropolitan systems, Padeiro (2016) studied the relation between land-use changes and municipal management plans in the Lisbon Metropolitan Area. The author found that the distance to the capital and former urban dynamics were more significant drivers of land-use change than the land-use plans.

Due to the complexity and highly changing features of peri-urban landscapes, planning of these regions might therefore have to shift away from traditional land-use design and act as a flexible platform to imbalance the activity of public policies and private initiatives, where trade-offs between different uses of the land can be negotiated (Moreira et al. 2016).

4.2 Planning Instruments, Spatial Governance, and Transferability of Approaches

The planning of peri-urban landscapes requires changes in the relations among different administration levels (e.g., local, metropolitan, or regional level), making necessary new political arrangements within the metropolitan areas where they are located. Overall, a combination of local and supra-local schemes (in planning, governance, regulations, and agreements) is viewed as necessary (Webber and Hanna 2014). Furthermore, an effective coordination between different levels of land-use planning is considered crucially important (Carruthers and Vias 2005).

To achieve this coordination, more complex and advanced planning schemes and instruments are needed. In some cases, the integration of different planning levels also means an integration between traditional planning based on spatial administrative units and more innovative tools. To this end, links among different planning levels should be revised and strengthened, involving all levels of the planning process, from the master plan to the subdivision plan and zoning (Lörzing 2006). Many authors suggest a possible combination of different approaches (e.g., traditional and innovative, top-down and bottom-up) that can work at different and integrated scales (e.g., the more strategic at the regional and metropolitan levels with the more operational at the very local level).

Classic planning approaches can be critical to apply in peri-urban and not effective in achieving sustainability. One example is the traditional classic land-use zoning that is viewed as acting as a barrier to sustainable development in peri-urban areas because it fails to consider their complex and dynamic features (Haller 2014) and the multifunctional use of spaces that support socially and environmentally sustainable practices, for example, the use of vacant residential lots for new forms of agriculture (Hara et al. 2013). Other planning instruments, such as master plans or local land-use plans, appear unable to consider peri-urban as part of larger metropolitan regions and are hence inadequate to act as effective solutions (Roose et al. 2013).

Some authors also report concerns about the lack of binding norms to protect peri-urban land that is considered valuable and strategic for sustainable development from urban development. This happens not only with farmlands in peri-urban areas but also with unmanaged open spaces and vacant lands in peripheral neighborhoods, where ecological auto-determination and unplanned but socially relevant land uses can flourish (Foster 2014). Additionally, to a certain degree, regulations are seen as a necessary legal frame to ensure more effective integration of planning choices in metropolitan systems (Carruthers and Vias 2005).

A recent work by Moreira et al. (2016) proposed alternative administrative units than traditional municipalities to better target sectorial policies at local scale within peri-urban contexts. For the Metropolitan Area of Lisbon, authors mapped different peri-urban areas and associated dynamics of landscape change through a set of landscape indicators to identify seven different units where to adopt inter-municipal planning policies and regulations adaptable to manage the urban and non-urban land uses, as well as promoting market tools to regulate land-use change initiatives in the desired directions. Such an approach might be able to avoid planning choices about future land-use and urban functions in metropolitan areas that have been traditionally based on the neat separation of spaces, administrative units, and related spatial policies.

The issue of appropriate schemes of governance for peri-urban landscapes relates to the ongoing debate about alternative modes of spatial governance. Governance is acknowledged as a key issue for these areas, which are frequently divided into different jurisdictions but also characterized by administrative overlaps (Korthals Altes and van Rij 2013) or by political marginality (Cash 2014). These uncertainties may lead to informality in urban development. Multilevel governance (MLG) plays a crucial role to effectively supporting the coordination of planning instruments. MLG is defined as the interplay of institutions, mechanisms, and processes through which political and administrative authority is exercised across different levels (Goldthau 2014). MLG is categorized into two types depending on its orientation toward particular administrative areas or particular policy problems (Hooghe and

Marks 2003). In the first type, bundled MLG, jurisdictional boundaries are separated and not intersecting or overlapping, where each level is assigned distinctive functions and clear lines of responsibility (Smith 2007). Here, the authorities and powers are bundled together within a jurisdiction, with those jurisdictions at the lower level "nested" into higher ones. Type I follows a rather traditional hierarchy of different levels of governance documents. However, its deficit is that it does not react properly to spillover effects, for example, while analyzing ecosystem services benefitting and provisioning areas. In the second type, the flexible jurisdictions form a complex pattern of formal and informal institutions and networks that often overlap with each other. They are no longer related to a jurisdiction but focus on specific policy sectors with task-specific institutions (Hooghe and Marks 2003). Implementing the second type of MLG produces a rich pattern of both formal, statutory spaces and "soft spaces" as more functional, fluid, and governance arrangements. Soft spaces involve creating new functional spaces inconsistent with political territorial boundaries (Allmendinger et al. 2015), which may result in "inefficiencies, spatial externalities, and spillovers" (Moss and Newig 2010).

An example of flexible governance for peri-urban contexts is proposed by Hedblom et al. (2017) with reference to Swedish examples: in Stockholm, the system of green wedges, a landscape previously unrecognized as environmental relevant, has become acknowledged and incorporated in multilevel landscape governance among the municipality, regional authorities, and NGOs. These partners established a long-term commitment and finally formalized a local-level governance structure at local level allowing the conservation of multiple functionalities the wedges provided to the peri-urban population.

Transferability of planning approaches to other geographical contexts is a key issue to evaluate their real flexibility and robustness. Geneletti et al. (2017) found a high level of uncertainty about the real transferability of successful planning approaches to contexts, in which physical, environmental, and socioeconomic conditions are different than the ones where these approaches have been developed (Ryan and Throgmorton 2003). One of the most important issues that make approaches difficult to transfer relies on the institutional variability of planning schemes and related legislation. Many papers highlighted that specific contexts call for specific approaches to incorporate sustainability in several respects of peri-urban planning processes (Todes 2004) and to fit the local specificities of spatial legislation (Harman and Choy 2011). For example, this is the case with approaches of performance-based planning, whose implementation represents a major challenge even for administrations with high institutional capacity (Baker et al. 2006): the possibility of its adoption in other contexts with different planning and governance systems should not be taken for granted and requires further investigation.

4.3 Challenges and Raised Trade-Offs in Planning Approaches

Several challenges can be identified when evaluating the effectiveness of sustainable planning to peri-urban landscapes. One of the most relevant is the aforementioned possibility to use traditional approaches (land-use Euclidean zoning, master plans, spatial regulations), mainly because these instruments may be not able to address the fast-changing features of peri-urban areas. Furthermore, such stand-alone instruments could be vulnerable to formal and informal pressures on planning processes (Mason and Nigmatullina 2011).

Another limitation highlighted by current research raised concerns of the real socio-environmental effectiveness of sustainable planning approaches and the measurability of the environmental effects of the sustainable development of peri-urban areas (Zimmerman 2001). Uncertainties about the short-/long-term environmental sustainability and the potential unwanted outcomes generated from the application of (presumed) sustainable planning approaches have been highlighted. Sustainable models for peri-urban areas have been unable to show whether the solution proposed is ecologically sound and even whether it can be considered livable. An important example of socio-environmental effects of planning is the positive correlation between population growth and the close proximity of peri-urban green spaces in the cities of Antwerp and Ghent (Van Herzele and Wiedemann 2003). This implies that, indirectly, development of peri-urban green spaces can generate more requests for urban development for people wishing to live close to greenery and thus can generate more urban sprawl.

A big challenge is related to the economic resources needed to implement any planning decision. This is a crucial issue, in times when many local authorities (e.g., municipalities) are experiencing a continuous decrease of available budget to be used for the acquisition of land needed to develop new public green spaces or other forms of public service. To this end, alternative sources of funding should be sought, such as grants or incentive schemes, by which landowners could be economically encouraged to directly create or manage new green spaces. Such mechanisms can also produce more effective results if linked to engagement of stakeholders who might provide additional economic support. For example, local communities might be willing to pay a limited fee to have access to green spaces that can be planned by local governments in private lands. Through the budget coming by these fees, the management costs could be covered. Other forms of land acquisition for public spaces include Transfer of Development Rights mechanisms. Landowners and developers exchange a right to build on concentrated portions of property with the obligation to transfer to the municipality the remaining area, zoned for public use (e.g., green spaces). This can increase the overall provision of public green spaces at reduced costs for the municipality (Martinico et al. 2014).

A recent study by Geneletti et al. (2017) reviewed approaches of sustainable planning for urban peripheries and peri-urban areas in particular, revealing challenges and trade-offs that emerge from existing planning research on peri-urban systems. An important category of trade-offs concerns the relation between peri-urbanization processes and the landscapes produced by these processes. For example, Haller (2014) argued that, even if the process of urbanization and peri-urbanization cannot be claimed as positive or negative per se but can produce both positive and negative outcomes, these need to be balanced considering the local socioeconomic and environmental characteristics of the context.

In fact, the possibility of generating sustainable and eco-compatible development can clash with socio-environmental and, particularly, equity issues. New peri-urban developments or retrofitting may generate inequalities by providing opportunities for particular social groups to get preferential access to environmental amenities and therefore allowing an unequal access to ecological/environmental benefits at the cost of low-density urban development (Leichenko and Solecki 2008).

Focusing more on planning approaches, some trade-offs may result from the application of particular spatial planning concepts. An example is the concept of the "compact city," where urban densification and consolidation can generate trade-offs with the condition of urban livability (Westerink et al. (2013). In developing countries, the increase of density as a solution to low-density developments located in peri-urban areas poses issues of availability of green spaces for the growing population (Ramos-Santiago et al. 2014). On the contrary, the request of having pleasant peri-urban environments may clash with the need for water conservation and sustainable management (Carruthers and Vias 2005).

5   Conclusions and Perspectives for Further Research

The previous sections have revealed how peri-urban landscapes are particular socio-ecological systems where it is challenging to find consolidated, easy-to-replicate planning approaches to enhance their level of sustainability. This is mainly due to their dynamic characters, fast-changing nature, and the many pressures that they have to face, especially from humans that tend to want more land for their different activities.

Different land-use and land-cover compositions and configurations as well as different and quick changing socioeconomic structures produce very diverse types of peri-urban landscapes, which are difficult to reduce and to classify. As a direct consequence, research on planning approaches of these systems is still limited and scattered, and they are more focused on solving context-specific issues than on providing comprehensive frameworks for sustainable planning.

According to the ongoing research, one of the most relevant approaches and topics for the planning of peri-urban landscapes is related to providing equal and facilitated access to green spaces for the different social subjects living nearby. Inclusive planning approaches to peri-urban greenery contexts should maximize the social benefits of woodlands (accessibility and qualities of goods and services, culturally appropriate development and fulfillment, self-reliance, etc.). The consideration of factors of equity and possible disparity within the current peri-urban population and between the present and future generations are crucial issues to be taken into account in making planning decisions (van Herzele et al. 2005).

Ecosystem services-based planning is an emerging field of research but still rarely applied in peri-urban contexts. New scenarios can be planned in order to conserve and/or maximize the overall provision of ES by peri-urban landscapes. The management and protection of ecosystem services by agro-ecosystems located in peri-urban and other peripheral areas is a possible way to fight against sprawl urban developments and reduce their negative impacts.

Nature-based solutions and green infrastructure provide sustainable, multipurpose, and flexible alternatives for various planning objectives. Particularly for peri-urban landscapes, green infrastructure may be able to achieve a multiple set of planning objectives such as environmental protection, the development of greenery with new distributions of public parks and gardens, the enhancement of the accessibility of current public green spaces, and the integration of peri-urban agriculture.

With reference to this last point, the possibility to readdress existing farmlands and—much importantly—abandoned farms to New Forms of Urban Agriculture is a fundamental planning strategy for peri-urban landscapes that fulfills multifunctional objectives including food safety, landscape conservation, and ES provision. Planning scenarios of NFUA have to be designed according to the specific features of geographical contexts and particularly evaluating variables such as accessibility by local residents. NFUA can be part of municipal programs and investments for public greenery and environmental conservation. However, due to the large size of these areas, their planning requires synergies and coordination among different administrative levels (e.g., for the creation of large agricultural parks).

This presence of many different public bodies and administrations raises the crucial issue of the choice of the most effective spatial governance instrument and mechanism that should be used to apply the sustainable planning approaches discussed in this chapter. Peri-urban landscapes have to be included in wide metropolitan systems, presenting complex relations with both the main cities and the rural surroundings. It is therefore important that new types of flexible metropolitan governance and related planning instruments are established and that they can integrate different planning levels (municipalities, provinces, regions) and sectors.

According to these considerations, Table 1 reports the main characteristics of peri-urban landscape together with planning recommendations and possible approaches (with reference to the scale of application).

Some future directions can be envisaged for new research on planning of peri-urban landscapes. First, it is essential to further explore to which extent some approaches that performed well in a particular context could be reused in other contexts with similar characteristics. This is probably the most relevant issue, as many examples from current literature have revealed uncertainties with regard to the transferability of successful case studies to other geographical contexts.

Another important research improvement includes the evaluation of the socio-environmental effects and outcomes of planning approaches that are adopted. In

Table 1 Synthesis of characteristics, planning recommendations, and planning approaches for peri-urban landscapes

Characteristic of peri-urban landscapes	Planning recommendations	Suitable planning approaches	Scale
High proximity to urban areas	Ensure equal accessibility to resources/services	Planning and design of peri-urban greenery Nature-based solutions	Local Local
Mix of land uses	Avoid rigid zoning Support/allow the multifunctional use of the land and the reuse of vacant/abandoned lots	Planning new forms of urban agriculture Ecosystem-based planning Green infrastructure planning	Regional Regional/metropolitan Local
Presence of ecological and agricultural values	Develop binding norms/regulations to protect land from urban development	Planning and design of peri-urban greenery Planning new forms of urban agriculture Nature-based solutions	Regional Local Local
High pressure for further urban development	Develop binding norms/regulations to mitigate/avoid urban sprawl	Strategic planning Planning and design of peri-urban greenery Nature-based solutions	Regional Regional/metropolitan Local
Part of wider metropolitan contexts	Co-development of metropolitan plans integrated with lower planning levels (municipal/local) by institutional and local stakeholders New planning instruments (strategic plans, landscape metropolitan plan)	Multilevel spatial governance Strategic planning	Regional Regional

some cases, specific approaches may eventually result in unsustainable outcomes, instead of a higher level of sustainability. This could happen if the effects on different sectors or at different scales are not adequately addressed. For example, urban planning aimed at densification to reduce sprawl can generate problem of green space availability for the growing peri-urban population. Methods and monitoring programs are needed to provide quantitative evidence on the extent to which a proposed solution can be considered sustainable and livable.

Finally, the same characteristics that make peri-urban landscapes challenging contexts for sustainable planning offer, on the other way, interesting and unique opportunities for current planning approaches. In particular, these opportunities are based on the local resources of peri-urban areas, including both environmental resources (e.g., through ecosystem services-based planning) and socioeconomic resources (e.g., through the integration of bottom-up processes into top-down approaches). Examples include the possibility of planning peri-urban landscapes by mixed configuration of new housing and different types of highly accessible green spaces and other spaces for local food production. Abandoned spaces and vacant lands can be turned into positive resources that increase socio-ecological opportunities and offer more sustainable food production for the peri-urban population.

To this aim, a shift in the conceptualization of the peri-urban contexts from the traditional urban-centric approaches (e.g., including zoning and vertical land-use planning) to an environmental and ecosystem-based interpretation is crucial. This will also allow a better understanding—and consequent regulation—of the social and economic consequences of the peri-urbanization processes (e.g., in terms of environmental intra- and intergenerational equity) and increase the overall sustainability level of these complex and dynamic systems.

References

Aguiar MR, Sala OE (1999) Patch structure, dynamics and implications for the functioning of arid ecosystems. Trends Ecosyst Evol 14:273–277

Ahaiblame L, Engel B, Chaubey I (2012) Effectiveness of low impact development practices: literature review and suggestions for future research. Water Air Soil Pollut 223(7):4253–4273

Albert C, Aronson J, Fürst C, Opdam P (2014) Integrating ecosystem services in landscape planning: requirements, approaches, and impacts. Landsc Ecol 29:1277–1285

Allen A (2003) Environmental planning and management of the peri-urban interface: perspectives and emerging field. Environ Urban 15:135–147

Allmendinger P, Haughton G, Knieling J, Othengrafen F (2015) Soft spaces, planning and emerging practices of territorial governance. In: Allmendinger P, Haughton G, Knieling J, Othengrafen F (eds) Soft spaces in Europe. Re-negotiating governance, boundaries and borders. Routledge, London, pp 3–21

Aquino FL, Gainza X (2014) Understanding density in an uneven city, Santiago de Chile: implications for social and environmental sustainability. Sustain Basel 6:5876–5897

Aubry C, Ramamonjiso J, Dabat MH, Rakotoariso J, Rakotondraib J, Rabehariso L (2012) Urban agriculture and land use in cities: an approach with the multi-functionality and sustainability concepts in the case of Antananarivo (Madagascar). Land Use Policy 29:429–439

Baker DC, Sipe NG, Gleeson BJ (2006) Performance-based planning: perspectives from the United States, Australia, and New Zealand. J Plan Educ Res 25(4):396–409

Barbedo J, Miguez JM, van der Horst D, Marins M (2014) Enhancing ecosystem services for flood mitigation: a conservation strategy for peri-urban landscapes? Ecol Soc 19(2):54

Benis K, Ferrão P (2017) Potential mitigation of the environmental impacts of food systems through urban and peri-urban agriculture (UPA)–a life cycle assessment approach. J Clean Prod 140:784–795

Benjamin K, Bouchard A, Domon G (2007) Abandoned farmlands as components of rural landscapes: an analysis of perceptions and representations. Land Urban Plan 83:228–244

Berke PR, Conroy MM (2000) Are we planning for sustainable development? J Am Plann Assoc 66(1):21–33

Bourne LS (2000) Living on the edge: conditions of marginality in the Canadian urban system. In: Lithwick H, Gradus Y (eds) Developing frontier cities, the GeoJournal Library. Springer, Netherlands, pp 77–97

Carruthers J, Vias AC (2005) Urban, suburban, and exurban sprawl in the Rocky Mountain West: evidence from regional adjustment models. J Reg Sci 45(1):21–48

Cash C (2014) Towards achieving resilience at the rural–urban fringe: the case of Jamestown, South Africa. Urban Forum 25(1):125–141

Clark JK, Munroe DK (2013) The relational geography of peri-urban farmer adaptation. J Rural Community Dev 8(3):15–28

Cohen-Shacham E, Walters G, Janzen C, Maginnis S (2016) Nature-based solutions to address global societal challenges. IUCN, Gland

Commission of the European Communities (2006) Thematic strategy for soil protection. www.ec.europa.eu/environment/soil. Accessed 10 May 2017

Conedera M, Del Biaggio M, Seeland K, Moretti M, Home R (2015) Residents' preferences and use of urban and peri-urban green spaces in a Swiss mountainous region of the Southern Alps. Urban For Urban Gree 14:139–147

Coppola A (2004) An economic perspective on land abandonment processes. Paper presented at the AVEC Workshop on Effects of land abandonment and global change on plant and animal communities, 11–13 Oct 2004, Anacapri, Italy

Dai D (2011) Racial/ethnic and socioeconomic disparities in urban green space accessibility: where to intervene? Landsc Urban Plan 102:234–244

de Groot RS, Alkemade R, Braat L, Hein L, Willemen L (2010) Challenges in integrating the concept of ecosystem services and values in landscape planning, management and decision making. Ecol Complex 7:260–272

Deelstra T, Girardet H (2000) Urban agriculture and sustainable cities. In: Bakker N, Dubbeling M, Gundel S, Sabel-Koschela U, de Zeeuw H (eds) Growing cities, growing food: urban agriculture on the policy agenda. Deutsche Stiftung fur Internationale Entwicklung (DSE), Feldafing, pp 43–65

Dijkstra L, Poelman H (2008) Remote rural regions: how proximity to a city influences the performance of rural regions. In: COM—Commission of the European Communities (ed) Regional policy 1/2008. European Commission, Brussels, pp 1–7

Donadieu P (1998) Les Campagnes Urbaines. Actes Sud, Arles

Dorning MA, Koch J, Shoemaker DA, Meentemeyer RK (2015) Simulating urbanization scenarios reveals tradeoffs between conservation planning strategies. Landsc Urban Plan 136:28–39

DTLR (Department for Transport Local Government and the Regions) (2002) Improving urban parks, play areas and green spaces. Improving Urban Parks, Play Areas, London

EEA (European Environmental Agency) (2006) Urban sprawl in Europe The ignored challenge. Report 10. EEA, Copenhagen

EEA (European Environmental Agency) (2015) State of the environment report. EEA, Copenhagen

EUROSTAT (2010) Regional yearbook 2010. Chapter 15: a revised urban–rural typology. EUROSTAT, Luxemburg, pp 240–253

FAO (2002) Urban and peri-urban forestry sub-programme: strategic framework for the Biennium 2002–2003 and mid- term 2002–2007. FAO FORC, Rome

Fletcher TD, Shuster W, Hunt WF, Ashley R, Butler D, Scott A, Trowsdale S, Barraud S, Semadeni-Daves A, Bertrand-Krajewski JL, Mikkelsen PS, Rivard G, Uhl M, Dagenais D, Viklander M (2014) SUDS, LIDS, BMs, WUDS and more – the evolution and application of terminology surrounding urban drainage. Urban Water J 12(7):525–542

Forman RTT (1995) Land mosaics. The ecology of landscapes and regions. University Press, Cambridge

Forrest M, Konijnendijk CC, Randrup TB (eds) (1999) COST action E12-Research and Development in urban forestry in Europe. Official Printing Office of the European Communities, Luxembourg

Foster J (2014) Hiding in plain view: vacancy and prospect in Paris' Petite Ceinture. Cities 40:124–132

Friedmann J (1987) Planning in the public domain: from knowledge to action. Princeton University Press, Princeton

Fürst C, Opdam P, Inostroza L, Luque S (2014) A balance score card tool for assessing how successful the ecosystem services concept is applied in participatory land use planning. Landsc Ecol 29:1435–1446

Gallent N, Shaw D (2007) Spatial planning, area action plans and the rural next urban fringe. J Environ Plan Manag 50:617–638

Gallent N, Andersson J, Bianconi M (2006) Planning on the edge : the context for planning at the rural-urban fringe. Routledge, Abingdon

Gant RL, Robinson GM, Shahab Fazal S (2011) Land-use change in the 'edgelands': policies and pressures in London's rural-urban fringe. Land Use Policy 28:266–279

Geneletti D (2013) Ecosystem services in environmental impact assessment and strategic environmental assessment. Environ Impact Assess Rev 40:1–2

Geneletti D, La Rosa D, Spyra M, Cortinovis C (2017) A review of approaches and challenges for sustainable planning in urban peripheries. Landsc Urban Plan 165:231. https://doi.org/10.1016/j.landurbplan.2017.01.013

Gill SE, Handley JF, Ennos AR, Pauleit S, Theuray N, Lindley SJ (2008) Characterising the urban environment of UK cities and towns: a template for landscape planning. Landsc Urban Plan 87:210–222

Goldthau A (2014) Rethinking the governance of energy infrastructure: scale, decentralization and polycentrism. Energy Res Soc Sci 1/0:134–140

Goodling E, Green J, McClintock N (2015) Uneven development of the sustainable city: shifting capital in Portland, Oregon. Urban Geogr 36(4):504–527

Gret-Regamey A, Altwegg J, Sirén EA, van Strien MJ, Weibel B (2016) Integrating ecosystem services into spatial planning—a spatial decision support tool. Landsc Urban Plann 165:206. https://doi.org/10.1016/j.landurbplan.2016.05.003

Haase D, Larondelle N, Andersson E, Artmann M, Borgström S, Breuste J, Gomez-Baggethun E, Gren A, Hamstead Z, Hansen R, Kabisch K, Kremer P, Langemeyer J, Rall E, McPhearson T, Pauleit S, Qureshi S, Schwarz N, Voigt A, Wurster D, Elmqvist T (2014) A quantitative review of urban ecosystem service assessments: concepts, models, and implementation. Ambio 43:413–433

Haller A (2014) The "sowing of concrete": peri-urban smallholder perceptions of rural–urban land change in the Central Peruvian Andes. Land Use Policy 38:239–247

Hara Y, Murakami A, Tsuchiya K, Palijon AM, Yokohari M (2013) A quantitative assessment of vegetable farming on vacant lots in an urban fringe area in Metro Manila: can it sustain long-term local vegetable demand? Appl Geogr 41:195–206

Harman BP, Choy DL (2011) Perspectives on tradable development rights for ecosystem service protection: lessons from an Australian peri-urban region. J Environ Plann Manag 54(5):617–635

Hedblom M, Andersson E, Borgströmc S (2017) Flexible land-use and undefined governance: from threats to potentials in peri-urban landscape planning. Land Use Policy 63:523–527

Heimlich RE (1989) Metropolitan agriculture: farming in the city's shadow. J Am Plann Assoc 55:457–466

Hooghe L, Marks G (2003) Unraveling the central state, but how? Types of multi-level governance. Am Polit Sci Rev 97(2):233–243

Hough M (2004) Cities and natural process: a basis for sustainability. Routledge, New York

Johnson MP (2001) Environmental impacts of urban sprawl: a survey of the literature and proposed research agenda. Environ Plann A 33:717–735

Kline JD, Thiers P, Ozawa CP, Alan Yeakley J, Gordon SN (2014) How well has land-use planning worked under different governance regimes? A case study in the Portland, OR-Vancouver, WA metropolitan area, USA. Landsc Urban Plann 131:51–63

Konijnendijk CC (2003) A decade of urban forestry in Europe. Forest Policy Econ 5:173–186

Korthals Altes WK, van Rij E (2013) Planning the horticultural sector. Managing greenhouse sprawl in the Netherlands. Land Use Policy 31:486–497

La Greca P, La Rosa D, Martinico F, Privitera R (2011) Agricultural and green infrastructures: the role of non-urbanised areas for eco-sustainable planning in a metropolitan region. Environ Pollut 159:2193–2202

La Rosa D (2014) Accessibility to greenspaces: GIS based indicators for sustainable planning in a dense urban context. Ecol Indic 42:122–134

La Rosa D, Privitera R (2013) Characterization of non-urbanized areas for land-use planning of agricultural and green infrastructure in urban context. Landsc Urban Plann 109:94–106

La Rosa D, Barbarossa L, Privitera R, Martinico F (2014) Agriculture and the city: a method for sustainable planning of new forms of agriculture in urban contexts. Land Use Policy 41:290–303

Lee Y-C, Ahern J, Chia-Tsung Yeh C-T (2015) Ecosystem services in peri-urban landscapes: the effects of agricultural landscape change on ecosystem services in Taiwan's western coastal plain. Landsc Urban Plann 139:137–148

Leichenko RM, Solecki WD (2008) Consumption, inequity, and environmental justice: the making of new metropolitan landscapes in developing countries. Soc Nat Resour 21:611–624

Lindley SJ, Handley JF, Theuray N, Peet E, Mcevoy D (2006) Adaptation strategies for climate change in the urban environment: assessing climate change related risk in UK urban areas. J Risk Res 9:543–568

Lörzing H (2006) Reinventing suburbia in The Netherlands. Built Environ 32(3):298–310. https://doi.org/10.2148/benv.32.3.298

Loupa Ramos I, Ferreiro M, Colaço C, Santos S (2013) Peri-urban landscapes in metropolitan areas: using transdisciplinary research to move towards an improved conceptual and geographical understanding. In: proceeding of the AESOP-ACSP joint congress, 15–19 July 2013, Dublin, p 1145. Available at http://aesop-acspdublin2013.com/uploads/files/AESOP_Programme_final.pdf. Accessed 10 May 2017

Martinico F, La Rosa D, Privitera R (2014) Green oriented urban development for urban ecosystem services provision in a medium sized city in southern Italy. iForest 7:385–395. https://doi.org/10.3832/ifor1171-007

Mason RJ, Nigmatullina L (2011) Suburbanization and sustainability in metropolitan Moscow. Geogr Rev 101(3):316–333

Meeus S, Gulinck H (2008) Semi-urban areas in landscape research: a review. Living Rev Landsc Res 2:1–45

Miguez MG, Mascarenhas F, Canedo de Magalhães L, D'Alterio C (2009) Planning and design of urban flood control measures: assessing effects combination. J Urban Plann Dev 135(3):100–109

Miller RW (1997) Urban forestry: planning and managing urban green spaces, 2nd edn. Prentice Hall, New Jersey

Miller JD, Kim H, Kjeldsen TR, Packman J, Grebby S, Dearden R (2014) Assessing the impact of urbanization on storm runoff in a peri-urban catchment using historical change in impervious cover. J Hydrol 515:59–70

Moreira F, Fontes I, Dias S, Batista e Silva J, Loupa-Ramos I (2016) Contrasting static versus dynamic-based typologies of land cover patterns in the Lisbon metropolitan area: towards a better understanding of peri-urban areas. Appl Geogr 75:49–59

Moss T, Newig J (2010) Multilevel water governance and problems of scale: setting the stage for a broader debate. Environ Manag 46(1):1–6

Munoz F (2003) Lock living: urban sprawl in Mediterranean cities. Cities 20(6):381–385

Neuvonen M, Sievänen T, Tönnes S, Koskela T (2007) Access to green areas and the frequency of visits—a case study in Helsinki. Urban For Urban Gree 6:235–247

Nowak DJ, Crane DE, Stevens JC (2006) Air pollution removal by urban trees and shrubs in the United States. Urban For Urban Gree 4:115–123

O'Brien K, Sygna L, Haugen JE (2004) Vulnerable or resilient a multi-scale assessment of climate impacts and vulnerability in Norway. Clim Change 64:193–225

O'Brien L, DeVreese R, Kern M, Sievanen T, Stojanova B, Atmis E (2017) Cultural ecosystem benefits of urban and peri-urban green infrastructure across different European countries. Urban For Urban Gree 24:236. https://doi.org/10.1016/j.ufug.2017.03.002

OECD (2002) Redefining territories: the functional regions. Organisation for economic co-operation and development (OECD), Paris

Opdam P, Albert C, Fürst C, Gret-Regamey A, Kleemann J, Parker DC, La Rosa D, Schmidt K, Villamor GB, Walz A (2015) Ecosystem services for connectingactors – lessons from a symposium. CASES – Change Adapt Soc Ecol Syst 2:1–7

Padeiro M (2016) Conformance in land-use planning: the determinants of decision, conversion and transgression. Land Use Policy 55:285–299

Padgham J, Jabbour J, Dietrich K (2015) Managing change and building resilience: a multi-stressor analysis of urban and peri-urban agriculture in Africa and Asia. Urban Climate 12:183–204

Pappalardo V, Campisano A, Martinico F, Modica C (2016) Supporting urban development master plans by hydraulic invariance concept: the case study of Acquicella catchment. Proceedings of the 9th International Conference Novatech, Lyon, July 2016

Pappalardo V, La Rosa D, La Greca P, Campisano A (2017) The potential of GI application in urban runoff control for land use management: a preliminary evaluation from a southern Italy case study. Ecosyst Serv., https://doi.org/10.1016/j.ecoser.2017.04.015 26:345

Patz JA, Campbell-Lendrum D, Holloway T, Foley JA (2005) Impact of regional climate change on human health. Nature 438(7066):310–317

Paül V, McKenzie FH (2013) Peri-urban farmland conservation and development of alternative food networks: insights from a case-study area in metropolitan Barcelona (Catalonia, Spain). Land Use Policy 30:94–105

Piorr A, Ravetz J, Tosics I (2011) Peri-urbanisation in Europe: towards a European policy to sustain urban-rural futures. University of Copenhagen – Academic Books Life Sciences, Copenhagen

Provè C, Dessein J, de Krom M (2016) Taking context into account in urban agriculture governance: case studies of Warsaw (Poland) and Ghent (Belgium). Land Use Policy 56:16–26

Pugnare FI, Lazaro R (2000) Seed bank and understorey species composition in a semi-arid environment: the effect of shrub age and rainfall. Ann Bot London 86:807–813

Pulighe G, Fava F, Lupia F (2016) Insights and opportunities from mapping ecosystem services of urban green spaces and potentials in planning. Ecosyst Serv 22:1–10

Ramos-Santiago LE, Villanueva-Cubero L, Santiago-Acevedo LE, Rodriguez-Melendez YN (2014) Green area loss in San Juan's inner-ring suburban neighborhoods: a multidisciplinary approach to analyzing green/gray area dynamics. Ecol Soc 19(2)

Rauws W, de Roo G (2011) Exploring transitions in the peri-urban area. Plan Theory Pract 12(2):269–284

Roose A, Kull A, Gauk M, Tali T (2013) Land use policy shocks in the post-communist urban fringe: a case study of Estonia. Land Use Policy 30(1):76–83. https://doi.org/10.1016/j.landusepol.2012.02.008

Ros-Tonen M, Pouw N, Bavinck M (2015) Governing beyond cities: the urban-rural interface. In: Gupta J, Pfeffer K, Verrest H, Ros-Tonen M (eds) Geographies of urban governance. Advanced theories, methods and practices. Springer International Publishing, Cham, pp 85–105

Rupprecht CD, Byrne JA, Ueda H, Lo AY (2015) It's real, not fake like a park': residents' perception and use of informal urban green-space in Brisbane, Australia and Sapporo, Japan. Landsc Urban Plann 143:205–218

Ryan S, Throgmorton J (2003) Sustainable transportation and land development on the periphery: a case study of Freiburg, Germany and Chula Vista, California. Transportation Res D-Tr E 8:37–52

Salet W, Savini F (2015) The political governance of urban peripheries. Environ Plann C 33(3):448–456

Salet W, Vermeulen R, Savini F, Dembski S, Thierstein A, Nears P, Vink B, Healey P, Stein U, Schultz H, Salet W, Vermeulen R, Savini F, Dembski F (2015) Planning for the new European metropolis: functions, politics, and symbols/metropolitan regions: functional relations between the core and the periphery/business investment decisions and spatial planning policy/metropolitan challenges, political responsibilities/spatial imaginaries, urban dynamics and political community/capacity-building in the city region: creating common spaces/which challenges for today's European metropolitan spaces? Plann Theory Pract 16:251–275. https://doi.org/10.10 80/14649357.2015.1021574

Schipperijn J, Ekholm O, Stigsdotter UK, Toftager M, Bentsen P, Kamper-Jørgensen F, Randrup TB (2010) Factors influencing the use of green space: results from a Danish national representative survey. Landsc Urban Plann 95:130–137

Shkaruba A, Kireyeu V, Likhacheva O (2017) Rural–urban peripheries under socioeconomic transitions: changing planning contexts, lasting legacies, and growing pressure. Landsc Urban Plann.: dx.doi.org 165:244. https://doi.org/10.1016/j.landurbplan.2016.05.006

Smith A (2007) Emerging in between: the multi-level governance of renewable energy in the English regions. Energy Policy 35(12):6266–6280

Sugiyama T, Leslie E, Giles-Corti B, Owen N (2008) Association of neighbourhood greenness with physical and mental health: do walking, social coherence and local social interaction explain the relationship? J Epidemiol Commun H 62(5):e9

Swanwick C (2009) Society's attitudes to and preferences for land and landscape. Land Use Policy 26:S62–S75

Swinton SM, Lupi F, Robertson GP, Hamilton SK (2007) Ecosystem services and agriculture: cultivating agricultural ecosystems for diverse benefits. Ecol Econ 64:245–252

Tan Z, Zhang F, Rotunno R, Snyder C (2004) Mesoscale predictability of moist baroclinic waves: experiments with parameterized convection. J Atmos Sci 61:1794–1804

Taylor Lovell S (2010) Multifunctional urban agriculture for sustainable land use planning in the United States. Sustain 2:2499–2522

Terres JM, Nisini Scacchiafichi L, Wania A, Ambar M, Anguiano E, Buckwell E, Coppola A, Gocht A, Nordström Källström H, Pointereau P, Strijker D, Visek L, Vranken L, Zobena A (2015) Farmland abandonment in Europe: identification of drivers andindicators, and development of a composite indicator of risk. Land Use Policy 49:20–34

The Sierra Club (1999) The dark side of the American dream: The costs and consequences of suburban sprawl. Available at http:// http://www.sierraclub.org/sprawl/report98/report.asp

Thompson A, Stalker Prokopy L (2009) Tracking urban sprawl: using spatial data to inform farmland preservation policy. Land Use Policy 26:194–202

Todes A (2004) Regional planning and sustainability: limits and potentials of South Africa's integrated development plans. J Environ Plann Manag 47(6):843–861

Tu G, Abildtrup J, Serge Garcia S (2016) Preferences for urban green spaces and peri-urban forests: an analysis of stated residential choices. Landsc Urban Plann 148:120–131

Tzoulas K, Korpela K, Venn S, Yli-Pelkonen V, Kazmierczak A, Niemela J, James P (2007) Promoting ecosystems and human health using green infrastructure: a literature review. Landsc Urban Plann 81:167–178

Van Herzele A, Wiedemann T (2003) A monitoring tool for the provision of accessible and attractive urban green spaces. Landsc Urban Plann 63:109–126

Van Herzele A, De Clercq EM, Wiedemann T (2005) Strategic planning for new woodlands in the urban periphery: through the lens of social inclusiveness. Urban For Urban Gree 3:177–188

Vejre H, Søndergaard Jensen F, Jellesmark Thorsen B (2010) Demonstrating the importance of intangible ecosystem services from peri-urban landscapes. Ecol Complex 7:338–348

Webber S, Hanna K (2014) Sustainability and suburban housing in the Toronto region: the case of the Oak Ridges Moraine conservation plan. J Urban Int Res Placemaking Urban Sustain 7(3):245–260

Westerink J, Haase D, Bauer A, Perpar A, Grochowski M, Ravetz J, Jarrige F, Aalbers C (2013) Dealing with sustainability trade-offs of the compact city in peri-urban planning across European city regions. Eur Plan Stud 21:473–497

Whitehand JWR (1988) Urban fringe belts: development of an idea. Plan Perspect 3:47–58

Woodruff SC, BenDor TK (2016) Ecosystem services in urban planning: comparative paradigms and guidelines for high quality plans. Landsc Urban Plann 152:90–100

Yang Z, Hao P, Liu W, Cai J (2016) Peri-urban agricultural development in Beijing: varied forms, innovative practices and policy implications. Habitat Int 56:222–234

Zasada I (2011) Multifunctional peri-urban agriculture-a review of societal demands and the provision of goods and services by farming. Land Use Policy 28:639–648

Zasada I, Loibl W, Berges R, Steinnocher K, Koestl M, Piorr A, Werner A (2013) Rural-urban regions: a spatial approach to define urban-rural relationships in Europe. In: Nilsson K, Pauleit S, Bell S, Aalbers C, Nielsen TAS (eds) Peri-urban futures: scenarios ad models for land use change in Europe. Springer, Berlin Heidelberg, pp 45–68

Zezza A, Tasciotti L (2010) Urban agriculture, poverty, and food security: empirical evidence from a sample of developing countries. Food Policy 35:265–273

Zimmerman J (2001) The "nature" of urbanism on the new urbanist frontier: sustainable development, or defense of the suburban dream? Urban Geogr 22(3):249–267

Žlender V, Ward Thompson C (2017) Accessibility and use of peri-urban green space for inner-city dwellers: a comparative study. Landsc Urban Plann 165:193. https://doi.org/10.1016/j.landurbplan.2016.06.011

Barriers and Bridges for Landscape Stewardship and Knowledge Production to Sustain Functional Green Infrastructures

Per Angelstam, Marine Elbakidze, Anna Lawrence, Michael Manton, Viesturs Melecis, and Ajith H. Perera

1 Introduction

Ecosystems constitute the ultimate foundation for human well-being. With our unique ability to modify the environment, *Homo sapiens* can be viewed as a keystone species – one that has a disproportionately large effect on its environment in relation to its abundance (Power et al. 1996). Insights about the importance of securing long-term ecological sustainability led to establishment of ancient taboos and norms, as well as medieval legislations, and have been the scope of scientific publications for more than 300 years (e.g. von Carlowitz 1713; Marsh 1864; Odum 1959). Ecosystems provide material goods, services and immaterial values, benefit portfolios of which are stressed with different emphasis over time, as well as among

P. Angelstam (✉) · M. Elbakidze
School for Forest Management, Swedish University of Agricultural Sciences, Skinnskatteberg, Sweden
e-mail: per.angelstam@slu.se

A. Lawrence
Scottish School of Forestry, University of Highlands and Islands, Inverness, UK

M. Manton
Faculty of Forest Science and Ecology, Aleksandras Stulginskis University, Kaunas, Lithuania

School for Forest Management, Swedish University of Agricultural Sciences, Skinnskatteberg, Sweden

State Environmental Institution National Park "Braslavskie Ozera", Braslav, Belarus

V. Melecis
Institute of Biology, University of Latvia, Salaspils, Latvia

A. H. Perera
Ontario Forest Research Institute, Sault Ste. Marie, ON, Canada

© Springer International Publishing AG, part of Springer Nature 2018
A. H. Perera et al. (eds.), *Ecosystem Services from Forest Landscapes*,
https://doi.org/10.1007/978-3-319-74515-2_6

cultures and stakeholders. However, the link between supply and demand of nature's benefits is not straightforward. Ecosystems may incur both services and disservices; there are trade-offs among services, stakeholder experiences and perspectives as well as spatial scales; abiotic resources also need to be considered. Finally, human investments are often required to realise the potential of ecosystems to deliver human benefits (e.g. Lele et al. 2013; Huntsinger and Oviedo 2014).

Stewardship and management towards functional ecosystems require that their composition, structure and function are understood in time and space. This is captured by the biodiversity concept (e.g. Noss 1990), which highlights the intrinsic value of nature, as well as its benefits to humans (Millennium Ecosystem Assessment 2005). More recently, research and contemporary policies seeking to reduce anthropogenic pressure on ecosystems have adopted the concept of ecosystem services as a metaphor and means of advocacy (Norgaard 2010). While the biodiversity concept captures the potential supply of ecosystem services in terms of what can be derived from components as species, structures as habitats and functions linked to ecosystem processes (e.g. Brumelis et al. 2011), the ecosystem services concept focuses on the benefits to human well-being in terms of provisioning, regulating, supporting/habitat and cultural dimensions (Millennium Ecosystem Assessment 2005; Koschke et al. 2012).

While global assessments of biodiversity and ecosystem services are crucial as high-level advocacy tools to communicate the principal importance of functional ecosystems to humans (e.g. de Groot et al. 2012), place-based approaches are needed in tandem to ensure functional ecosystems and delivery of benefits in landscapes and regions on the ground (Angelstam et al. 2013a; Singh et al. 2013). This is captured by the policy concept of green infrastructure (GI), which emerged as an analogy to functional transport infrastructure networks. "GI is a strategically planned network of natural and semi-natural areas with other environmental features designed and managed to deliver a wide range of ecosystem services. It incorporates green spaces (or blue if aquatic ecosystems are concerned) and other physical features in terrestrial (including coastal) and marine areas, and is present in rural and urban settings" (European Commission 2013a).

Place-based implementation of GI policy on the ground requires two iterated aspects: (1) production of evidence-based knowledge about GI's states and trends and knowledge about ecological tipping points for assessment of sustainability and (2) cross-sectoral collaborative spatial planning and management at multiple levels of societal steering. This process is captured by the analogy to "compass and gyroscope" (Lee 1993) and is the core of concepts like ecosystem approach (e.g. Yaffee 1999), landscape approach (Axelsson et al. 2011; Sabogal et al. 2015) and landscape stewardship (Plieninger et al. 2015). Understanding landscapes as social-ecological systems and the contribution of stakeholder collaboration are common denominators for all these concepts.

This chapter is divided into two parts. First, we analyse barriers to integrate researchers from human and natural sciences with practitioners and stakeholders in knowledge production and learning about how to sustain human benefits from landscapes as integrated social-ecological systems. This is based on reviewing results from analyses of place-based initiatives belonging to several international concepts aiming at knowledge production and learning about how to sustain ecosystem services, as well as researchers' and stakeholders' experiences of practising integrative research. Second, we present an agenda about how to bridge barriers for knowledge production and learning towards maintaining and restoring functional GIs with a list of seven concrete key actions.

2   Barriers and Bridges Towards Functional Green Infrastructures (GI)

2.1   Long-Term Studies of Landscape Approach Initiatives

2.1.1   Initiatives Belonging to Several Concepts

Systematic studies of different landscape approach concepts, and what local initiatives using these concepts deliver on the ground, can encourage learning from experiences of place-based integrative research in social-ecological systems. Below we review experiences from six long-term place-based initiatives linked to different combinations of five landscape approach concepts (Table 1) aiming at knowledge production and learning about the supply and demand of ecosystem services in landscapes with different landscape histories and governance arrangements. These are the Kristianstad Vattenrike Biosphere Reserve (e.g. Manton et al. 2016) and Bergslagen Model Forest and LTSER platform (Axelsson et al. 2013) in Sweden, the Engure LTSER platform in Latvia (Melecis et al. 2014), the Roztochya Biosphere Reserve in Ukraine (Elbakidze et al. 2013a), and the Pskov

Table 1 Basic information of the suite of six long-term studies of landscape approach initiatives reviewed in this chapter

Case study	Kristianstad	Bergslagen	Engure	Roztochya	Pskov	Komi
Country	Sweden	Sweden	Latvia	Ukraine	Russia	Russia
Area (km²)	1043	14,500[a]	644	744	184	8000
Number of local administrative units	1	18[a]	5	21	1	1
Biosphere Reserve (Est. year)	2005			2014		
Ecomuseum (Est. year)	1989	1986				
Long-term socio-ecological research (Est. year)		2011	2011			
Model Forest (Est. year/ duration)	2016	2008			2000– 2008	1996– 2006
Ramsar (Est. year)	1974		1995			

[a]The informal Bergslagen region has many definitions (Angelstam et al. 2013b); this refers to the area studied in Andersson et al. (2013a)

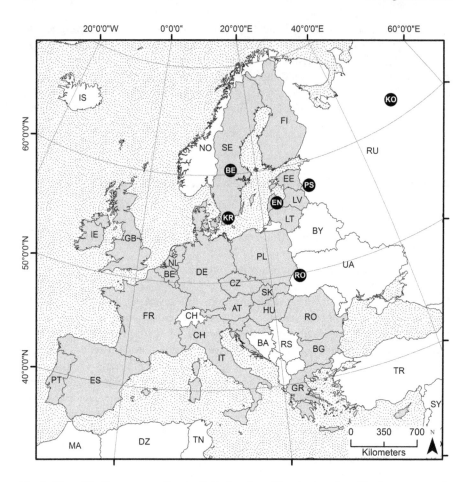

Fig. 1 Map of the European continent with the location of the six place-based landscape approach initiatives reviewed in this chapter (Kristianstad (KR) and Bergslagen (BE) in Sweden, Engure (EN) in Latvia, Roztochya (RO) in Ukraine and Pskov (PS) and Komi (KO) in Russia). They form three pairs, each representing a long history of democratic governance (Sweden), top-down steering (Russia) and countries in transition (Latvia and Ukraine). Countries with grey colour are EU Member States

and Komi Model Forest initiatives in Russia (Elbakidze et al. 2010) (Fig. 1). All case studies are reviewed with respect to (1) social-ecological context, (2) activities, as well as (3) reflections on process, outputs and consequences (sensu Rauschmayer et al. 2009).

2.1.2 Kristianstad Vattenrike Biosphere Reserve

The Kristianstad Vattenrike Biosphere Reserve (KVBR) is located in southernmost Sweden. The Biosphere Reserve concept is an approach aiming at reconciling and promoting conservation of natural and cultural diversity, environmentally and

Fig. 2 The potential natural vegetation in the Kristianstad area is beech (*Fagus sylvatica*) forests. However, a very long history of forest clearing to create agricultural fields and meadows has left only 10–20% of forest cover (Manton et al. 2016). Much of this is Norway spruce (*Picea abies*), which is not native to this area (Photo by Per Angelstam)

socioculturally sustainable development, and research. The Kristianstad area has a long history of human-induced forest clearing and land cover change dating back to the Stone Age (Fig. 2). This long land use history in combination with particular biophysical conditions has created two of southern Sweden's most valuable land covers for biodiversity conservation, namely wet grasslands and the xeric sand steppe (Manton 2016). However, being severely modified and fragmented, such key land covers are subject to a diversity of conservation, management and restoration measures (e.g. Dawson et al. 2017). Additionally, ecosystem processes have been altered, e.g. brownification of water as a downstream effect of changing landscapes (Tuvendal and Elmqvist 2011), predator-prey interactions (Manton et al. 2016) and invasive species (Tuvendal and Elmberg 2015).

Stakeholders in the Kristianstad landscape, and later the entire river catchment of Helgeå (Richnau et al. 2013) of which the KVBR is a part, have been a member of different international landscape stewardship concept for >40 years. Firstly, the wet grasslands were pronounced as a Ramsar Convention wetlands area in 1974 but did not yield the desired results (Walker and Salt 2006). Subsequently, a bridging organisation, the Kristianstad Water Kingdom Ecomuseum, was founded in 1989 as a local community response to deal with degradation and management issues of the wet grassland landscape (Folke et al. 2005). This set the foundations to expand the protected area network and develop governance and management systems through the establishment of the KVBR in 2005. In 2016, the KVBR joined another landscape approach concept aimed at creating partnerships for sustainability, namely, Model Forest (Besseau et al. 2002). The key tasks of the KVBR include improving the conservation efforts for biodiversity of wet grasslands, as well as other land covers (Magnusson et al. 2004). The ultimate aim of the actors and stakeholder of the Kristianstad landscape is thus clearly to use the Ramsar, Ecomuseum, Biosphere Reserve and Model Forest concepts to contribute to the implementation of both social and ecological sustainability.

Despite these efforts to develop landscape governance and being deemed a social system success (e.g. Schultz et al. 2007, 2015), the long-lasting effect towards improving the states of the focal ecosystems' composition, structure and function is largely unknown (Manton 2016). Neither the maintenance of the key land cover patterns as functional green infrastructure nor the regulation of ecosystem processes have been successful (Manton et al. 2016). This should be highlighted as an area of concern for the KVBR as these focal ecosystems have also been prioritised as the most important ecosystems for the local community (Lindström et al. 2006; Johansson and Henningsson 2011). However, on a positive note, the new infrastructure developments (e.g. the Kristianstad Nature Centre, nature trails and recreation facilities) within the KVBR aimed towards nature tourism have delivered both inspiration and access to nature (Beery and Jönsson 2015) and thus promoted human well-being. However, many issues and questions are still unaccounted for; does the development of recreational infrastructure benefit the restoration and maintenance of functional GI for both human well-being and species?

2.1.3 Bergslagen LTSER and Model Forest

Bergslagen in south central Sweden is an informal region with a long history of intensive use of its landscapes (Fig. 1, Table 1). The legacies of more than 2000 years of integrated use of ore, forests and water of major national and international economic importance now involve several challenges for the maintenance of sustainable landscapes (Andersson et al. 2013a, b; Angelstam et al. 2013b, 2014). This includes sustainability of rural communities as well as functional green infrastructures for natural capital and human well-being in forests, cultural landscapes and streams (Angelstam et al. 2014). To cope with such challenges at the local level, the Foundation Säfsen Forests was initiated in 1999 (Elbakidze et al. 2010). To gain momentum from international landscape approach concepts, the emerging network and NGO Sustainable Bergslagen was established and facilitated the move to join both the International Model Forest network and the network for Long-Term Socio-Ecological Research (LTSER) (Axelsson et al. 2013). Key motives were that stakeholders in Bergslagen can learn from regional sustainability assessments and sustainable development processes from other regions and to make Bergslagen more visible internationally.

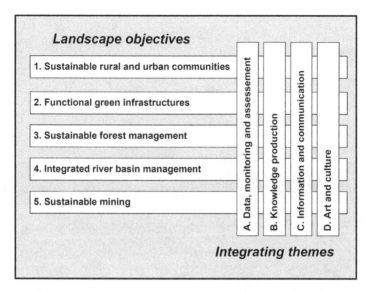

Fig. 3 The strategy to promote resilient networking, collaboration and learning as a continuous process towards a sustainable Bergslagen region was to divide different sectors work into five goals (1–5) and four cross-cutting themes (A–D) (Angelstam and Axelsson 2014). At the interface between objective and themes place-based work occurs in local hubs in Bergslagen

Aiming at cross-sectoral integration at multiple levels of public, private and civil sectors, as well as research, short-term regional, national and EU-level funding resulted in several local and regional projects. These have focused on (1) monitoring and assessment of different aspects of sustainability; (2) knowledge production about material and immaterial landscape values relevant for the management of ecological, economic, social and cultural dimensions; (3) information and communication using both traditional media; as well as (4) through art and culture (Fig. 3). An example of the first item is that the county of Dalarna and researchers in Bergslagen have collaborated to make assessments of the extent to which old Scots pine and Norway spruce forest patches function as two different green infrastructures for biodiversity conservation. The second item is illustrated by the collaboration between a company that produces hydroelectricity, a county administration, a forest company and researchers (Fig. 4). This resulted in joint production of knowledge about restoration of blue infrastructure by integrating water regulation, stream restoration and evidence-based knowledge about the requirements of focal aquatic species (Törnblom et al. 2017). Finally, art and culture is represented by the theatre company "Teatermaskinen" which highlights rural development challenges through performances, workshops and interaction with stakeholders.

Axelsson et al. (2013) evaluated the extent to which learning at multiple governance levels for sustainable landscapes occurred in 18 local development initiatives in the network of Sustainable Bergslagen. Activities at multiple levels were mapped during 10 years, and key actors in the network were interviewed.

Fig. 4 Knowledge-based stakeholder participation in spatial planning across sectors is a way forward for landscape restoration where the history of intensive forest management is long, such as in the Swedish Bergslagen region. Collaborative learning among representatives from a county administrative board, a forest company and a NGO and an archaeologist together with researchers and students can build capacity for evidence-based landscape restoration (Photo by Per Angelstam)

While activities resulted in exchange of experiences, innovations and some local solutions, a major challenge was to secure systematic learning and make new knowledge explicit at multiple levels. None of the development initiatives used a systematic approach to secure learning, and sustainability assessments were not made systematically. Nevertheless, the attempt towards a landscape approach based on opportunities for integration of different sectors on the one hand, and knowledge production, communication and awareness-rising, has empowered stakeholders. However, while several concrete effects have been delivered, understanding the long-term consequences of attempts towards collaborative learning in social and ecological systems require long-term core funding.

2.1.4   Engure LTSER

Latvia's Engure LTSER platform encompasses the drainage basin of the Lake Engure and the adjacent coastal areas including the eastern sand dunes of the Engure Spit and the adjacent Gulf of Riga in Latvia (Figs. 1 and 5, Table 1). To protect nature values, a nature reserve was established already in 1957 (Viksne 1997). To strengthen its status and stress its international significance, Lake Engure was

Fig. 5 The Engure Lake nature park was established in 1998. It contains the Engure Lake, both deciduous and Scots pine (*Pinus sylvestris*) forests and also the seashore of the Gulf of Riga. Being a cultural landscape, management by grazing and mowing is important management measures. The area is popular for bird watching, fishing and recreation (Photo by Per Angelstam)

declared an Important Bird Areas of Europe in 1994, and in 1995 it was included as a Ramsar site. In 1998 the Lake Engure Nature Park was established, and in 2004, it became listed as a Natura 2000 site (Viksne et al. 2011). The large islands and coastal habitats are particularly favourable for water birds (Viksne 1997), and also hosts 844 species of vascular plants (Gavrilova et al. 2000).

An important step towards habitat management was the introduction of regular ecological monitoring and long-term research. For more than 50 years, ecological monitoring of water birds within the region has been undertaken (Viksne 2000). In addition, research programmes on the sustainability of Lake Engure and its adjacent terrestrial ecosystems began in 1995 (Melecis 2000). With a long history of anthropogenic management linked to animal husbandry, mowing of grasslands and reeds and manipulation of wetlands to increase suitable nesting places are needed to maintain bird species diversity. A grassland management project was thus initiated by the Latvian Fund for Nature in the frame of an EU LIFE-Nature project in 2001. This project included "rewilding" by the introduction of konik horses and heck cows for grassland habitat management. In 2010, a national cooperative project was started with the aim to develop the Engure Long-Term Socio-Ecological Research (LTSER) platform (Melecis 2011). Integrated studies were performed on ecological and social components of Lake Engure LTSER, and a conceptual model was worked out (Melecis et al. 2014).

The areas for which mowing or pasturing is necessary to maintain focal bird species are much larger than those presently mowed, or grazed by konik horses and heck cows. Grassland management is partly performed by local land owners under compensation agreements. Reeds also are regularly removed from the lake by a local company. Except for collaboration between the national park and researchers, the inclusion of the social system research has turned out to be difficult. Local stakeholders view efforts towards biodiversity conservation as competing with production of tangible provisioning ecosystem services. The main problem to sustain the Engure LTSER lies in securing funding for necessary activities including scientific research, monitoring and ecological management, as well as stakeholder collaboration. During the 2008 economic crisis in Latvia, monitoring programmes were interrupted and funding for scientific research was significantly reduced. However, until today the burden of environmental problems tends to be solved mostly on funding available from EU programmes.

2.1.5   Roztochya Biosphere Reserve

Roztochya in Western Ukraine and Eastern Poland is a corridor of upland forest and cultural landscape that crosses the Eastern border of the European Union (Figs. 1 and 6, Table 1). The Ukrainian part of Roztochya was approved as a Biosphere

Fig. 6 The Ukrainian countryside, like in the Roztochya area near Lviv in Western Ukraine, is a mixture of intensively managed agricultural land, abandoned fields where forest is coming back and forest patches (either beech or Scots pine). Old trees are an important cultural heritage, often with mistletoe (*Viscum album*) (Photo by Per Angelstam)

Reserve in 2011. During the Soviet period of today's Ukraine (1939–1991), sulphur mining was the main industry in the area, and collective agricultural enterprises were the main employers in rural areas. Following the disintegration of the Soviet Union in 1991, the mining industry was closed and the collective agricultural enterprises were reorganised into small-scale private farms or abandoned. A large portion of both the urban and rural populations lost their jobs. Currently, this region is still facing high levels of unemployment, a poor health care system, lack of access to suitable markets for local products, insufficient road infrastructure and depopulation of rural areas (Elbakidze et al. 2013a).

Motivations for establishment of the Roztochya Biosphere Reserve initiative were (i) to protect biodiversity, (ii) to address issues associated to the heritage of local sulphur mining industry and (iii) to encourage regional economic development driven by regional and international tourism. The proposed Biosphere Reserve was also considered as an attractive tool for generating interest and investments from international and national sources (Elbakidze et al. 2013a).

Elbakidze et al. (2013b) studied the legal recognition of the Biosphere Reserve concept as a tool for sustainable development in Ukraine and what impact legislation has had on implementation in the Roztochya Biosphere Reserve. The Biosphere Reserve concept has been incorporated into Ukrainian nature conservation legislation. The implementation process on the ground in Roztochya was socially constrained because the legislative domain of the Biosphere Reserve concept, being linked to nature conservation, impacted the different stakeholders' perceptions of what the Biosphere Reserve concept means. Interviews with locals engaged with the Roztochya Biosphere Reserve initiative thus revealed that the aim to promote sustainability through stakeholder collaboration was poorly implemented (Elbakidze et al. 2013b). Biosphere Reserve implementation may thus be improved by (1) choosing national terminology describing the concept carefully, because this affects stakeholder perceptions, (2) ensuring that legislation for BRs has a multisectoral character and (3) ensuring that those who implement Biosphere Reserve initiatives have the understanding, knowledge and will to lead and facilitate sustainable development as a collaborative social learning process towards ecological, economic, social and cultural sustainability. However, with the current economic and political challenges in Ukraine, the Roztochya Biosphere Reserve is facing multiple obstacles.

2.1.6 Pskov Model Forest

The Pskov Model Forest project was carried out in the westernmost of part Russia's boreal forest biome (Fig. 1, Table 1). Intensive wood mining during the Soviet period (e.g. Naumov et al. 2016), a lack of silvicultural programmes (Angelstam et al. 2017a) and abandonment of agricultural land (Prishchepov et al. 2012) have resulted in a high proportion of deciduous trees and large volumes of dead wood compared to Fennoscandian managed forests (Fig. 7) (Angelstam and Dönz-Breuss 2004). The area of Pskov Model Forest is state-owned and was a leasehold territory

Fig. 7 Russian forest policy aims at transitioning from Soviet wood mining to sustainable forest management by intensification of forestry (Angelstam et al. 2017a). The Pskov Model Forest project demonstrated that this is indeed feasible by introducing pre-commercial thinning. The red berries are rowan (*Sorbus aucuparia*); that rowans are not browsed indicates that there is no need to limit the abundance of large browsing herbivores like in Sweden (Angelstam et al. 2017c) (Photo by Per Angelstam)

of an international forest company interested in increased sustained yield wood production.

The motivation for creating Pskov Model Forest was to create new regional forestry norms for intensification of forest management to sustain the wood resource base by employing the Nordic intensive sustained yield approach, primarily for international forest companies (Elbakidze et al. 2010). Large forest industry companies in Sweden and Finland, which were using Russian timber and pulpwood, experienced problems with a reduced supply starting in the early 1990s after the collapse of the Soviet Union. In the 1990s, the Pskov region began to play an important role in the Baltic timber trade. However, the Nordic approach to intensification contradicted the existing Russian system of forestry norms and regulations. To improve economic efficiency, Stora Enso Co. initiated a project targeted at sustaining profits from timber industry on a long-term basis. At that time, harvesting operations by western companies in Russia aroused serious protests among the local population.

The Pskov Model Forest (Yablochkina et al. 2007) was a development and demonstration project, which appeared as a result of simultaneous interests of foreign donors for development of approaches to sustainable forest management in Russia, and the presence of local and regional champions able to act as brokers. This made it possible to promote and implement demonstration sites aimed at intensifying forest management by tree planting and pre-commercial and commercial thinning. The majority of the activities were initiated, facilitated and financed by foreign donors. The Pskov Model Forest operated in a specific governance domain on a national level, which enhanced the ability to develop adaptive capacities in the local Model Forest initiative. Stakeholders in Pskov Model Forest initiative began to develop a network-based type of governance system both locally and regionally, and the dissemination of project experiences is currently in progress. However, scaling-up is hampered by a fragmented governance system in Russia with poorly functioning institutions, insufficient education, corruption and low levels of social capital (e.g. Naumov et al. 2016; Angelstam et al. 2017a). Cultural, political and language barriers currently hinder learning based on comparisons of what portfolios of ecosystem services that different approaches to forest management deliver.

2.1.7 Komi Model Forest

Protecting pristine forests from logging was the original motive for developing what became the Komi Model Forest initiative. In the beginning of the 1990s, several foreign forest companies began logging operations in the naturally dynamic forests adjacent to the Pechora-Ilych Reserve in the eastern Komi Republic (Elbakidze and Angelstam 2008). To prevent exploitation of these last large intact forest landscapes, researchers from Russia and Sweden prepared a project with the aim to elaborate approaches to sustainable forest landscape management and submitted it to World Wildlife Fund. The project idea was accepted, and it began in 1996. The Swiss Agency for Development and Cooperation, which supported implementation of sustainable forest management policy in countries in transition in the mid-1990s, funded the project. In 1999, this donor decided to shift the focus of the project to southwest Komi and to use the term Model Forest, despite its departure from the Canadian Model Forest concept (Elbakidze et al. 2010). Criteria for selecting a Model Forest were formulated, and the Priluzje state forest enterprise in south-westernmost Komi's local administrative unit Obyachevo was chosen (Figs. 1 and 8, Table 1).

A specially created non-governmental organisation named Silver Taiga (http://silvertaiga.ru/) facilitated the Komi Model Forest activities, including identification of problems in forest use or management through consultations with a working group or a coordinating board. The issues were evaluated and solutions developed. These were discussed with stakeholders, especially with governmental organisations, and then with donor representatives. An action plan targeted at implementation of sustainable forest management policy that matched regional and local conditions and interests of the stakeholders was initiated by collaboration between managers and stakeholders. A working group conducted ten brainstorming sessions up to 1 week in length during a 6-month period. The major difficulties involved were (1) the partner's capabilities of being open and honest during discussions; (2)

Fig. 8 Forestry in remote areas in Russia, such as in the Komi Republic, can be described as wood mining. A key challenge is to maintain a road network that can be used not only to transport the harvested wood to the industry but also to carry our different silvicultural treatments (Elbakidze et al. 2013c). Still, however, wood mining continues (Naumov et al. 2017) (Photo by Per Angelstam)

overcoming professional stereotypes and widening the view the issues of forest use; (3) reaching equality between partners with different professional and social status, and in the course of discussions and decision-making processes; (4) in the development of teamwork. Gradually, these difficulties were overcome; and (5) a process of constructive and creative work resulted in support to the Model Forests by many stakeholders from local and regional levels.

After approval by the donor, decisions were implemented by project executives who worked with stakeholders. The transparency of the governance system was ensured by the work of the public relations group via mass media and publication of various materials. Local people participated in the decision-making and implementation processes through (1) public hearings, (2) formation of forest clubs as neutral platforms and (3) provision of grants for different activities in the Model Forest, such as forest clubs discussions, ecological festivals and creating ecological trails. Libraries, local schools and cultural establishments were the primary recipients of grants. Educational activities were key components in the governance systems of the Komi Model Forest. These activities created an open and transparent environment, attracting public attention to issues of forest management and use. Finally, to promote the principles of sustainable development on different levels, new specialists were trained with emphasis on solving problems related to sustainable forest management and to propose management approaches for wood production and establishment of protected areas (e.g. Mariev et al. 2005). This training was intended to target young professionals as potential future leaders in society. Nevertheless, sustained yield forestry lags behind (Elbakidze et al. 2013c). While the Komi Model Forest project ended in 2006, Silver Taiga has developed into a successful facilitator of sustainable forest management in NW Russia.

2.2   Researchers' and Stakeholders' Experiences

All six case studies reviewed aimed at collaborative learning towards sustainability in concrete landscapes and regions. This is consistent with Lee's (1993) notion of compass in terms of evidence-based knowledge about states and trends and gyroscope in terms of stakeholders' social and collaboration learning, such as towards spatial planning for functional green infrastructure (Fig. 9). The six case studies reveal several barriers for transdisciplinary knowledge production and learning (Table 2) towards functional green infrastructure or simply to sustainable provision of multiple ecosystem services. Overall, the major challenges were linked to the social system in terms of different stakeholders' desired benefits from landscapes. Several innovations were developed and applied towards maintenance of green

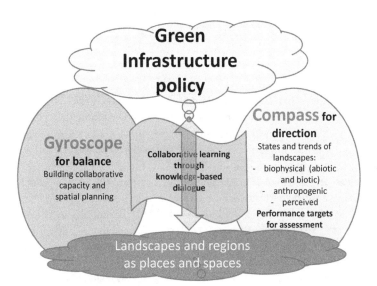

Fig. 9 The term landscape approach as a means of linking green infrastructure policy and landscapes and regions as places and space is captured by combination of "compass and gyroscope" (Lee 1993), and collaborative learning through knowledge-based dialogue that links them. Model Forest, Biosphere Reserve and Long-Term Socio-Ecological Research platform are three examples of the landscape approach (Axelsson et al. 2011), see also Table 4

Table 2 Key to successes and major challenges analysis of integrative/transdisciplinary (TD) knowledge production and learning – a prerequisite for evidence-based landscape stewardship. A comparison of the five case studies reviewed in Crow et al. (2007: 208ff) and the six case studies in this chapter

Topic	Issue	Crow et al. (2007)	This chapter
Major challenges	Inadequate technological infrastructure	x	
	Lack of common language	x	x
	Organisational cultures, resource constraints	x	x
	Land ownership complexity	x	x
	Difficulties coordinating among numerous and dispersed partners	x	x
	Shifting organisational priorities	x	x
	Complexities associated with multiple, diverse audiences	x	x
	Shifting organisational priorities	x	x
	Bureaucratic barriers	x	x
	Institutional and organisational barriers	x	x
	Lack of long-term funding for team building		x
Keys to success	Shared vision of the outcome and commitment of all participants	x	x
	Political will to adapt local capacity to generate knowledge and enabling technological and personnel infrastructure	x	x
	Demonstrating an ability to solve critical issues	x	x
	Outspoken policy towards transdisciplinarity		x
	Champions able to act as brokers		x
	Ability to integrate funding from different sources		x

infrastructures as providers of ecosystem services. The use of demonstration sites, spatial modelling of habitat functionality and presence of brokers that can bridge different stakeholder interests were crucial. However, Axelsson et al. (2013) found that building up the necessary trust and trustworthiness takes a long time. A key factor for successful implementation of a place-based landscape approach was therefore availability of core funding. This was achieved in two ways. Experienced funding agencies provided long-term funding knowing that establishment of collaborative learning takes time. One example is the donor's support of the Komi Model Forest during 10 years (Elbakidze et al. 2010). Eventually there was a transition from a long-term project to sustained knowledge production and learning as an enterprise (Komi Model Forest and Silver Taiga, respectively). The other was through a bottom-up approach based on a suite of researcher and stakeholder-driven projects with or without public sector support (e.g. Kristianstad Vattenrike Biosphere Reserve, and Sustainable Bergslagen).

This is consistent with previous experiences. For example, Angelstam et al. (2013c) used group modelling to map perceived barriers and bridges for researchers' and practitioners' joint knowledge production and learning towards transdisciplinary work. The analysis indicated that this process is influenced by (1) the amount of traditional disciplinary formal and informal control, (2) adaptation of project applications to fill the transdisciplinary research agenda, (3) stakeholder participation and (4) functional team building based on self-reflection and experienced leadership.

Stakeholders and practitioners have similar experiences. A study of municipal comprehensive planning observations in Bergslagen showed that it is difficult for planners to engage stakeholders in municipal planning, even for large municipalities with financial and human resources (Elbakidze et al. 2015). In Sweden a concrete example of GI development led by practitioners' realities is the approach to landscape zoning developed by Sveaskog Co. (Angelstam and Bergman 2004; European Commission 2013b). These are issues that characterise the challenge of multifunctional, multi-stakeholder governance. The issue has been well-studied in the USA, where the public US Forest Service is obligated to involve stakeholders in forest planning but struggles to attract participants, particularly in the implementation stages (Cheng and Mattor 2006). Urban forestry is also an area where both planners and public engagement are necessarily involved and where the quality of participation affects outcomes (Lawrence et al. 2013).

The six case studies also indicate the role of societal context represented by the governance gradient between Sweden with stable democratic institutions, countries in transition from planned to market economy such as Latvia and Ukraine and finally in the Russian Federation with hybrid system of governance with public institutions that exercise top-down governmental control and private institutions that have been established in transition to a market economy. In countries formerly part of the past Soviet Union, such as Latvia and Ukraine, there is a legacy of private land seizure by the state and control of natural resources that contributes to mistrust or suspicion of the government. Later, during independence after the end of the Soviet Union in 1990, some of these lands were returned to previous owners. In this sense, land ownership is a source of pride (Elbakidze and Angelstam 2007). However, local people do not yet feel fully secure with their land ownership and are afraid that the government could take their property again. This history, in combination with current social and economic insecurity, contributes to local stakeholder distrust towards initiatives such as Biosphere Reserve and LTSER platform that originate outside of their own community. Thus, in the Ukrainian Roztochya Biosphere Reserve where local livelihoods depend directly on use of natural resources, the legislative misplacement of Biosphere Reserves threatens its implementation towards collaboration to satisfy all dimensions of sustainability. This notion is also supported by other studies (e.g. Kušová et al. 2008; Wallner et al. 2007). A poor society would be stopped from devastating use of natural resources only by ecological catastrophe or strong political pressure from countries they are dependent from. In general, there is thus a paradox in terms of a low level of ecological knowledge but high demands for ecosystem services. Exchange of knowledge and experience from different social-ecological contexts regarding how to maintain functional green infrastructure is one approach (e.g. Angelstam et al. 2013b; Nordberg et al. 2013; Naumov et al. 2017).

3 Towards Functional Green Infrastructures

3.1 The Social-Ecological Context Matters

Generic and sectorial sustainability policies call for maintenance and improvement of landscapes' GIs as a tool that maintains biodiversity as a base for providing ecosystem services for human well-being. This is expressed in the EU forest strategy (European Commission 2013c) and in policies on the conservation of cultural landscapes (Council of Europe 2000) and water (European Commission 2000). At the same time, the European forest-based sector has the vision that by 2030 it will be a key contributor to a sustainable European society (www.forestplatform.org). In a new, bio-based and customer-driven European economy, forestry is expected to make significant societal contributions. These policy trends imply increased conflicts between intensified economic use of forests and woodlands in rural and urban landscapes on the one hand, and the maintenance of these forest and woodland ecosystems' composition, structure and function as green infrastructure on the other. For example, the continued intensification of forestry in Sweden will lead to a division into production and biodiversity conservation functions and reduced social functions in rural landscapes (Angelstam and Elbakidze 2017).

Given that landscape histories and governance contexts among regions and countries are very diverse, policies regarding GI are likely to be comprehended and implemented differently. The comparison of applications of different concepts in this chapter illustrates that development of successful approaches to landscape stewardship for functional GI needs to acknowledge and address the diversity of social-ecological and political culture contexts. The proliferation of landscape approach concepts and initiatives on the ground provides promising opportunities for multilevel learning based on comprehensive meta-analyses (Axelsson et al. 2013). As suggested by Price et al. (2010) and Reed and Egunyu (2013), sharing of quality-assured practices across the international network of BRs, and across other international networks (e.g. Ramsar, Model Forest, Ecomuseum), provides opportunity for learning by evaluation, which ultimately can improve governance, planning and management towards functional GI.

3.2 Transdisciplinary Knowledge Production and Learning

Advocates of a landscape approach emphasise the importance of transitioning from land management as site-level monitoring or technical interventions towards learning within and among landscapes as social-ecological systems (Axelsson et al. 2011; Laestadius et al. 2015; Sabogal et al. 2015; Lawrence 2016). Focusing on understanding how public forest administrations in the UK can develop a capacity for adaptive forest governance, Lawrence (2016) studied how forest managers are using their own knowledge to establish more diverse tree species choices, particularly in response to high impact tree health disasters. Their knowledge and practice is supported or constrained by their organisations, e.g. the politics within and between devolved forest administrations, the personalities in senior management positions, reduced budgets and staff for forest management. Practitioners' social capital and collaborative capacity can nevertheless support communication and learning. The new circumstances are stimulating a creative approach grounded in silvicultural knowledge and experience in the forest, rather than hierarchical science-led approaches.

These initiatives are important in terms of adaptation, but they are limited unless the public administrations can more explicitly acknowledge the role of practitioners' knowledge in a more transdisciplinary approach to learning. Additionally, as described above, both academic as well as non-academic participants perceive barriers for co-production of knowledge based on monitoring and assessment of the sustainability of forest landscapes.

We argue that much can be learned from regionally adapted traditional governance systems that have evolved over long times, even before the sustainable development discourse was invented, such as traditional village systems (Angelstam and Elbakidze 2017). The challenge with switching forest management paradigms can be eased, at least temporarily, by resorting to expert knowledge (Perera et al. 2012) among academics, land management professionals and stakeholders to understand the challenges as well as solutions associated with GI.

Increasing focus on sustaining ecosystem services from landscapes via GI is a paradigm shift in land and water management policies and practices. As a part of this transition, policy-makers and land managers demand a new and reliable body of knowledge from researchers to help develop and implement land management plans with a broader content than in the past when the focus was only on provisioning ecosystem services. If a relevant body of knowledge is not built rapidly and rigorously, that would pose a significant barrier to adoption and popularisation of the GI approach. Consequently, novel approaches to land management require a paradigm shift not just in policies and practices but also in how research and practice is carried out. Exploration of additional sources knowledge, as alternatives to that which is typically built by empirical research is one such paradigm shift in ecological research, especially at landscape scale. Three such sources becoming increasingly popular are citizen science (Dickinson et al. 2010), expert knowledge (Perera et al. 2012) and comparative studies of places with different landscape histories and governance arrangements (Angelstam et al. 2013b).

Citizen science is based on the involvement of non-scientist volunteers to collect data of ecological phenomena that are readily measured. This tactic is not new; rather it has been in practice for a very long time in some ecological fields such as ornithology. Dickinson et al. (2010) observed that citizen science is a tool for knowledge production in many areas of ecology where data gathering over larger geographic areas are involved. Lawrence (2010) has also reviewed, across examples from the north and the south as well as post-industrial and developing countries, the ways in which such knowledge can contribute effectively to decision-making. Seeking a universal conclusion across very diverse contexts, from birdwatchers in

England to community forest groups in Nepal, she concluded that different stake-holders can have different objectives, knowledge, information needs, cultures and power relations, as well as methods for collecting and sharing information. These different positions and needs can be accommodated through partnerships, which can provide distinct but complementary and mutually rewarding outcomes.

The typically implicit (latent and tacit) knowledge of ecological experts such as practitioners (and sometimes scientists) can be elicited and analysed to compose an explicit body of expert knowledge. When performed well, expert knowledge can contribute in several ways to the traditional bodies of empirical knowledge, espe-cially under demands for rapid solutions (Table 3), from specific instances to synop-tic syntheses. The methodology of eliciting, analysing and using expert knowledge is extensive and well founded, and designed to identify as well as minimise the biases and uncertainties, with explicit and repeatable methods (e.g. Ericsson et al. 2006; O'Hagan et al. 2006). Expert knowledge methods are steadily evolving towards the rigour of traditional empirical research. Furthermore, expert knowledge development process presents an ideal venue for multiparty interactions (policy developers, land use planners, scientists and stakeholders) and thereby promotes communication and information exchange, especially during the early stages of plan formulation. This is a distinct advantage in highly interactive and multiparty ventures such as planning for GI. Not surprisingly, expert knowledge is beginning to be used in various aspects of sustaining ecosystem services, e.g. in identifying (Koschke et al. 2012; Quijas et al. 2012), mapping (Grêt-Regamey et al. 2013; Jacobs et al. 2015), planning (Kopperoinen et al. 2014) and decision-making (de Groot et al. 2010; Zavrl and Zeren 2010).

After a long and gradual domestication of naturally dynamic forests into resilient traditional cultural landscapes, the contemporary focus on economic development based on maximum sustained yield of food, feed, wood, fibre and biomass has been

Table 3 Various avenues of expert knowledge contributions in ecology when rapid solutions for applications are sought

Motivation	Contribution of expert knowledge	Example applications
Absence of empirical knowledge	Qualitative information	Exploring and identifying knowledge gaps, research questions and prioritising topics
	Quantitative surrogates	Formulating conceptual models, parameterising simulation models, building data bases and developing DSS
Major gaps and uncertainties in empirical knowledge	Quantitative complements	Identifying priorities for land management, predicting ecological functions, refining and parameterising simulation models
	Quantitative supplements	Predicting spatial patterns, refining databases, fine-tuning model predictions
Complications and disagreements in empirical knowledge	Insights and decisions	Informing strategic and tactical land management decisions, simulating scenarios and generating research hypotheses

Source: Summarised and adapted from Drescher et al. (2013)

successful but has also led to effects that are not desired by society. This involves human migration from rural to urban areas, loss of natural forest and cultural woodlands, modified natural and anthropogenic processes as well as extirpation and extinction of species. In response to these issues, policies about rural development, biodiversity (i.e. species, habitats and processes) as natural capital, landscapes as integrated social-ecological systems and ecosystem services have emerged. The GI concept recently appeared as an implementation tool for coordinated actions among different sectors. However, to change undesired trajectories in both ecological and social systems on the ground is not easy. This calls for landscape stewardship approaches that involve understanding the states and trends of landscapes as social-ecological systems and which can foster collaborative spatial planning that is regionally adapted. Angelstam and Elbakidze (2017) reviewed results from place-based diagnoses of the current states of both ecological and social systems in gradients of landscape history and governance to government.

The development of sustainability science (e.g. Komiyama and Takeuchi 2006; Kates 2011) is a response to the need for holistic knowledge production and learning towards sustainable landscapes on the ground. Moving from natural science or human science research to transdisciplinary knowledge production by researchers, practitioners and citizens means a radical change in the way knowledge production is carried out and how infrastructure for this is built. Although addressing multiple spatial scales from land cover patches to regions and engaging stakeholders at different level problem-oriented research is highly topical (Durham et al. 2014), there is epistemological and methodical friction (Furman and Peltola 2013). The next section presents seven concrete proposals towards knowledge production and learning.

4 Key Actions

4.1 Comparative Studies

Comparison of several different social-ecological contexts is one avenue towards natural experimentation to improve knowledge production and learning towards sustainable landscapes. With policy and management of different representative types of GI often being similar or the same within a region or country, relationships between land cover patterns and processes in local landscape show limited variation. International comparisons can contribute (Angelstam et al. 2013a; Manton 2016). As illustrated in this chapter by six landscape approach initiatives using five different concepts, the European continent hosts a variety of different landscape histories and different governance arrangements. The resulting diversity of contexts provides different opportunities for supply of ecosystem services and different human well-being demands. Comparative studies can thus help to understand the effects humans have on the functionality of GI and what ecosystem services are important for biodiversity conservation and human well-being in different contexts.

Using different European countries' landscapes as a time machine provides unique opportunities to learn from the consequences of the past and present (Angelstam et al. 2011a; Elbakidze et al. this book), as well as understand the role of past legacies for societal steering (Elbakidze et al. 2010; 2013b).

In doing so, however, it is important to think about two aspects of methodology. First, those describing individual case studies are often embedded in the case. It can be challenging to think about the example objectively or to separate subjective and objective aspects (both of which are important). This can be greatly helped by using an interactive approach which develops the comparisons, gradually developing a comparative framework in an iterative manner. Nevertheless, flexibility is needed in the framework. As Lawrence et al. (2013) reflected on the process of developing a framework for comparative study of urban forest governance, the process of finding language to communicate succinctly each other's experiences of, for example, public engagement, to accommodate all of this in one framework and to insist that every case study should use exactly the same language, would be to suffocate the emergent value of such a framework.

4.2 Learning by Evaluation

Collaborative learning based on monitoring and assessment is crucial in the quest towards attractive places and regions. Policy implementation research through the evaluation of (1) policy processes, (2) governance and stewardship outputs and (3) management consequences on the ground in landscapes as social-ecological systems is a key step in the progress towards agreed policy goals about sustainability through the maintenance of functional GI (Lundquist 1987; Rauschmayer et al. 2009). Evaluation of policy processes involves the assessment of what constitutes good governance, including elements such as improved information management and learning, legitimate processes and the normative aims of transparency and participation. Outcomes of policy processes can be divided into two parts. Firstly, the outputs in terms of implementation of policy tools, norms and rules to be applied by governors, planners and managers at multiple levels; pronouncements of criteria and indicators; short-term and long-term performance targets; and finally tactical planning and operational management approaches. Secondly, the consequences of operational implementation of strategic and tactical plans on the ground need to be assessed. However, this final step in the policy implementation process is often poorly studied (e.g. Popescu et al. 2014).

4.3 Landscape Approach as Tool

The ecosystem services framework aims to integrate natural capital into political and economic decisions at multiple governance levels. However, this framework fails to capture the complexity of interactions between social and ecological systems, leading to fragmented policy and governance, thus hindering multifunctional land management and spatial planning (Garrido et al. 2017a, b). Applying a landscape approach can be used as a tool for integration of ecological and social systems.

The term landscape captures the manifold dimensions of places where people live and work (Angelstam et al. 2013a). Consideration of landscapes' biophysical, anthropogenic and perceived dimensions at multiple scales represents a holistic approach to implement green infrastructure policy through spatial planning and integrated land use management. Climate, terrain, soil and the flow of water determine the particular types of natural ecosystems and form the biophysical checkerboard that underpins the delivery of ecosystem services. These range from tangible goods and ecological functions to habitat for species and cultural values. Human land use has modified once natural ecosystems and resulted in cultural landscapes, agricultural fields, managed forests and built infrastructure. Finally, landscapes' different land covers provide intangible cultural values, including sense of place to people. When landscapes have been intensively used to deliver one kind of ecosystem service, other ones may not be satisfied or disservices may occur.

Therefore, to maintain natural capital and enhance human well-being, modified landscapes often require capacity-building in social systems to support maintenance by protection, management and restoration of landscapes as social-ecological systems (Dawson et al. 2017). This involves both place-based modification of the biophysical environment, coordination of human management of land and water and motivation of stakeholders and actors to act sustainably. The term landscape approach captures this (Axelsson et al. 2011; Sabogal et al. 2015; Sayer et al. 2013). Model Forest, Biosphere Reserve and Long-Term Socio-Ecological Research platform are three examples of the landscape approach (see Table 4).

4.4   Reflective Practitioner

In recent decades there have been many calls for students and professionals in forestry, and natural resource management more generally, to learn new social-science skills, and yet social sciences continue to be at best marginal in many higher education programmes (Innes 2009; Stummann and Gamborg 2014). In particular, there is a lack of training on reflective practice. Natural resource practitioners are often characterised as practical, can-do types, with little inclination to waste time on reflection; indeed they characterise themselves in that way (Lawrence and Gillett 2011). Yet recent work in the UK shows how foresters are actively experimenting to adapt to new constraints (tree health and climate) and developing new ways of sharing their findings, through field based discussions and innovation (Lawrence 2016). What is most challenging for practitioners but perhaps to a lesser degree for students is to step back and reflect on the wider system and different levels of learning. Reflexivity is greatly strengthened if theoretical frameworks can be used, such as the learning theories of Argyris and Schön (1978) who distinguish between single loop learning (are we doing things right?)

Table 4 Comparison of three landscape approach concepts using Lee's (1993) idea of compass and gyroscope

Criterion	Model Forest (IMFN 2008)	Biosphere Reserve (UNESCO 1995, 2008)	LTSER platform (Grove et al. 2013; Mirtl et al. 2013)
Area (landscape, region, catchment)	A large-scale biophysical area representing a broad range of landscape values, including social, cultural, economic and environmental concerns	Biosphere Reserves are organised into three management zones, known as core areas, buffer zones and transition areas	Siting in a large area
Infrastructure	A neutral forum that welcomes voluntary participation of representatives of stakeholder interests and values on the landscape	Logistical support for research, education, monitoring, etc.	Construction of coordination capacity (data bases, laboratories, monitoring schemes such as LTER sites)
Gyroscope (sustainable development as a societal process)	The process is representative, participatory, transparent and accountable and promotes collaborative work among stakeholder The activities are reflective of stakeholder needs, values and management challenges Builds on stakeholder capacity to engage in the sustainable management of natural resources and collaborate and share results and lessons learned through networking	Different stakeholders are informed and should participate on the procedure for the elaboration and review of integrated management policy Environmental education: (1) respect natural and cultural heritage, (2) favour responsible relationships with the environment and better land management, (3) create citizens who are aware of their responsibilities to future generations Institutional structure for local sustainable development	Stakeholder engagement for regional/local development involving decision-makers at different levels of governance, land use stakeholders, the public Collaborative learning builds on both quantitative and qualitative research and stakeholders' skills and experiences
Compass (sustainability as consequences)	Stakeholders are committed to the use, conservation and sustainable management of natural resources and the landscape	Development of integrated management policy Conservation of biological diversity (genetic variation, species, ecosystems, landscapes) and sustainable use of natural resources	Long-term monitoring of social and ecological systems Biological conservation and sustainable provision of ecosystem services to nature and people

and double loop learning (are we doing the right things?). The researcher-practitioner can be so embedded in the system that he or she cannot describe it in a balanced way. Some discourses, rules and institutional frameworks are so obvious to the observer that they are overlooked or are so culturally embedded that they stay invisible. However, they can be essential to comparing with other cases in other geographical regions and/or other cultural or institutional settings (Lawrence 2009; Lawrence et al. 2013; Brukas 2015).

Beyond the everyday carrying out of forest management duties, there are wider partnerships and systems that do not normally accommodate reflexivity but can be made to do so. Monitoring and evaluation systems within partnerships, for example, within National Biodiversity Strategies and Action Plans, are usually tick-box approaches but can be adapted to invite reflection on barriers and bridges. The UK Biodiversity Action Plan is an example of a bureaucratic approach that evolved to provide scope for reflexivity and holistic thinking by avoiding this tick the box method to artificially defined species and their habitats (Lawrence and Molteno 2012). To conclude, there is a need to enhance integrative knowledge production in research, schools, university education, vocational training and collaboration among stakeholders and actors with a range of sectors relevant for forest landscapes ecosystem services at multiple levels of governance.

4.5   Integrated Spatial Planning and Zoning

Sustainable forest management goals are often rival and hard to achieve at the same time in the same area. Implementation of a specific mix of forest management regimes is therefore a fundamental step towards maintaining representative green infrastructures in forest landscapes (e.g. Duncker et al. 2012). Forest management systems can be described in economical or ecological terms. Economically, the focus in on consolidation of all production factors including soil, machinery, energy, financial capital and manpower to get the highest economic return. Ecologically, one can describe the degree of anthropogenic transformation caused by forest management operations aimed at wood production. Forest management approaches can thus be grouped by the degree of management intensity (e.g. Duncker et al. 2012) and the extent to which management regimes emulate natural disturbances (e.g. Angelstam 1998). Additionally, different spatial planning extents need to be considered when designing the portfolio of forest management systems.

Integrated strategic, tactical and operational spatial planning at different scales from stands and forest management units to landscapes and regions can segregate intensive wood production, multiple forest use and forest protection (e.g. the triad approach proposed by Seymour and Hunter 1992). Three examples are the Ekopark developed and implemented by Sveaskog Co. in Sweden (Angelstam and Bergman 2004), the Latvian Ecoforest (Elmars Peterhofs, personal communication) and the Russian forest zoning system (Lazdinis and Angelstam 2005; Angelstam and Lazdinis 2017; Naumov et al. 2017). However, where land ownership is fragmented, both spatial planning and zoning can be difficult to realise. Key issues include what area proportions at different scales should be devoted to different forest functions, and how different functions should be configured spatially. Here evidence-based knowledge and negotiated targets about what biodiversity conservation requires are clearly diverging (Tear et al. 2005; Svancara et al. 2005; see next section).

4.6 Performance Targets for Green Infrastructure Functionality

Human behaviour is largely regulated by emotions. A person will often be more affected by a novel of a talented writer or presentation by an artist than a lecture given by one of the most erudite scholars. In an ideal case, narratives would combine both good theatrics and knowledge. Assessments of GI should therefore be presented as narrative and visual policy messages that are supported by evidence-based analyses. Given the complexity of assessing green infrastructure functionality, this is a major challenge. Protected area development is a good example.

As a base for discussing what a certain proportion of protected area actually constitutes for biodiversity conservation, Tear et al. (2005) used the terms representation, redundancy and resilience. Representation means capturing all ecological elements or target of interest (e.g. a population, species, biotope, landscape type or ecoregion). Redundancy (i.e. to protect more than is required for a specific ambition level) is necessary to reduce the risk of losing representative examples of these targets. Resilience, often referred to as the quality or health of an ecological element, is the ability of the element to persist through severe hardships.

Biodiversity can be maintained with different levels of ambition (Angelstam et al. 2004). A population's persistence in a landscape or region depends on how much habitat there is, whether individuals or propagules can move between different patches of suitable habitat and how the habitat network is maintained over time. Additionally, the role of the matrix among habitat patches need to be understood. While the term biotope refers to an environmentally uniform area, i.e. a natural or anthropogenic land cover with fine thematic resolution, a habitat is defined by the requirements of particular species or populations and often includes several biotopes (Udvardy 1959). Thus, a habitat often consists of a number of biotopes (i.e. land covers), such as for feeding, resting and breeding. Therefore, there is a need to identify and assess the quality, size and spatial distribution of biotopes that form habitats. However, habitats are more than just biotopes or land covers; also predators, competitors or micro- and macroclimate affect the function of biotopes. The combination of decreasing amounts of habitat, which decreases the number of individuals that can be supported, and increased fragmentation, which makes it harder for individuals to move in the landscape, are the most common reasons why species disappear locally and regionally and finally completely (Fahrig 2003).

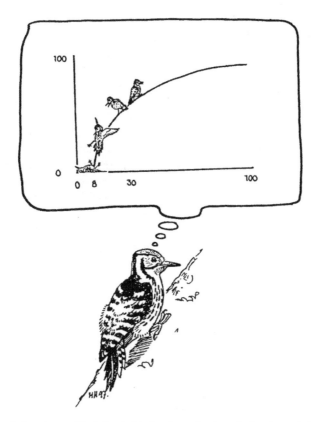

Fig. 10 Knowledge about critical thresholds for the extinction of a local population is a crucial component of analyses of the functionality of green infrastructure for biodiversity conservation. The white-backed woodpecker (*Dendrocopus leucotos*) is a species which is both endangered by the transformation of the natural dynamic forest landscapes and which is well studied (e.g. Edman et al. 2011; Stighäll et al. 2011). This drawing illustrates that when the amount of habitat decreases in the landscape, the likelihood that the species will remain is reduced, at first slowly, then faster and faster. Three important things need to be pointed out: (1) even in an ideal white-backed woodpecker habitat is not 100% sure that there is are white-backed woodpeckers, (2) when only a little of the original amount of habitat remains the probability of survival rapidly declines and (3) if the proportion of habitat is less than 8%, the probability of the local population survival is zero (Drawing by Martin Holmer)

There is both theoretical and empirical evidence for the existence of thresholds for extirpation of a population as the amount of available habitat is reduced. The threshold refers to the fact that the risk for population extinction shift from low to high within a limited range of further loss of habitat (Fig. 10). That there are limits to how much of different forest habitats that may disappear without threatening the viability of populations of all naturally occurring species forms the basis for the formulation of long-term evidence-based performance targets for how much of different forest habitats are needed (e.g. Svancara et al. 2005; Tear et al. 2005).

There are two key thresholds or tipping points. The first is when contiguous habitat is broken up into patches, thus no longer permitting percolation of individuals or propagules of different species through an un-fragmented habitat (With and Christ 1995). There is a threshold of 40% remaining forest land as habitat for the potential occurrence of species that need contiguous forest (Fahrig 2003). The second key threshold is when fragments begin increasing in inter-patch distance and thus isolation. As a minimum proportion of necessary habitat for the potential occurrence of species that can use local habitat patches, but are sensitive to forest landscape fragmentation, a target value of 20% can be used (Angelstam et al. 2011b).

These components can be illustrated by the different interpretations of how much forest has been set aside for provisioning of habitat for biodiversity conservation in Sweden (e.g. Angelstam et al. 2011b). At the national level across all five ecoregions in Sweden (Fig. 11), a total of 4.7% of productive forests were formally protected by the end of 2015 (Naturvårdsverket and Skogsstyrelsen 2017). Additionally, 5.8% productive forest were voluntarily set aside as stands within the framework of forest certification schemes, and 7.0% were set aside within the framework of reten-

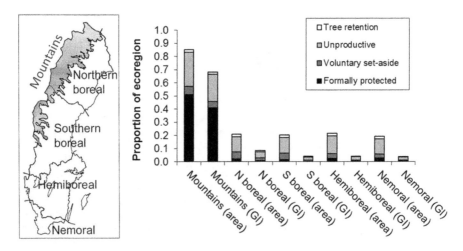

Fig. 11 Sweden extends 1572 km from south to north and hosts four main forest ecoregions plus mountains that include subalpine coniferous and deciduous forests. There are three categories of set-asides of forest and other wooded land at the scale of stands in landscapes (formally protected, voluntary set-aside, unproductive; cf. Aichi target 11) plus retention trees in stands (retention forestry, cf. Aichi target 7). These are not subject to forest management with the main aim to produce wood and biomass. The figure illustrates two aspects of these two groups of set-asides as components of a functional green infrastructure. The first is the large difference between the mountain forests of little economic interest versus all other regions focusing on maximum sustained yield forestry. The other is the difference between the total area of different kind set-asides (area) and an estimate of the functionality as green infrastructure (GI) for biodiversity conservation by considering the role of connectivity of formally protected and voluntary set-aside areas (see Angelstam et al. 2011a, b: 1124 with 40% functionality for north boreal, 20% functionality for south boreal, hemiboreal and nemoral forest ecoregions; for mountain forests, functionality was set at 80%). Data about the different kinds of set-asides from the Swedish Forest Agency and the Swedish Environmental Protection agency 2012–2017

tion forestry (Claesson et al. 2015). Summing up all those three categories yields a figure of 17.5% for Sweden; adding unproductive forests and woodlands yields about 27%. This can be compared with the Aichi biodiversity target 11 (CBD 2010), which states:

> By 2020, at least 17 per cent of terrestrial and inland water areas, and 10 per cent of coastal and marine areas, especially areas of particular importance for biodiversity and ecosystem services, are conserved through effectively and equitably managed, ecologically representative and well connected systems of protected areas and other effective area-based conservation measures, and integrated into the wider landscapes and seascapes.

However, there is a very large difference in the amount of forests not used for wood production between the generally unproductive subalpine mountain forests in NW Sweden with a total of 83% of formally protected, voluntarily set aside, retention set-asides and non-productive forests, respectively, and 17–20% in the four other ecoregions (Fig. 11). Moreover, voluntarily set-aside forest have lower quality (Elbakidze et al. 2016) and do not always host species dependent on natural forest components. Also the role of edge effects, decline of natural disturbances as well as trophic interactions needs to be understood and taken into account. Additionally, formally protected forest areas are most often neither representative (Nilsson and Götmark 1992) nor with sufficient connectivity (Angelstam et al. 2011b; Elbakidze et al. 2011). Based on the role of functional connectivity (e.g. Angelstam et al. 2011b; i.e. Aichi target number 11s "well connected systems") for the categories of forest not managed for wood and biomass production, the effective area that can be reported is considerably lower (see details in Figs. 10 and 11). The estimates show that when different ecoregions are assessed separately, only the mountain forests with 67%, with limited forestry due to low rates of biomass production, reach and exceed the Aichi target 11s 17%. All other forest ecoregions, which make up 93% of all forests in Sweden, reach only 3–8%-units of the 17% target (Fig. 11). This clearly suggests that the functionality for biodiversity conservation of the protected forest and woodland patches not managed for wood and biomass production is severely overestimated. At the same time, the transformation of near-natural remnants to intensively managed forest continues (e.g. Jonsson et al. 2016). This stresses the need for assessing the net consequences for biodiversity conservation of formal and voluntary forest protection on the one hand, and consequences of continued transformation of near-natural forest to forests managed for maximum sustained yield on the other.

Also the Aichi target number 7 (CBD 2010) should be considered. It states that by 2020 areas under agriculture, aquaculture and forestry are managed sustainably, ensuring conservation of biodiversity. In Sweden this includes voluntary set-asides, areas with low productivity and retention forestry. These three methods can be viewed as tools to both create habitat and to make the matrix around formally protected areas more permeable for dispersal of individuals of different species. The total functionality of (1) formally protected areas, (2) voluntary set-asides, (3) areas not used for forestry and (4) retention forestry would mirror the sum of Aichi targets 11 and 7. However, even then Sweden does not reach the 17% target (see Fig. 11, which NB! includes both targets 7 and 11).

At the same time, as there are positive effects of operational management supporting biodiversity conservation at multiple spatial scales, there are also negative effects in terms of continued gradual loss and transformation of forest stands never subject to clear-felling (e.g. Angelstam and Andersson 2013; Degteva et al. 2015; Naumov et al. 2017). Additionally, the effects of intensified forestry to improve wood and bioenergy yields need to be understood. This emphasises the need to estimate the net effect of continued transformation of near-natural forest remnants and the accumulated effect on biodiversity of establishing protected forests areas with or without conservation management, active habitat restoration and what can be achieved by increased nature conservation in the managed matrix. Resolving this challenge could bridge barriers of rhetoric among stakeholder groups with different stakes.

4.7 Building a Multilevel Infrastructure of Landscape Approach Initiatives

Our review highlights the experiences of a suite of place-based initiatives representing different landscape approach concepts in Europe's West and East. Early advocates of the landscape approach emphasised the key importance of transitioning from land management as site-level technical interventions towards collaborative learning within and among social-ecological systems (see Besseau et al. 2002; Axelsson et al. 2011; Laestadius et al. 2015; Sabogal et al. 2015). The proliferation of landscape approach concepts, and initiatives realising them on the ground, provides excellent opportunities for multilevel learning based on comprehensive analyses, such as presented in this chapter. Thus, the sharing of quality-assured practices among initiatives on the ground across the international networks of different landscape approach concepts could improve practices for learning by evaluation and collaboration. Ultimately this could improve governance, planning and management towards functional GI and thus support the development of sustainable landscapes.

Given sufficient time for developing collaborative capacity ("gyroscope" sensu Lee (1993)), long-term data about ecological and social systems ("compass" sensu Lee (1993)) and coordination within and among these initiatives, there is potential for developing an infrastructure for transdisciplinary research that represents different social-ecological contexts (Angelstam and Elbakidze 2017). At the national level, the French network of place-based so-called Zone Atelier initiatives is a good example (Mauz et al. 2012). The former Soviet Union network for environmental monitoring is another example (Spetich et al. 2009; Sokolov 1981; Zhulidov et al. 2000) but which needs to be complemented with stakeholder engagement.

To implement the United Nations' Sustainable Development Goals (e.g. Folke et al. 2016; Mbow et al. 2015) as well as ecological and green infrastructure policy (Angelstam et al. 2017b), different landscape approach concepts and local initia-

- UN Sustainable Development goals
  - Sustainable development: societal process
  - Sustainability: consequences in social-ecological systems (states and trends of assets)
- Landscape approach
  - Biophysical, anthropogenic and percieved dimensions
  - Collaboration among stakeholders at multiple levels

Strategic (Policy)

- Place-based research infrastructure networks for knowledge production and learning jointly by researchers and stakeholders
  - Siting
  - Constructing
  - Maintaining

Tactical (Planning, management)

- Green infrastructure policy as case study: supply and demand of ecosystem services
  - Green infrastructure functionality: state and trends of the potential supply of ecosystem services
  - Human demands: states and trends of socio-economic benefits
  - How is functionality of green infrastructure related to socio-econoomic benefits?
- Landscape approach infrastructure delivers sustainable development and sustainability (=UN goals achieved!!)

Operational (Consequences)

Fig. 12 Illustration of links between (1) United Nations' Sustainable Development Goals as strategic level policy and landscape approach as a principle for linking this to knowledge production, (2) place-based clusters of collaborating stakeholders aiming a sustainability in a social-ecological system through spatial planning and (3) consequences of operational management and monitoring of states and trends of sustainability

tives ought to collaborate strategically at the policy level (Fig. 12). Next, at the tactical level, different place-based initiatives that have similar aims and employ similar methods (see Table 4) should share knowledge and approaches to stimulate collaborative learning. Finally, operational work needs to be subject to evaluations of progress, ideally through active adaptive management (Shea et al. 2002). The extent to which green infrastructure functionality and socio-economic states and trends are monitored in the long-term, as well as how they are related to each other in terms of the supply and delivery of ecosystem services, is a prime example. However, there is so far insufficient cohesion, neither among different concepts nor among place-based initiatives on the ground.

The European Strategy Forum on Research Infrastructures (ESFRI), which encourages a coherent, strategy-led approach across Europe, is one option that could improve cohesion among place-based initiatives aimed at transdisciplinary research. ESFRI was established in 2002, with a mandate from the European Union Council to support policy-making on research infrastructures. ESFRI (2016) declares that the future prosperity of landscapes and regions in an increasingly competitive, globalised and knowledge-based economy relies on the potential of scientific and technological innovation. This requires high-quality educational and research institutions, a strong focus on skills and high-quality facilities for research that provide evidence-based knowledge. To facilitate multi-lateral initiatives leading to the better use and development of research infrastructures, ESFRI publishes roadmaps for the construction and development of the next generation of pan-European research infrastructures across a broad range of scientific fields (ESFRI 2016).

European Union's Horizon 2020 funding for establishment of a research infrastructure based on Long-Term Socio-Ecological Research (LTSER) platforms (Mirtl et al. 2013) is an attempt to create a research infrastructure in the European Union (see http://www.lter-europe.net/elter). At the international level, several networks focus on landscape restoration (IUCN and WRI 2014) and sustainable landscapes in the tropics (Denier et al. 2015). Global level examples are UNESCO's Biosphere Reserves, the International Model Forest Network (www.imfn.net) and the Global Landscapes Forum (www.landscapes.org). There is thus potential for integration among different landscape approach concepts and initiatives as a research infrastructure that can support implementation on the ground (Figs. 12 and 13). However, the high-level praise of landscape approach as a tool (e.g. World

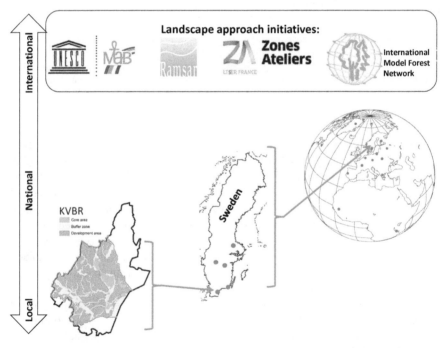

Fig. 13 Illustration of a Biosphere Reserve's (Kristianstad Vattenrike in Sweden (KVBR)) logistic function that monitors progress in the BR's conservation and development functions based on multilevel learning towards conservation and use of biodiversity. Undertaking multilevel learning by evaluation should be applied (1) within local initiatives, among them within the BR concept networks both (2) nationally and (3) internationally and ideally also (4) among different concept networks with similar ambitions (e.g. UNESCO's man and biosphere, Ramsar, the French network of LTSER platforms (Zone Ateliers) and the international Model Forest network

Forestry Congress 2009; Sayer et al. 2013) needs to be matched by effective bridging of barriers in terms of competition between organisations and concepts that love their own version of landscape approach.

5   Conclusions

A key task to sustain the provision of broad portfolios of ecosystem services in forest landscapes is to bridge barriers in social systems. Stakeholders have different benefit profiles as well as different opportunities to participate in landscape stewardship. Intensified economic use of forests and woodlands in rural and urban landscapes on the one hand, and the maintenance of these forest and woodland ecosystems' composition, structure and function (i.e. biodiversity) on the other cannot be achieved without integrated spatial planning. When comparing different countries, we see a paradox in terms of insufficient ecological knowledge, or limited interest, but high demands for provisioning ecosystem services.

The task of evidence-based landscape approach initiatives and other landscape stewardship platforms inspired by concepts like Biosphere Reserve, Model Forest and Long-Term Socio-Ecological Research platforms and their networks is to enhance long-term supply of data and stakeholder collaboration at multiple levels that support the sustainability of landscapes as social-ecological systems. This must be founded on the understanding of (1) landscapes' supply of material goods as well as nonmaterial services and values and (2) desired benefits of stakeholders from different sectors at multiple governance levels. This supply-demand interaction requires integration of quantitative and qualitative research methods including both expert and traditional/local knowledge. Thus, transdisciplinary place-based research is crucial with three integrated components: (1) collaboration among academic units with different portfolios of expertise, (2) engagement of stakeholders at multiple levels within landscapes and regions, including practitioners' knowledge, and (3) exchange of knowledge about both barriers and bridges among place-based researcher-stakeholder clusters (see Elbakidze et al. this book).

Maintenance of multiple place-based landscape approach initiatives as a distributed social-ecological research infrastructure requires high-level collaboration among different concepts with similar objectives, comprehensive long-term funding to sustain individual landscape approach initiatives and multilevel collaborative learning among both initiatives on the ground and among concepts internationally (Fig. 13).

Acknowledgements This work was supported by the Swedish Research Council Formas (grant number 2011-1737) to Per Angelstam and by the Swedish Institute (grant number 10976/2013) to Marine Elbakidze. We thank Peter Besseau, Lars Laestadius and the editors for constructive and stimulating reviews.

References

Andersson K, Angelstam P, Axelsson R, Elbakidze M, Törnblom J (2013a) Connecting municipal and regional level planning: analysis and visualization of sustainability indicators in Bergslagen, Sweden. Eur Plan Stud 21(8):1210–1234

Andersson K, Angelstam P, Elbakidze M, Axelsson R, Degerman E (2013b) Green infrastructures and intensive forestry: need and opportunity for spatial planning in a Swedish rural–urban gradient. Scand J For Res 28(2):143–165

Angelstam P (1998) Maintaining and restoring biodiversity by developing natural disturbance regimes in European boreal forest. J Veg Sci 9(4):593–602

Angelstam P, Andersson K (2013) Grön infrastruktur för biologisk mångfald i Dalarna. Har habitatnätverk för barrskogsarter förändrats 2002–2012? Länsstyrelsen Dalarnas län, Rapport 24

Angelstam P, Bergman P (2004) Assessing actual landscapes for the maintenance of forest biodiversity – a pilot study using forest management data. Ecol Bull 51:413–425

Angelstam P, Dönz-Breuss M (2004) Measuring forest biodiversity at the stand scale: an evaluation of indicators in European forest history gradients. Ecol Bull 51:305–332

Angelstam P, Elbakidze M (2017) Forest landscape stewardship for functional green infrastructures in Europe's West and East: diagnosing and treating social-ecological systems. In: Bieling C, Plieninger T (eds) The science and practice of landscape stewardship. Cambridge University Press, Cambridge, UK, pp 124–144

Angelstam P, Lazdinis M (2017) Tall herb sites as a guide for the protection and restoration of riparian continuous forest cover. Ecol Eng 103(B):470–477

Angelstam P, Boutin S, Schmiegelow F, Villard M-A, Drapeau P, Host G, Innes J, Isachenko G, Kuuluvainen M, Mönkkönen M, Niemelä J, Niemi G, Roberge J-M, Spence J, Stone D (2004) Targets for boreal forest biodiversity conservation – a rationale for macroecological research and adaptive management. Ecol Bull 51:487–509

Angelstam P, Axelsson R, Elbakidze M, Laestadius L, Lazdinis M, Nordberg M, Pătru-Stupariu I, Smith M (2011a) Knowledge production and learning for sustainable forest management: European regions as a time machine. Forestry 84(5):581–596

Angelstam P, Axelsson R. (2014) Sustainable Bergslagen - a landscape approach initiative in Sweden. Eur Rep 2014:4

Angelstam P, Andersson K, Axelsson R, Elbakidze M, Jonsson B-G, Roberge JM (2011b) Protecting forest areas for biodiversity in Sweden 1991-2010: policy implementation process and outcomes on the ground. Silva Fenn 45(5):1111–1133

Angelstam P, Grodzynskyi M, Andersson K, Axelsson R, Elbakidze M, Khoroshev A, Kruhlov I, Naumov V (2013a) Measurement, collaborative learning and research for sustainable use of ecosystem services: landscape concepts and Europe as laboratory. AMBIO 42(2):129–145

Angelstam P, Andersson K, Isacson M, Gavrilov DV, Axelsson R, Bäckström M, Degerman E, Elbakidze M, Kazakova-Apkarimova EY, Sartz L, Sädbom S, Törnblom J (2013b) Learning about the history of landscape use for the future: consequences for ecological and social systems in Swedish Bergslagen. AMBIO 42(2):150–163

Angelstam P, Andersson K, Annerstedt M, Axelsson R, Elbakidze M, Garrido P, Grahn P, Jönsson KI, Pedersen S, Schlyter P, Skärbäck E, Smith M, Stjernquist I (2013c) Solving problems in social-ecological systems: definition, practice and barriers of transdisciplinary research. AMBIO 42(2):254–265

Angelstam P, Andersson K, Axelsson R, Degerman E, Elbakidze M, Sjölander P, Törnblom J (2015) Barriers and bridges for Sustainable Forest Management: the role of landscape history in Swedish Bergslagen. In: Kirby KJ, Watkins D (eds) Europe's changing woods and forests: from wildwood to cultural landscapes. CABI, Wallingford, pp 290–305

Angelstam P, Naumov V, Elbakidze M (2017a) Transitioning from Soviet wood mining to sustainable forest management by intensification: are tree growth rates different in northwest Russia and Sweden? Forestry 90(2):292–303

Angelstam P, Barnes G, Elbakidze M, Marsh A, Marais C, Mills A, Polonsky S, Richardson DM, Rivers N, Shackleton R, Stafford W (2017b) Collaborative learning to unlock investments for functional ecological infrastructure: bridging barriers in social-ecological systems in South Africa. Ecosyst Serv 27:291–304

Angelstam P, Manton M, Pedersen S, Elbakidze M (2017c) Disrupted trophic interactions affect recruitment of boreal deciduous and coniferous trees in northern Europe. Ecol Appl 27(4):1108–1123

Argyris C, Schön DA (1978) Organizational learning: a theory of action perspective. Addison-Wesley, Reading

Axelsson R, Angelstam P, Elbakidze M, Stryamets N, Johansson K-E (2011) Sustainable development and sustainability: landscape approach as a practical interpretation of principles and implementation concepts. J Landsc Ecol 4(3):5–30

Axelsson R, Angelstam P, Myhrman L, Sädbom S, Ivarsson M, Elbakidze M, Andersson K, Cupa P, Diry C, Doyon F, Drotz MK, Hjorth A, Hermansson JO, Kullberg T, Lickers FH, McTaggart J, Olsson A, Pautov Y, Svensson L, Törnblom J (2013) Evaluation of multi-level social learning for sustainable landscapes: perspective of a development initiative in Bergslagen, Sweden. AMBIO 42(2):241–253

Beery T, Jönsson KI (2015) Inspiring the outdoor experience: does the path through a nature center lead out the door? J Interpret Res 20:67–85

Besseau P, Dansou K, Johnson F (2002) The International Model Forest Network (IMFN): elements of success. For Chron 78(5):648–654

Brukas V (2015) New world, old ideas – a narrative of the Lithuanian forestry transition. J Environ Policy Plan 17:495–515

Brumelis G, Jonsson BG, Kouki J, Kuuluvainen T, Shorohova E (2011) Forest naturalness in northern Europe: perspectives on processes, structures and species diversity. Silva Fenn 45:807–821

CBD (2010) The strategic plan for biodiversity 2011–2020 and the aichi biodiversity target. Available at https://www.cbd.int/doc/decisions/COP-10/cop-10-dec-02-en.pdf

Cheng AS, Mattor KM (2006) Why won't they come? Stakeholder perspectives on collaborative national forest planning by participation level. Environ Manag 38:545–561

Claesson S, Duvemo K, Lundström A, Wikberg P-E (2015) Skogliga konsekvensanalyser 2015. Skogsstyrelsen Rapport 2015, p 10

Council of Europe (2000) European landscape convention european treaty series no. 176. Council of Europe, Florence

Crow RR, Perera AH, Buse LJ (2007) Synthesis: what are the lessons for landscape ecologists? In: Perera AH, Buse LJ, Crow RR (eds) Forest landscape ecology. Transferring knowledge to practice. Springer, Dordrecht, pp 205–214

Dawson L, Elbakidze M, Angelstam P, Gordon J (2017) Governance and management dynamics of landscape restoration at multiple scales: learning from successful environmental managers in Sweden. J Environ Manag 197:24–40

de Groot RS, Alkemade R, Braat L, Hein L, Willemen L (2010) Challenges in integrating the concept of ecosystem services and values in landscape planning, management and decision making. Ecol Complex 7:260–272

de Groot R, Brander S, van der Ploeg S, Constanza R, Bernard F, Braat L, Christie N, Crossman N, Ghermandi A, Hein L, Hussain S, Kumar P, McVittie A, Portela R, Rodriguez LC, ten Brink P, van Beukering P (2012) Global estimates of the value of ecosystems and their services in monetary unites. Ecosyst Serv 1:50–61

Degteva SV, Ponomarev VI, Eisenman SW, Dushenkov V (2015) Striking the balance: challenges and perspectives for the protected areas network in northeastern European Russia. AMBIO 44(6):473–490

Denier L, Scherr S, Shames S, Chatterton P, Hovani L, Stam N (2015) The little sustainable landscapes book. Global Canopy Programme, Oxford, UK

Dickinson JL, Zuckerberg B, Bonter DN (2010) Citizen science as an ecological research tool: challenges and benefits. Annu Rev Ecol Evol Syst 41:149–172

Drescher M, Perera AH, Johnson CJ, Buse LJ, Drew CA, Burgman MA (2013) Toward rigorous use of expert knowledge in ecological research. Ecosphere 4(7):1–26

Duncker PS, Barreiro SM, Hengeveld GM, Lind T, Mason WL, Ambrozy S, Spiecker H (2012) Classification of forest management approaches: a new conceptual framework and its applicability to European forestry. Ecol Soc 17(4):51

Durham E, Baker H, Smith M, Moore E, Morgan V (2014) The BiodivERsA stakeholder engagement handbook. BiodivERsA, Paris

Edman T, Angelstam P, Mikusinski G, Roberge J-M, Sikora A (2011) Spatial planning for biodiversity conservation: assessment of forest landscapes' conservation value using umbrella species requirements in Poland. Landsc Urban Plan 102:16–23

Elbakidze M, Angelstam P (2007) Implementing sustainable forest management in Ukraine's Carpathian Mountains: the role of traditional village systems. For Ecol Manag 249:28–38

Elbakidze M, Angelstam P (2008) Model Forest in the north west of the Russian Federation: view from outside. Ustoychivoe Lesopolzovanie 1(17):39–47. (In Russian)

Elbakidze M, Angelstam P, Sandström C, Axelsson R (2010) Multi-stakeholder collaboration in Russian and Swedish Model Forest initiatives: adaptive governance towards sustainable forest management? Ecol Soc 15(2):14

Elbakidze M, Angelstam P, Andersson K, Nordberg M, Pautov Y (2011) How does forest certification contribute to boreal biodiversity conservation? Standards and outcomes in Sweden and NW Russia? For Ecol Manag 262(11):1983–1995

Elbakidze M, Angelstam P, Sandström C, Stryamets N, Crow S, Axelsson R, Stryamets G, Yamelynets T (2013a) The Biosphere Reserves for conservation and development in Ukraine? Legal recognition and establishment of the Roztochya initiative. Environ Conserv 40(2):157–166

Elbakidze M, Hahn T, Mauerhofer V, Angelstam P, Axelsson R (2013b) Legal framework for biosphere reserves as learning sites for sustainable development: a comparative analysis of Ukraine and Sweden. AMBIO 42(2):174–187

Elbakidze M, Andersson K, Angelstam P, Armstrong GW, Axelsson R, Doyon F, Hermansson M, Jacobsson J, Pautov Y (2013c) Sustained yield forestry in Sweden and Russia: how does it correspond to sustainable forest management policy? AMBIO 42(2):160–173

Elbakidze M, Dawson L, Andersson K, Axelsson R, Angelstam P, Stjernquist I, Teitelbaum S, Schlyter P, Thellbro C (2015) Is spatial planning a collaborative learning process? A case study from a rural–urban gradient in Sweden. Land Use Policy 48:270–285

Elbakidze M, Ražauskaite R, Manton M, Angelstam P, Mozgeris G, Brumelis G, Brazaitis G, Vogt P (2016) The role of forest certification for biodiversity conservation: Lithuania as a case study. Eur J For Res 135(2):361–376

Ericsson KA, Charness N, Feltovitch PJ, Hoffman RR (eds) (2006) The Cambridge handbook of expertise and expert performance. Cambridge University Press, New York

ESFRI (European Strategy Forum on Research Infrastructures) (2016) Strategy report on research infrastructures. Science and Technology Facilities Council. Available at http://ec.europa.eu/research/infrastructures

European Commission (2000) Directive, 2000/60/EC of the European parliament and of the council of 23rd October 2000. Establishing a framework for community action in the field of water policy. European Commission, Brussels

European Commission (2013a) Green Infrastructure (GI) — enhancing Europe's natural capital. COM 249. European Commission, Brussels

European Commission (2013b) Commission staff working document. Technical information on Green Infrastructure (GI) SWD 155. European Commission, Brussels

European Commission (2013c) A new EU Forest Strategy: for forests and the forest-based sector. COM 659

Fahrig L (2003) Effects of habitat fragmentation on biodiversity. Annu Rev Ecol Evol Syst 34:487–515

Folke C, Hahn T, Olsson P, Norberg J (2005) Adaptive governance of social-ecological systems. Annu Rev Environ Resour 30(1):441–473

Folke C, Biggs R, Norström A, Reyers B, Rockström J (2016) Social-ecological resilience and biosphere-based sustainability science. Ecol Soc 21(3):41

Furman E, Peltola T (2013) Developing socio-ecological research in Finland: challenges and progress towards a thriving LTSER network. In: Singh SJ, Haberl H, Chertow M, Mirtl M, Schmid M (eds) Long term socio-ecological research. Springer, Dordrecht, pp 443–459

Garrido P, Elbakidze M, Angelstam P (2017a) Stakeholders' perceptions on ecosystem services in Östergötland's (Sweden) threatened oak wood-pasture landscapes. Landsc Urban Plan 157:96–104

Garrido P, Elbakidze M, Angelstam P, Plieninger T, Pulido F, Moreno G (2017b) Stakeholder perspectives of wood pasture ecosystem services: a case study from Iberian dehesas. Land Use Policy 60:324–333

Gavrilova G, Baroniņa V, Sulcs V (2000) Vascular plant flora of the Lake Engures (Engure) drainage basin, Latvia, and the coastal zone of the Gulf of Riga. Proc Latvian Acad Sci B 54(5/6):177–179

Grêt-Regamey A, Brunner SH, Altwegg J, Christen M, Bebi P (2013) Integrating expert knowledge into mapping ecosystem services tradeoffs for sustainable forest management. Ecol Soc 18(3):34–50

Grove JM, Pickett STA, Whitmer A, Cadenasso ML (2013) Building and urban LTSER: the case of the Baltimore ecosystem study and the D.C./B.C., ULTRA-Ex project. In: Singh JS, Haberl H, Schmid M, Mirtl M, Chertow M (eds) Long term socio-ecological research studies in society nature interactions across temporal and spatial scales. Springer, Dordrecht, pp 369–408

Huntsinger L, Oviedo JL (2014) Ecosystem services are social-ecological services in a traditional pastoral system: the case of California's Mediterranean rangelands. Ecol Soc 19(1):8

IMFN (2008) Model Forest development guide. International Model Forest Network Secretariat, Ottawa

Innes J (2009) Is forestry a social sciences? J Trop For Sci 21:V–VI

IUCN and WRI (2014) A guide to the Restoration Opportunities Assessment Methodology (ROAM). IUCN, Gland

Jacobs S, Burkhard B, Van Daele T, Staes J, Schneiders A (2015) "The matrix reloaded": a review of expert knowledge use for mapping ecosystem services. Ecol Model 295:21–30

Johansson M, Henningsson M (2011) Social-psychological factors in public support for local biodiversity conservation. Soc Nat Resour 24:717–733

Jonsson BG, Ekström M, Esseen PA, Grafström A, Ståhl G, Westerlund B (2016) Dead wood availability in managed Swedish forests – policy outcomes and implications for biodiversity. For Ecol Manag 376:174–182

Kates RW (2011) What kind of science is sustainability science? Proc Natl Acad Sci U S A 108:19449–19450

Komiyama H, Takeuchi K (2006) Sustainability science: building a new discipline. Sustain Sci 1:1–6

Kopperoinen L, Itkonen P, Niemelä J (2014) Using expert knowledge in combining green infrastructure and ecosystem services in land use planning: an insight into a new place-based methodology. Landsc Ecol 29:1361–1375

Koschke L, Fürst C, Frank S, Makeschin F (2012) A multi-criteria approach for an integrated land-cover-based assessment of ecosystem services provision to support landscape planning. Ecol Indic 21:45–66

Kušová D, Těšitel J, Matějka K, Bartoš M (2008) Biosphere reserves - an attempt to form sustainable landscapes: a case study of three biosphere reserves in the Czech Republic. Landsc Urban Plan 84:38–51

Laestadius L, Buckingham K, Maginnis S, Saint-Laurent C (2015) Before Bonn and beyond: the history and future of landscape restoration. Unasylva 66:11–18

Lawrence A (2009) Forestry in transition: imperial legacy and negotiated expertise in Romania and Poland. Forest Policy Econ 11(5):429–436

Lawrence A (2010) Introduction: learning from experiences of participatory biodiversity assessment. In: Lawrence A (ed) Taking stock of nature: participatory biodiversity assessment for policy and planning. Cambridge University Press, Cambridge, UK, pp 1–29

Lawrence A (2016) Adapting through practice: silviculture, innovation and forest governance for the age of extreme uncertainty. Forest Policy Econ 2017(79C):50–60

Lawrence A, Gillett S (2011) Human dimensions of adaptive forest management and climate change: a review of international experience. Forestry Commission Research Report. Edinburgh, Forestry Commission, p 52

Lawrence A, Molteno S (2012) From rationalism to reflexivity? Reflections on change in the UK Biodiversity Action Plan. In: Brousseau E, Dedeurwaerdere T, Siebenhüner B (eds) Reflexive governance for global public goods. MIT Press, Cambridge, MA, pp 283–298

Lawrence A, De Vreese R, Johnston M, van den Bosch CCK, Sanesi G (2013) Urban forest governance: towards a framework for comparing approaches. Urban For Urban Green 12(4):464–473

Lazdinis M, Angelstam P (2005) Functionality of riparian forest ecotones in the context of former Soviet Union and Swedish forest management histories. Forest Policy Econ 7(3):321–332

Lee KN (1993) Compass and gyroscope: integrating science and politics for the environment. Island Press, Covelo

Lele S, Springate-Baginski O, Lakerveld R, Deb D, Dash P (2013) Ecosystem services: origins, contributions, pitfalls, and alternatives. Conserv Soc 11(4):343–358

Lindström M, Johansson M, Herrmann J, Johnsson O (2006) Attitudes towards the conservation of biological diversity. A case study in Kristianstad Municipality, Sweden. J Environ Plan Manag 49:495–513

Lundquist L (1987) Implementation steering. An actor-structure approach. Studentlitteratur, Lund

Magnusson S-E, Magntorn K, Wallsten E, Cronert H, Thelaus M (2004) Kristianstads Vattenrike biosphere reserve nomination form. Kristianstad Kommun, Sweden

Manton M (2016) Functionality of wet grasslands as green infrastructure. Waders, avian predators and land covers in Northern Europe. Acta Univ Agric Sueciae 2016:119

Manton M, Angelstam P, Milberg P, Elbakidze M (2016) Governance and management of green infrastructures for ecological sustainability: wader bird conservation in Southern Sweden as a case study. Sustainability 8(4):340

Mariev AN, Kutepov DZ, Mikheev RB, Poroshin EA (2005) Recommendations on final felling operations with focus on biodiversity conservation in pristine forests of the Komi Republic. Agency of Forest Management in Komi Republic, Syktyvkar. (in Russian)

Marsh GP (1864) Man and nature; or, physical geography as modified by human action. Charles Scribner, New York

Mauz I, Peltola T, Granjou C, Van Bommel S, Buijs A (2012) How scientific visions matter: insights from three long-term socio-ecological research (LTSER) platforms under construction in Europe. Environ Sci Pol 19:90–99

Mbow C, Neely C, Dobie P (2015) How can an integrated landscape approach contribute to the implementation of the Sustainable Development Goals (SDGs) and advance climate-smart objectives? In: Minang PA, van Noordwijk M, Freeman OE, Mbow C, de Leeuw J, Catacutan D (eds) Climate-smart landscapes: multifunctionality in practice. World Agroforestry Centre (ICRAF), Nairobi, pp 103–117

Melecis V (2000) Integrated research: renaissance of ecology in Latvia. Proc Latvian Acad Sci B 54(5/6):221–225

Melecis V (2011) Project on development of a conceptual integrated model of socioeconomic biodiversity pressures, drivers and impacts for the long-term socioecological research platform of Latvia. Proc Latvian Acad Sci B 65(5/6):206–212

Melecis V, Klavins M, Laivins M, Rusina S, Springe G, Viksne J, Krisjane Z, Strake S (2014) Conceptual model of the long-term socio-ecological research platform of Engure ecoregion, Latvia. Proc Latvian Acad Sci B 68(1/2 (688/689)):1–19

Millennium Ecosystem Assessment (2005) Ecosystems and human well-being: synthesis. Island Press, Washington, DC

Mirtl M, Orenstein DE, Wildenberg M, Peterseil J, Frenzel M (2013) Development of LTSER platforms in LTER-Europe: challenges and experiences in implementing place-based long-term socio-ecological research in selected regions. In: Singh SJ, Haberl H, Chertow M, Mirtl M, Schmid M (eds) Long term socio-ecological research. Springer, Dordrecht, pp 409–442

Naturvårdsverket and Skogsstyrelsen (2017) Värdefulla skogar. Redovisning av regeringsuppdrag. Skrivelse 2017-01-31. Available at http://www.naturvardsverket.se/Miljoarbete-i-samhallet/Miljoarbete-i-Sverige/Regeringsuppdrag/Redovisade-2017/Ny-nationell-strategi-for-formellt-skydd-av-vardefulla-skogar/

Naumov V, Angelstam P, Elbakidze M (2016) Barriers and bridges for intensified wood production in Russia: insights from the environmental history of a regional logging frontier. Forest Policy Econ 66:1–10

Naumov V, Angelstam P, Elbakidze M (2017) Satisfying rival objectives in forestry in the Komi Republic: effects of Russian zoning policy change on forestry intensification and riparian forest conservation. Can J For Res 47:1339–1349

Nilsson C, Götmark F (1992) Protected areas in Sweden: is natural variety adequately represented? Conserv Biol 6(2):232–242

Nordberg M, Angelstam P, Elbakidze M, Axelsson R (2013) From logging frontier towards sustainable forest management: experiences from boreal regions of North-West Russia and North Sweden. Scand J For Res 28(8):797–810

Norgaard RB (2010) Ecosystem services: from eye-opening metaphor to complexity blinder. Ecol Econ 69(6):1219–1227

Noss RF (1990) Assessing and monitoring forest biodiversity: a suggested framework and indicators. For Ecol Manag 115(2):135–146

O'Hagan A, Buck CE, Daneshkhah A, Eiser JR, Garthwaite PH, Jenkinson DJ, Oakley JE, Rakow T (2006) Uncertain judgements: eliciting experts' probabilities. Wiley, Chichester

Odum EP (1959) Fundamentals of ecology, 2nd edn. W.B. Saunders, Philadelphia

Perera AH, Drew CA, Johnson CJ (eds) (2012) Expert knowledge and its application in landscape ecology. Springer, New York

Plieninger T, Kizos T, Bieling C, Le Dû-Blayo L, Budniok M-A, Bürgi M, Crumley CL, Girod G, Howard P, Kolen J (2015) Exploring ecosystem-change and society through a landscape lens: recent progress in European landscape research. Ecol Soc 20(2):5

Popescu VD, Rozylowicz L, Niculae IM, Cucu AL, Hartel T (2014) Species, habitats, society: an evaluation of research supporting EU's Natura 2000 network. PLoS One 9(11):e113648

Power ME, Tilman D, Estes JA, Menge BA, Bond WJ, Mills LS, Daily G, Castilla JC, Lubchenco J, Paine RT (1996) Challenges in the quest for keystones. Bioscience 46(8):609–620

Price MF, Park JJ, Bouamrane M (2010) Reporting progress on internationally designated sites: the periodic review of biosphere reserves. Environ Sci Pol 13(6):549–557

Prishchepov AV, Radeloff VC, Dubinin M, Alcantara C (2012) The effect of Landsat ETM/ETM+ image acquisition dates on the detection of agricultural land abandonment in Eastern Europe. Remote Sens Environ 126:195–209

Quijas S, Jackson LE, Maass M, Schmid B, Raffaelli D, Balvanera P (2012) Plant diversity and generation of ecosystem services at the landscape scale: expert knowledge assessment. J Appl Ecol 49:929–940

Rauschmayer F, Berghöfer A, Omann I, Zikos D (2009) Examining processes or/and outcomes? Evaluation concepts in European governance of natural resources. Environ Policy Gov 19(3):159–173

Reed MG, Egunyu F (2013) Management effectiveness in UNESCO Biosphere Reserves: learning from Canadian periodic reviews. Environ Sci Pol 25:107–117

Richnau G, Angelstam P, Valasiuk S, Zahvoyska L, Axelsson R, Elbakidze M, Farley J, Jönsson I, Soloviy I (2013) Multi-faceted value profiles of forest owner categories in South Sweden: the River Helge å catchment as a case study. AMBIO 42(2):188–200

Sabogal C, Besacier C, McGuire D (2015) Forest and landscape restoration: concepts, approaches and challenges for implementation. Unasylva 66(3):3–10

Sayer J, Sunderland T, Ghazoul J, Pfund JL, Sheil D, Meijaard E, Venter M, Boedhihartono AK, Day M, Garcia C, van Oosten C (2013) Ten principles for a landscape approach to reconciling agriculture, conservation, and other competing land uses. Proc Natl Acad Sci 110(21):8349–8356

Schultz L, Folke C, Olsson P (2007) Enhancing ecosystem management through social-ecological inventories: lessons from Kristianstads Vattenrike, Sweden. Environ Conserv 34(2):140–152

Schultz L, Folke C, Österblom H, Olsson P (2015) Adaptive governance, ecosystem management, and natural capital. Proc Natl Acad Sci 112(24):7369–7374

Seymour RS, Hunter ML (1992) New forestry in eastern spruce-fir forests: principles and applications to Maine. College of Forest Resources, University of Maine, Orono

Shea K, Possingham HP, Murdoch WW, Roush R (2002) Active adaptive management in insect pest and weed control: intervention with a plan for learning. Ecol Appl 12(3):927–936

Singh SJ, Haberl H, Chertow M, Mirtl M, Schmid M (eds) (2013) Long-term socio-ecological research. Springer, Dordrecht

Sokolov V (1981) The biosphere reserve concept in the USSR. AMBIO 2–3:97–101

Spetich MA, Kvashnina AE, Nukhimovskya YD, Rhodes OE Jr (2009) History, administration, goals, value, and long-term data of Russia's strictly protected scientific nature reserves. Nat Areas J 29(1):71–78

Stighäll K, Roberge J-M, Andersson K, Angelstam P (2011) Usefulness of biophysical proxy data for modelling habitat of a threatened forest species: the white-backed woodpecker. Scand J For 26(6):576–585

Stummann CB, Gamborg C (2014) Reconsidering social science theories in natural resource management continuing professional education. Environ Educ Res 20:496–525

Svancara LK, Brannon R, Scott JM, Groves CR, Noss RF, Pressey RL (2005) Policy-driven versus evidence-based conservation: a review of political targets and biological needs. Bioscience 55:989–995

Tear TH, Kareiva P, Angermeier PL, Comer P, Czech B, Kautz R, Landon L, Mehlman D, Murphy K, Ruckelshaus M, Scott JM, Wilhere G (2005) How much is enough? The recurrent problem of setting measurable objectives in conservation. Bioscience 55(10):836–849

Törnblom J, Angelstam P, Degerman E, Tamario C (2017) Prioritizing dam removal and stream restoration using critical habitat patch threshold for brown trout (*Salmo trutta* L.): a catchment case study from Sweden. Écoscience. https://doi.org/10.1080/11956860.2017.1386523

Tuvendal M, Elmberg J (2015) A handshake between markets and hierarchies: geese as an example of successful collaborative management of ecosystem services. Sustainability 7(12):15937–15954

Tuvendal M, Elmqvist T (2011) Ecosystem services linking social and ecological systems: river brownification and the response of downstream stakeholders. Ecol Soc 16(4):21

Udvardy MF (1959) Notes on the ecological concepts of habitat, biotope and niche. Ecology 40(4):725–728

UNESCO (1995) The Seville strategy and the statutory framework of the world network of biosphere reserves. UNESCO, Paris

UNESCO (2008) Madrid action plan for biosphere reserves (2008–2013). UNESCO, Paris

Viksne J (1997) The bird lake Engure. Publ. House Jana Seta, Riga

Viksne J (2000) Changes of nesting bird fauna at the Engure Ramsar site, Latvia, during the last 50 years. Proc Latvian Acad Sci Sect B 54(5/6):213–220

Viksne J, Janaus M, Mednis A (2011) Factors influencing the number of breeding water birds in Lake Engure. Latvia Sect B 65(5/6):190–196

von Carlowitz HC (1713) Sylvicultura oeconomica [Economic forest management] oder Hausswirthliche Nachricht und naturmässige Anweisung zur wilden Baumzucht (reprint from 2000 edited by K. Irmer). Technische Universität Bergakademie Freiburg, Freiburg

Walker B, Salt D (2006) Resilience thinking: sustaining ecosystems and people in a changing world. Island Press, Washington, DC

Wallner A, Bauer N, Hunziker M (2007) Perceptions and evaluations of biosphere reserves by local residents in Switzerland and Ukraine. Landsc Urban Plan 83:104–114

With KA, Christ TO (1995) Critical threshold in species' responses to landscape structure. Ecology 76:2446–2459

World Forestry Congress (2009) Forest development: a vital balance, findings and strategic actions. Findings and Strategic Actions. http://foris.fao.org/meetings/download/_2009/xiii_th_world_forestry_congress/misc_documents/wfc_declaration.pdf

Yablochkina EM, Romanyuk BD, Chernenkova EA (2007) The WWF project "Pskov Model Forest". WWF Russia, Moscow and Strugy-Krasnye

Yaffee SL (1999) Three faces of ecosystem management. Conserv Biol 13(4):713–725

Zavrl MS, Zeren MT (2010) Sustainability of urban infrastructures. Sustainability 2:2950–2964

Zhulidov AV, Khlobystov VV, Robarts RD, Pavlov DF (2000) Critical analysis of water quality monitoring in the Russian Federation and former Soviet Union. Can J Fish Aquat Sci 57(9):1932–1939

Solving Conflicts among Conservation, Economic, and Social Objectives in Boreal Production Forest Landscapes: Fennoscandian Perspectives

Mikko Mönkkönen, Daniel Burgas, Kyle Eyvindson, Eric Le Tortorec, Maiju Peura, Tähti Pohjanmies, Anna Repo, and María Triviño

1 Introduction

The boreal biome encompasses almost one-third of the world's forests (UNEP et al. 2009). Unlike tropical and temperate forests, boreal forests have remained relatively stable in area in recent decades (UNEP et al. 2009; FAO 2010). Overall, the region is characterized by a net gain in growing forest stock (FAO 2010). However, extensive tracts of boreal forests are actively managed and harvested for timber production, with changes to the structure and tree species composition of the forests and impacts on wildlife and ecosystem functioning (Östlund et al. 1997; Bradshaw et al. 2009; Kuuluvainen et al. 2012). Moreover, there is ongoing pressure to increase forest biomass use to reduce CO_2 emissions and increase renewable energy use according to set targets (Stupak et al. 2007).

Even though forest cover is still extensive, forestry has caused profound changes in the landscape and stand structure in the boreal zone (Bryant et al. 1997; Esseen et al. 1997). Boreal forests also harbor unique biodiversity (UNEP et al. 2009). However, habitat availability and resources for species associated with processes and structures characteristic for pristine forests have severely declined. Consequently, forest management is the most prominent threat for a number of species in the north. For a large proportion of the red-listed species, for example, in Finland and Sweden (totally about 1700 and 3500 species, respectively), forestry is a main threat, and invertebrates and fungi are particularly common among forest-associated red-listed species (Rassi et al. 2010; Westling 2015).

Throughout the boreal region, intact forests are concentrated in the northernmost or otherwise inaccessible regions and, even still, are not extensively protected

M. Mönkkönen (✉) · D. Burgas · K. Eyvindson · E. Le Tortorec · M. Peura · T. Pohjanmies · A. Repo · M. Triviño
Department of Biological and Environmental Sciences, University of Jyvaskyla, Jyvaskyla, FL, Finland
e-mail: mikko.monkkonen@jyu.fi

© Springer International Publishing AG, part of Springer Nature 2018
A. H. Perera et al. (eds.), *Ecosystem Services from Forest Landscapes*,
https://doi.org/10.1007/978-3-319-74515-2_7

(Potapov et al. 2008). Protected areas cover only less than 5% of productive forest land in the Nordic countries. There is also a strong bias concerning protected areas so that most protected areas are situated in the north, and they consist, for large part, of forest types that are less productive than average (e.g., Angelstam and Pettersson 1997).

The multifunctional role of forests is widely recognized from scientific (Harrison et al. 2010) and policy points of view (e.g., the EU Forestry Strategy[1]). Forests are a major source of timber, but also provide a range of other goods and services that are essential to human societies (Vanhanen et al. 2012). In addition to providing jobs, income, and raw material to industry, forests, for example, regulate and purify fresh water, prevent soil erosion, conserve biodiversity, protect against landslides, provide recreation and non-timber products, and act as carbon sinks thus contributing to climate change mitigation (EC 2012). For both biodiversity and ecosystem services, it is crucially important how the forests outside the protected areas are managed. Forest exploitation has a long history in the Nordic countries dating back to the seventeenth century (Esseen et al. 1997). Intensive forestry practices such as clear-cut harvesting, soil preparation, and planting new trees were developed in the mid-twentieth century to maximize the production of timber and pulpwood. Forest industry constitutes a large proportion of national gross production of Nordic countries and is very important to many local economies. This imposes a great challenge: is it possible to preserve biological diversity and the diversity of ecosystem services in the boreal zone and yet at the same time maintain intensive timber extraction from Fennoscandian boreal forests? In this chapter, we aim at providing an answer to this question based on the understanding provided by the research on the natural dynamics of the boreal biome, on management effects on the structure and dynamics of the forest at different spatial, and on multifunctionality of Fennoscandian forests. Our focus is on Fennoscandian forest landscapes where we scrutinize sustainability of forest management from economic, social, and ecological perspectives.

We first provide a general overview of boreal forest ecosystems. We introduce the main abiotic and biotic factors affecting the distribution and causing variation in boreal forest, to lay a basis for understanding factors affecting both biodiversity and ecosystem services. We provide an overview of the history of the biome, describe the natural dynamics of boreal forests, and provide an overview of biodiversity patterns. Further, we describe the main ecosystem services from boreal forest and identify ecosystem services that are typical for boreal forests, emphasizing differences with warm temperate forests. We provide insights on how important these ecosystem services are globally, regionally, and locally. We particularly discuss the so-called everyman's right tradition in Fennoscandia and how this influences possibilities to benefit from forest ecosystem services.

Secondly, we describe the traditional and current forest management practices in Fennoscandia and go on by describing how management mechanistically connects with biodiversity and the flow of ecosystem services. We review synergies and conflicts among alternative ecosystem services as well as biodiversity. Third, we

---

[1] https://ec.europa.eu/agriculture/forest/strategy_en.

provide an overview of the forestry practices and policies that aim to ensure multifunctionality of Fennoscandian forests, i.e., diversity of efforts on sustaining biodiversity, timber production, and other ecosystem services from forest landscapes. Finally, we present alternative methods for assessing conflicts among different ecosystem services and for finding solutions for them. We conclude the chapter by providing insights for future management aiming at sustainability from economic, ecological, and social perspectives.

It becomes obvious from our overview of boreal forest ecosystems that Fennoscandian forests are probably the simplest of all forest ecosystems on Earth. Because of long traditions in natural history and forest ecological research in all Nordic countries, Fennoscandian boreal forests are relatively well-known ecosystems. From global perspective, national economies of the Nordic countries are affluent, stable, and predictable. We assert therefore that in comparative terms, Fennoscandia is an ideal test laboratory to find out if sustainable forest management, in ecological, economic, and social terms, is an achievable goal (Mönkkönen 1999).

2 Boreal Forest Ecosystem

2.1 Distribution of Boreal Forests

The boreal biome is characterized by a relatively cold climate, large differences between summer and winter temperatures, and a persistent snow cover in winter. In some regions of the boreal, precipitation may fall mainly as snow. This combination, along with nutrient-poor soils, favors the preponderance of conifer species (*Pinus* spp., *Picea* spp., *Abies* spp., *Larix* spp.), although species of broad-leaved deciduous trees, particularly *Betula* spp., *Populus* spp., and *Salix* spp., are also common. Compared to lower latitudes, forests in the boreal biome are home to relatively few tree species. The boreal region meets the tundra to the north and at high altitudes and the temperate forest to the south (Fig. 1). The transition zone between the boreal and temperate regions, called the hemiboreal (or sometimes "boreonemoral") zone, is characterized by a mixture of boreal and temperate elements (Nilsson 1997).

Olson et al. (2001) recognized 29 ecoregions within the boreal zone, 18 in the Nearctic, and 11 in the Palearctic region. Boreal forests of Fennoscandia and western Russia, stretching a latitudinal range of ~57°–69° N, are one of the Palearctic boreal ecoregions. There are clear variations in climate at larger scales within the ecoregion, i.e., a gradient from a maritime climate in Norway to increasingly continental climates eastward (Tuhkanen 1980), as well as a latitudinal and altitudinal gradient from the southern boreal zone to the northern boreal zone (Moen 1999). The location of Fennoscandia at the western edge of the Eurasian boreal forest belt makes the regions climatically different from other boreal ecoregions. Because of

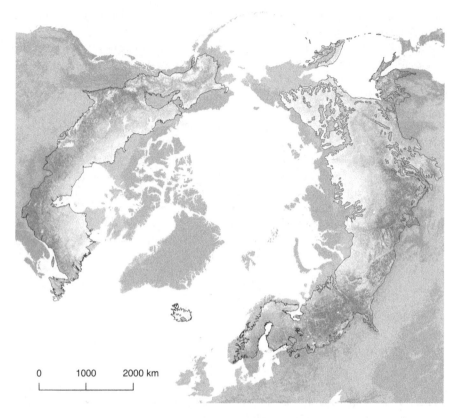

Fig. 1 Map showing the global extent of the boreal zone (outlined) and estimated cover of woody vegetation greater than 5 meters in height in 2010. Darker colors represent more dense forest cover (Data source: Global Land Cover Facility, Tree Canopy Cover 2010)

relatively maritime climate, annual variations in temperature and precipitation are less pronounced than in more continental areas in Siberia and the Nearctic. Also mean annual temperature is higher in the western Palearctic than at the same latitudes elsewhere, and consequently, the southern edge of the boreal zone in Fennoscandia is at the same latitude as the northern edge of the zone, for example, in Eastern North America.

The main gradients of variation influencing the local characteristics of natural boreal forests are (a) biogeographical patterns partly driven by climatic gradients, (b) the soil's nutrient and moisture gradients and their effects on tree growth and natural disturbances, and (c) time since disturbance (forest development). All these three gradients represent continuous variation rather than discreet classes. Nevertheless, classifying the gradients has large heuristic value in recognizing classes of forests and their developmental stages because they provide tools to handle underlying abiotic and biotic factors affecting both biodiversity and ecosystem services.

The biotic zonation of boreal forests in Fennoscandia has received much interest. Widely accepted subdivision scheme was created by Ahti et al. (1968), which identified four zones mainly based on the composition of plant communities in mesic sites. A main pattern is the increase of conifer tree dominance from hemiboreal zone toward the north boreal and concurrent decline of temperate broad-leaved trees. Likewise, the dominance of herbs in the field layer decreases and that of dwarf shrubs increases with increasing latitude. In the middle boreal zone, herbs are restricted to nutrient-rich sites and even more so in the north boreal zone where bryophytes and lichens also become more abundant (Esseen et al. 1997).

Variation in nutrient availability and moisture also causes variation in forest community composition, which is summarized in forest site type classification systems (e.g., Cajander 1949; Arnborg 1990). Generally, soil type, soil moisture, and vegetation in the European boreal forest are closely related. The vegetation of the understory of Fennoscandian boreal forest can be classified as follows. Very dry/nutrient-poor sites are lichen-dominated heaths. Dry-mesic and moderately rich sites are dominated by dwarf shrubs, e.g., *Calluna vulgaris*, *Empetrum hermaphroditum*, *Vaccinium vitis-idaea*, and *V. myrtillus*. Moist-wet, moderately rich sites are characterized by *V. uliginosum* and *Ledum palustre*, while mesic, moderately rich by pleurocarpous mosses. In moist and rich sites, herb layer with pteridophytes and e.g. *Maianthemum bifolium* is typical, and in moist-wet and very rich sites tall herb layer with e.g. *Filipendula ulmaria* prevails.

Time since disturbance is yet another gradient of variation because boreal forests can be considered disturbance-driven ecosystems (see below). This variation can be classified into forest successional stages based on stand structural characteristics (Shorohova et al. 2009).

2.2 History of Boreal Forests

Boreal forest biome is the youngest of the Earth's forest biomes. It has developed during gradual cooling that has characterized the global climate during the past 20 million years. During the early Oligocene, some 30 million years ago, tropical and subtropical forests thrived as far north as the Spitzbergen or Canadian (presently) artic. At the evolutionary time scale, the evolution of cold hardiness in trees has been critical. The origin of the present boreal tree taxa likely locates in East Asian mountain regions, where frost tolerance evolved in high-altitude conifer and mixed forest during the gradual cooling of the Miocene climate. During the Pleistocene, these cold-hardy tree taxa then spread across the Palearctic and Nearctic regions with further cooling of global climate (Latham and Ricklefs 1993). The current extent of the boreal zone is relatively recent phenomenon, developed during the past 2 million years when the global cooling culminated in the Pleistocene era. Thus, the boreal biome as we see it today is a child of the Ice Ages.

Pleistocene differs from the preceding era not only in terms of cooler global climates but also in terms of large climatic variations with recurrent glacial periods interrupted by warmer interglacial periods. Most of the area currently occupied by boreal forests, particularly in the western Palearctic and the Nearctic regions, was glaciated several times during the Pleistocene. Typically, tree species reacted to Pleistocene climatic oscillations in idiosyncratic ways, and thus forest zones or communities did not retract toward south during glaciations (Huntley et al. 1983). Consequently, forests in the glacial refugia lack modern equivalents in the present boreal tree communities both taxonomically and structurally. The latest glacial period ended approximately 10,000 years ago, after which species gradually (re) colonized the region. As a consequence, boreal species assemblages can be considered relatively "young" from a geological time perspective.

During the peak of glaciations, forests were almost completely wiped out from Europe except from the south- and sea-facing slopes of mountains in the Balkan, Apennine, and Iberian peninsulas. Further north the climate was too cold and lower down in the Mediterranean region too dry for forest growth. Of the boreal tree species, glacial refugia of the Scots pine (*Pinus sylvestris*) and birches were much larger than those of the Norwegian spruce (*Picea abies*) that only existed in the Balkans and further east. In the Nearctic and in eastern Palearctic, total forest area remained much more extensive than in the western Palearctic. In the Nearctic, both conifer and temperate deciduous zones extended across the entire continent as continuous belts (Cox and Moore 2010).

2.3   Natural Disturbances in Boreal Forests

Boreal forests are shaped by a range of disturbances varying in size, severity, and frequency, including fire, flooding, windthrow, snowbreak, avalanches, soil erosion, fungal diseases, outbreaks of defoliating insects, ungulate browsing, and the actions of beaver (*Castor fiber* in Eurasia) (Esseen et al. 1997). Disturbances are important natural drivers of forest ecosystem dynamics (Kuuluvainen and Aakala 2011) and have strong effects on the structure and functioning of forest ecosystems (Turner 2010). Particularly, fire has been one of the disturbances most studied because fires have been the most important factor shaping the structure and dynamics of natural boreal forests (Gromtsev 2002).

The boreal forest disturbance regimes range from natural succession following disturbances, such as severe stand-replacing fires and windstorms, to small-scale dynamics associated to gaps in the canopy created by the loss of individual trees (Angelstam and Kuuluvainen 2004). Furthermore, boreal forest disturbance regimes may vary considerably depending on the characteristics of the dominant tree species, landscape, local site conditions, and regional climate (Angelstam and Kuuluvainen 2004). According to Angelstam (1998), three main broad types of boreal forest can defined based on natural disturbance regimes in northern Europe: (1) Norway spruce dominated forest on wet and moist soils, characterized by internal gap dynamics and often forming more or less narrow elements in the landscape along depressions and water courses; (2) successional forest following stand-replacing disturbance such as fire (commonly on mesic soils), characterized by a gradual change from open conditions to closed forest and from more deciduous trees and pine shortly after disturbance to more Norway spruce after several decades; and (3) Scots pine dominated forest on drier sites subjected to frequent low-intensity fires, with different cohorts of trees having survived past fire events (Fig. 2). Evidence suggests that in Fennoscandian conditions gap dynamics (1) and cohort dynamics (3) were much more common than successional dynamics following stand-replacing fires (2) under natural disturbance dynamics (Angelstam and Kuuluvainen 2004). Also simulations suggest that old-growth forests dominated the natural landscapes of northern Europe, including the European part of the Russian boreal zone (Gromtsev 2002; Pennanen 2002). Similarly in northeastern North America, presettlement forests were dominated by relatively frequent, low-intensity disturbances that produced a heterogeneous mosaic, while large-scale stand-replacing disturbances were rare (Seymour et al. 2002). In more continental regions

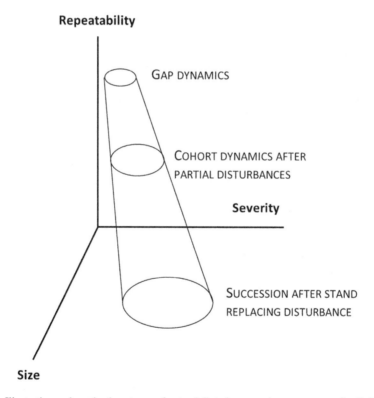

Fig. 2 Illustration on how the three types of natural disturbance regimes are not totally distinct but form a continuum in terms of size, severity, and repeatability of disturbance. Gap disturbances are frequent at landscape scale, small in size, and often low in severity leaving most of the vegetation alive. Partial disturbances in cohort dynamics are in the middle of the gradient. Succession is caused by severe disturbances such as fire or strong wind, which often cover large areas, but may occur relatively seldom (Figure taken from Angelstam and Kuuluvainen (2004))

such as Siberia and central Canada, stand-replacing fires and subsequent even-aged successional forests have been the dominating natural dynamics (Cogbill 1985).

Many studies showed that natural disturbances have a positive effect on biodiversity. Therefore, emulation of natural disturbances has been proposed as a sustainable forest management (Kuuluvainen and Grenfell 2012). In boreal forest ecosystems where fire is the major natural disturbance agent, it has been suggested the application of prescribed burning as a measure to restore natural forest conditions (Toivanen and Kotiaho 2007), whereas in wind- and bark beetle-dominated disturbance regimes, the creation of gaps of various sizes and shapes has been recommended to increase biodiversity (Kuuluvainen 2002).

In recent decades, forest disturbances have increased their frequencies and severity in many parts of the world (Chapin et al. 2000), and it is expected to increase even more in the future as a result of climate change (Seidl et al. 2014). As a consequence, it is very important to know the impacts of natural disturbances on boreal forest biodiversity and ecosystem services. A recent review by Thom and Seidl (2015) carried out in temperate and boreal forests to evaluate the impacts of three of the most relevant disturbance factors (fire, wind, and bark beetle) on 13 different ecosystem services and three indicators of forest biodiversity revealed a "disturbance paradox" reporting that disturbances can have negative effects on ecosystem services while simultaneously facilitating biodiversity. Thom and Seidl (2015) found that under a high-severity disturbance event, it was expected a loss of 38% of carbon storage while an increase in species richness by 36%. However, since negative effects on carbon are rapidly reduced with time, the positive effects on biodiversity do not substantially change over time. Therefore, managing for a low- to medium-frequency disturbance regime would result in a reduced negative impact on ecosystem services while still benefiting biodiversity. This emphasizes the need to put more attention on disturbance risk and resilience in ecosystem management, and new strategies to address the disturbance paradox in management are needed.

2.4   Biodiversity in Boreal Forests

The deep evolutionary history and Pleistocene glaciations are etched in the species diversity of current temperate forest regions. For example, the effects of Pleistocene glaciations have been most drastic in the western Palearctic region where several tree species and genera went extinct, mostly due to diminishing area of forests. Consequently, forests in the western Palearctic are less diverse in tree species than Nearctic and eastern Palearctic forests (Huntley 1993). Fennoscandian boreal forests contain only two common conifer tree species (Scots pine and Norway spruce), whereas in eastern Siberia there are eight common conifer species and in the eastern Nearctic typically 4–6 widespread conifers. Parallel differences among regions with the boreal zone are evident in forest-associated vertebrate species (Mönkkönen and Viro 1997). Compared to tropical forests where up to 300 tree species may occur in a single hectare (Gentry 1988) or temperate deciduous forest biome with almost

1200 tree species in total (Latham and Ricklefs 1993), boreal forests may appear dull. However, for some taxa such as beetles, the monotonous appearance is misleading, and consequently, Hanski and Hammond (1995) asserted that "if we do not soon recognize that boreal forests are not simple stands of pine and spruce, that is just what they may become."

Disturbances and processes following them are important drivers of biodiversity in boreal forests. Large-scale disturbances such as forest fires or storms create structural variation at the landscape scale, and boreal ecosystems under natural disturbance regimes are composed of forests representing different developmental stages (Syrjänen et al. 1994; Kuuluvainen and Aakala 2011) providing habitat and resources for species with varying preferences and requirements. Smaller scale disturbances resulting in replacing individual trees or group of trees by new individuals provide structural variation within forest stands, e.g., maintain mixed species tree communities and variation in tree ages, which is important from biodiversity perspective.

Tree deaths following disturbances influence a number of patterns and processes occurring in forest ecosystems (Franklin et al. 2002). Large quantities of decaying wood are one of the most characteristic features of natural, both young and old, boreal forests. Decaying wood is an important resource and habitat for a very large number of species. Both in Sweden and Finland, about 20%–25% of all multicellular forest species (20,000 in Finland, 25,000 in Sweden) are associated with deadwood, i.e., 5000 and 6500 saproxylic species in Finland and Sweden, respectively (Siitonen 2001; De Jong et al. 2004). Because of intensive forestry, deadwood volumes in managed Fennoscandian boreal forests have decreased by some 95% compared with forests under natural disturbance regime. This is a prime example of the fact that forestry-induced disturbances typically differ from natural disturbances in their effects on structures and processes important for biodiversity (Fig. 2).

The nonlinear relationship between the number of species and area is one of the few recurrent pervasive patterns in ecology (Rosenzweig 1995). When the area of a habitat type starts to diminish, for example, on old-growth forests, the changes in the species number are very small at first, but the rate of loss of species accelerates considerably with further habitat loss. Habitat loss is globally the main reason for increasing species extinction rates (MEA 2005). The proportion of old-growth forests in the present Fennoscandian landscape is getting so small in the modern landscape that we very likely operate on the steeply declining part of the species-area relationship for species strictly associated with old-growth forests. For example, in Finland, forests more than 120 years in age comprise about 4% and 18% of forest land in southern and northern parts, respectively (Peltola 2014); in a natural boreal landscape, this percentage is much higher, e.g., 70% in pristine NE European taiga in Russia (Syrjänen et al. 1994). Similarly, resulting from diminishing habitat availability for deadwood-associated species, they are disproportionally numerous among the threatened species in Fennoscandia: the reduction of large diameter deadwood has been identified as the principal threat factor for more than 50% of forest species (Berg et al. 1994; Siitonen 2001).

A simple calculation suggests that many species associated with virgin forests and their characteristic forest structures either have become extinct or will be lost in near future, as these natural forests comprise an ever-decreasing fraction of the total land area in Fennoscandia (Hanski 2000). This is a sheer challenge for sustainable forest management. Because of pressures to increase forest biomass harvesting, future boreal landscapes will likely face further reductions in the area of old, mature forests. The sustainability challenge actually is to break the species-area relationship with careful landscape level planning, i.e., maintain species and their populations when their habitat is decreasing. This seems very challenging indeed, because with low proportion of original habitat and resources in the landscape, the secondary consequences of habitat fragmentation such as edge effects tend to draw down the species number even more than predicted by the species-area relationship (Mönkkönen 1999).

2.5 Main Ecosystem Services from Boreal Forests

Ecosystem services can be divided into three main classes: *regulating, provisioning,* and *cultural* services according to the Common International Classification of Ecosystem Services (CICES) (Haines-Young and Potschin 2011; Haines-Young and Potschin 2013). Regulating services are the benefits obtained from the regulation of ecosystem processes such as climate regulation and pest control. Provisioning services are the products obtained from ecosystems such as food and fresh water. Cultural services are nonmaterial benefits that people obtain from ecosystems such as recreation and spiritual enrichment. Here, we provide an overview of the main ecosystem services in boreal forests for each class.

2.5.1 Regulating

Globally the most important *regulating* service in boreal forests is climate regulation (Burton et al. 2010). Carbon storage and sequestration by boreal forests is hugely important for global climate change mitigation. Boreal forests contain more than 30% of the global carbon storage and more than 20% of the global carbon sinks in forests (Pan et al. 2011). In boreal forests, 60% of carbon is stored in soil and 20% in woody biomass, whereas in other forests, such as tropical forests, the shares are opposite (56% in biomass and 32% in soil). In addition, especially old natural boreal forests store and also sequestrate carbon (Luyssaert et al. 2008), and contrary to temperate deciduous forests, some forested areas in the boreal zone still exist without human disturbance (Burton et al. 2010). Moreover, there are differences between forest biomes in climate regulation in terms of evapotranspiration and albedo effects (Snyder et al. 2004). In the boreal zone, albedo effect is more important for climate change mitigation than in tropical and temperate forests where cooling effects by evapotranspiration are more important (Bonan 2008).

Many regionally and locally important regulating services in forests are related to water and soil and are considered as public goods. Boreal forests are part of hydrological cycles and regulate water flows; they filter groundwater and act as buffer zones for adjacent waters by retaining nutrients (Saastamoinen et al. 2014).

Boreal regions have one of the largest freshwater supplies in the world (Schindler 1998). In addition, boreal forests retain nutrients and maintain soil productivity (Maynard et al. 2014) and provide resistance against natural disturbances such as pests, diseases, fire, wind, and floods, which may become even more important in the future if climate change increases disturbances (Thom and Seidl 2015). Furthermore, they provide habitats for many beneficial organisms, such as pollinators and decomposers, e.g., honeybees living in forests are pollinators of many commercially valuable crop species (Kettunen et al. 2012).

2.5.2 Provisioning

Timber production is the economically most important *provisioning* service in boreal forests (Burton et al. 2010; Vanhanen et al. 2012). Boreal forests contain approximately 45% of the world's stock of growing timber, and approximately 25% of global exports of forest industry derive from boreal forests. Slow tree growth produces strong, narrow-ringed wood with excellent properties as construction timber and fiber for papermaking. Logging residues and small diameter trees are used for bioenergy production to substitute for fossil fuels (Repo et al. 2011).

The intensity of forestry varies greatly across boreal forests. Most of Alaska's boreal forests are beyond timber production, whereas only 7% of Finland's forest-land is excluded from timber production. The share of the forest sector in the gross domestic product (GDP) is approximately 1% in Russia, 2% in Canada and Sweden, and even 4% in Finland (Vanhanen et al. 2012). Especially in Fennoscandia, many forests are privately owned and provide economic benefits from timber for private people. In addition, forest industry is an important employer and creates jobs for many local people.

In addition, non-timber forest products, such as mushrooms and berries, are valuable provisioning services in boreal forests (Burton et al. 2010; Saastamoinen et al. 2014). These products are economically and culturally important especially for rural and aboriginal communities (Turtiainen et al. 2012; Natural resources of Canada 2016). Everyman's right tradition in Fennoscandia allows all people to have access to forested areas and opportunity to collect products even in privately owned forests (Kettunen et al. 2012). This right makes berries and mushrooms a public good (Paassilta et al. 2009). The berries and mushrooms are mainly collected for household use, but some species, such as bilberries (*Vaccinium myrtillus*, *Vaccinium corymbosum*), cowberries (*Vaccinium vitis-idaea*), and mushrooms, particularly *Lactarius* spp. and *Boletus* spp., are also commercially harvested. In Fennoscandia, approximately 5%–10% of the berry and mushroom crops are harvested annually. Selling collectables is tax-free thus providing income for local people (Turtiainen et al. 2012).

Commercial harvesting of non-timber products of boreal forests and their international trade are increasing; Finland and Sweden are the main exporters of bilberries (Paassilta et al. 2009), and Canada is the main producer of maple products (Natural resources of Canada 2016). Wild mushrooms from boreal forests, such as cep (*Boletus edulis*) and chanterelle (*Cantharellus cibarius*) species, are exported to Central Europe and Asia (MARSI 2014; Natural resources of Canada 2016).

Moreover, hunting game species, such as moose (*Alces alces*), provide income for local people, and in Fennoscandian countries, the economic value of game meat is between 44 and 125 million euros per year (Kettunen et al. 2012). Nature tourism is both nationally and regionally important sector in the Nordic countries. For example, in Finland nature tourism provides employment for 32,000 persons and forestry for 25,000 persons (Ministry of Employment and the Economy 2014). In Finnish Lapland, nature tourism is the most important regional economy sector. Other regionally and locally important provisioning services distinct to boreal region are services such as reindeer herding and Christmas tree harvesting (Kettunen et al. 2012; Natural Resources of Canada 2016).

2.5.3 Cultural

Many non-timber forest products are also categorized as *cultural* ecosystem services because of their recreational and cultural value. Recreational berry and mushroom picking and hunting are popular activities for local people (Kettunen et al. 2012; Brandt et al. 2013). For example, more than half of the Finns annually participate picking berries and mushrooms (Turtiainen et al. 2012). Hunting has long traditions, and it still has importance to local communities also in terms of cultural identity.

Boreal forests are used also for other recreational outdoor activities, such as hiking, camping, and watching, e.g., bird species. For example, almost half of the Norwegians go hiking in forests or mountains more than twice a month (Kettunen et al. 2012). Also the scenic beauty of forest landscapes and national species have their own cultural and recreational values (Saastamoinen et al. 2014). Forests also provide possibilities to improve human health conditions. The results of a vast amount of research show that forest visits promote both physical and mental health by reducing stress (Karjalainen et al. 2010). In Fennoscandian countries, blueberry and cloudberry as well as birch sap are used for health-related products and cosmetics (Kettunen et al. 2012). Aboriginal communities in Canada have long traditions using medical plant species (Uprety et al. 2012). In general, a great potential exists in the development and utilization of health benefits deriving from boreal forests.

3   Past and Present Forest Use and Management in the Boreal Zone

3.1   Historical Use of Boreal Forests

From a global perspective, the boreal biome has sparse human population and relatively low anthropogenic impacts (Gauthier et al. 2015). In the boreal, only a minor proportion of forest has been converted to farmland, and land clearing for agriculture has mostly occurred in lowland areas along the seacoasts, near large lakes, and in river valleys. Forest land is a dominating land-use type across the entire boreal zone, but still, large parts of the European boreal forest have been influenced by a long history of shifting slash-and-burn agriculture (Myllyntaus et al. 2002). Importantly, most of the European boreal forest has been subjected to some form of logging or management for wood production (Bryant et al. 1997). In Fennoscandia, the history of logging is relatively long compared to other parts of the boreal biome. Already in the seventeenth to nineteenth centuries, forests were logged for charcoal production to supply the mining industry, and potash and tar production had large impacts on forests in some regions (Esseen et al. 1997; Östlund and Roturier 2010).

Extensive logging for saw timber – typically diameter-limit cutting targeting the largest trees – increased mostly from the nineteenth century, spreading gradually from the south into the north and inland of Fennoscandia (Imbeau et al. 2001). Also the demand for pulp wood considerably increased in the first half of the twentieth century. Since the 1950s, even-aged forestry involving clear felling and thinning has been the dominating management regime throughout the boreal zone. In Russia, forest management has thus far been less intensive than in the Nordic countries from a silvicultural perspective (Elbakidze et al. 2013), although logging has affected most parts of the Russian boreal forest (Potapov et al. 2008). In the Nearctic boreal zone, large-scale forest management and harvesting started in the mid-twentieth century, and harvesting for timber still largely considers pristine forests (Bryant et al. 1997; Imbeau et al. 2001). Moreover, forest fires have been suppressed very successfully in the Nordic countries since the later parts of the nineteenth century (Zackrisson 1977), whereas they still are a relatively more common disturbance agent in Russia and in Canada.

3.2 Forestry

In today's Fennoscandia, most of the productive forest area is under management for timber production, while low-productive areas are typically less affected by forestry. At a larger scale, there are still some relatively intact forest landscapes in the boreal region, unlike the situation in most of temperate Europe. However, these landscapes are confined to the northern parts of European Russia (Potapov et al. 2008) as well as to areas along the Finnish-Russian border and the Scandinavian and Ural mountain ranges.

Forest management is the conduct of human actions to extract resources and modify the growing potential of the forest. From a recently harvested stand (or bare ground), there are regional level recommendations, that is, "best practice" guidelines, that indicate which actions may be suitable to conduct on different forest structures (i.e., Äijälä et al. 2014). The set of management actions possible depend on the age and density of the stand and the access to the site. On recently harvested stands, the primary interest is often to ensure the rapid establishment of an appropriate tree species. In Fennoscandia, there is generally access to the forest site; silvicultural treatments may be applied to assist in the establishment and early development of the stand. For instance, preparatory actions may be conducted to aid in regeneration and planting, to sowing seeds or leaving a selection of seed trees to allow for natural regeneration. Once the stand has been established, and has grown for several years, the option to conduct a pre-commercial thinning could increase growth by reducing competition between trees. Once the forest has matured, there are possibilities to extract timber resources. Two different methods of extracting timber are possible: to conduct thinnings (selective removal of trees) or to conduct a clear felling (removal of most trees in the stand), returning the stand to a "bare ground" state. When access to the forest site is the key limitation (such as in Russia and Canada), road construction is required to do forest operations. This restricts the economically feasible management alternatives.

When access to site is not a limiting factor, the exact timing of the operations depends upon the ability of the forest to grow. The key input variables for growth models relate to conditions specific to the stand (Skovsgaard and Vanclay 2008). This includes the quantity of sunlight received and the soil conditions, and water availability determines the fertility of the site. For instance, forests located on sites with reduced soil nutrition or water availability will require a longer growing period, resulting in a delayed harvest when compared to a site with improved soil nutrition and improved water availability.

In Finland, legal restrictions were removed with an update of the forest law (Finnish Forest Act 2014); in 2014, continuous cover forestry (CCF) or uneven aged forestry is now allowed. Currently there is substantial research being conducted to find the most appropriate method of conducting CCF operations (Pukkala et al. 2011; Rämö and Tahvonen 2014; Lundmark et al. 2016). Although the legal restrictions have been relaxed, the primary method to extract timber resources is through a clear felling.

In Fennoscandia, predictions of future forest resources are conducted through forest management software. These tools integrate a large number of models to predict development of the forest. Models predict the growth, mortality, and ingrowth of the trees dependent on the specific site conditions (location, fertility, soil type). There is a wide variety of software options; for instance, in Finland there are three primary options, MELA (Redsven et al. 2012), MOTTI (Hynynen et al. 2005), and SIMO (Rasinmäki et al. 2009), in Sweden the Heureka system (Wikström et al. 2011), and AVVIRK in Norway (Eid and Hobbelstad 2000). In addition to generating predictions of future forest development, these software packages also include optimization tools (i.e., linear programming and heuristics; Kangas et al. 2015), which can be used to develop forest management alternatives.

4   Conflicts and Synergies among Ecosystem Services and Biodiversity

4.1   Mechanistic Pathways from Management to Biodiversity and Ecosystem Services

The structure of forests under natural disturbance regime vs. managed forests differs at several spatial scales (Fig. 3). These are important to acknowledge because species possess adaptations to structures that they have encountered during the evolutionary past and may therefore have difficulties in coping with changes in forest

Fig. 3 Forests under natural disturbance regime vs. under intensive management. Pictures on the left show early successional (top), middle-aged (appr. 80 years old), and old-growth boreal forest (bottom), and on the right, there are corresponding developmental stages in managed forests. Notice the higher level of structural diversity in forests under natural disturbance regime, particularly at early and late successional stages (All photos by Timo Kuuluvainen)

structures due to forest management. The basic pattern is that natural disturbances such as forest fires and gap formation create much structural variation and functional diversity in forest ecosystems that are missing, severely diminished, or altered in managed forests (Hansen et al. 1991; Esseen et al. 1997). These stem from differences in the disturbances: natural disturbances show much variation in their intensity and extent, whereas forest management actions usually follow standard sets of measures with little variation among stands (Table 1). At the stand scale, the most significant differences are low diversity of tree species, low amount of dead wood, and low abundance of very large tree individuals in managed stands compared with stands under natural dynamics. At the landscape scale, differences in size and age distribution of stands are large, managed forest landscapes having much less variation in the size of disturbed (managed) areas and more even age distribution. Managed landscapes typically are patchworks of stands at different developmental stages having more sharp edges between stands and less connectivity between patches of similar habitat types.

These structural differences have important repercussions to biodiversity and potentially also to ecosystem services in managed forests because of the tight links between the structure and function of ecosystems. Structural differences have direct effects on species' ability to find habitats, disperse in the landscape, and eventually persist as viable populations, and thereby indirect effects on the functioning of forest ecosystem. Indeed, in boreal forests where extensive areas have been actively managed and harvested primarily for timber production for decades, changes in forest structure and negative impacts on biodiversity and ecosystem functioning are now apparent (Bradshaw et al. 2009; Kuuluvainen et al. 2012). For example, the clear-cutting of even-aged forest stands commonly used in boreal timber production can radically change both biotic and abiotic conditions within a very short time, leading to negative impacts on biodiversity. In Finland, where production forests have been intensively managed with clear-cutting for almost 100 years, within-stand forest structure has become relatively even-aged, and the amount of deadwood has

Table 1 Comparison of disturbance dynamics between natural disturbance dynamics (natural forests) and modern intensive forestry (managed forests) (Modified from Kuuluvainen et al. (2004))

Disturbances	Natural forest	Managed forest
Number of different disturbance factors	Large	Small
Qualitative variation of disturbances	Large	Small
Proportion of trees dying during disturbance	0–100%	95–100%[a]
Proportion of dead trees remaining as deadwood on site	100%	0–5%
Recurrency of disturbances	Every 10–500 years	Every 70–130 years[b]
Occurrence of disturbances	Irregular	Regular
Area disturbed	0.001 →100.000 ha	0.5–10 ha

[a]Proportion of trees harvested at final felling
[b]Average forest rotation length in Fennoscandia

been considerably reduced (e.g., Vanha-Majamaa et al. 2007). As a consequence species requiring, e.g., deadwood and very large (overmature) trees have faced severe habitat loss.

The delivery of ecosystem services may be described as a process with four interacting stages: ecosystem structures, ecosystem functions, benefits experienced by humans, and finally the values assigned to these benefits (Haines-Young and Potschin 2010). Here, ecosystem structures refer to both the biotic and abiotic components of the ecosystem. As described above, in boreal forest ecosystems, these characteristics (Fig. 4) are heavily modified by forest management, and this has potential impacts on ecosystem functions. As these structures and functions are the basis of ecosystem services, the changes to the forest ecosystem following management interventions may extend to the various benefits required by different stakeholders (Fig. 4).

Species biodiversity, encompassing the diversity of species, the ecological functions that they possess, as well as their phylogenetic history, forms the basis of ecosystem functions, which in turn provide ecosystem services to humans (Duncan et al. 2015). Both experimental and correlational research into the links between biodiversity and ecosystem functioning have shown that losses in biodiversity have clear negative impacts on ecosystem functioning (Cardinale et al. 2012; Tilman et al. 2012). For example, Costanza et al. (2007) showed a positive link between vascular species diversity and net primary productivity in North America. Similarly, Paquette and Messier (2011) showed that increased functional diversity of shrubs and trees was associated with increased forest productivity in temperate and boreal forests in Canada. In addition, studies looking at the links between biodiversity and ecosystem services have shown a similar pattern. For example, Gamfeldt et al. (2013) showed in Sweden that forests with a higher diversity of tree species showed higher levels of ecosystem services, namely, biomass production, berry production, and soil carbon storage. Similarly, Vilà et al. (2007) showed that in the Mediterranean, tree species richness was positively associated with wood production. Maes et al.

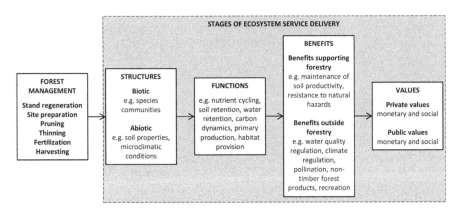

Fig. 4 Framework linking forest management activities via forest structures and functions to final benefits and values experienced by humans (Modified from Pohjanmies et al. (2017a))

(2012) found that on a European scale, increased levels of biodiversity had a general positive influence on forest-associated ecosystem services such as timber production, carbon storage, and recreation. Although previous research into the links between biodiversity and ecosystem functioning and services has focused on species richness, it has become increasingly clear that the diversity of functions processed by species plays the most important role (Díaz et al. 2007).

In general, the relation between biodiversity and ecosystem services is complex because biodiversity plays an important role at many levels of ecosystem service production (Mace et al. 2012). Multiple species are involved in producing ecosystem functions, and multiple ecosystem functions can be required to produce even a single ecosystem service. Biodiversity is also a multifaceted concept with several alternative elements (e.g., species) and attributes (e.g., amount or composition) whose role in providing any given ecosystem service can vary from indispensable to harmful. In addition, the links between biodiversity and ecosystem functioning and biodiversity and ecosystem services are not always linear. For example, even though Gamfeldt et al. (2013) showed a positive association between biodiversity and several ecosystem services, in half of the cases the relation was hump-shaped. In practice this means that sites with six or more tree species had diminishing levels of biomass and bilberry production. Spatial scale also plays an important role since positive relationships between biodiversity and ecosystem services have been found at the global scale (e.g., Strassburg et al. 2010), but this relationship weakens at national or regional scales (e.g., Thomas et al. 2013).

Despite the uncertainties outlined above, the majority of evidence points to a positive impact of increased biodiversity on ecosystem functioning and services. Thus, forest management practices, such as timber felling, that have negative impacts on biodiversity can also negatively impact both ecosystem functioning and ecosystem services. In addition to final harvesting, other forest management practices can also impact ecosystem services. For example, the frequency and intensity of thinning play very important roles in timber production and carbon sequestration, yet thinning also reduces structural diversity important to biodiversity. Indeed, simulation studies have shown that foregoing thinning altogether can have clear positive impacts on biodiversity and ecosystem functioning by resulting in 5–6 times more deadwood than green tree retention (Tikkanen et al. 2012).

4.2   Conflicts

As described above, forest management actions taken to increase timber yields affect also other forest functions and services. If stand management and harvesting cause losses in ecosystem services, these functions are in conflict, and there are trade-offs between timber production and other benefits provided by the forest. These situations have been found to be common in boreal forests. For example, Pohjanmies et al. (2017b) evaluated all of the pairwise conflicts among timber production and four other ecosystem services (bilberry production, carbon storage, pest regulation, and biodiversity conservation) using forest management simulations. They found strong conflicts between timber production and the other ecosystem services to be common, whereas conflicts among the four non-timber objectives were less severe (Fig. 5). For example, prioritizing timber production caused an average loss of 58% in the ecosystem service of pest regulation, while prioritizing carbon storage caused an average loss of only 9% in the same service. Similarly, prioritizing pest regulation caused average losses of 94% and 5% in timber production and carbon storage, respectively. According to these results, timber production and pest regulation are thus strongly conflicting, but carbon storage and pest

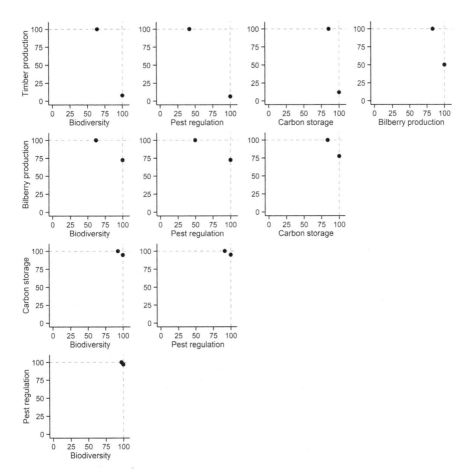

Fig. 5 Pairwise conflicts between five ecosystem services measured as the level of one service when forest management has been planned to maximize another service. The black points show the average value of the services across nearly 30,000 forest stands in Central Finland. These values are expressed as percentage of the maximal achievable value; units on all axes are thus percentages (%). Dashed gray lines have been added to all plots at $y = 100$ and $x = 100$ for graphical comparison. The closer to the point (100, 100) the two points are, the weaker is the conflict between the two services (Modified from Pohjanmies et al. (2017b))

regulation are highly compatible management objectives. However, Triviño et al. (2017) showed that targeting high levels of timber revenues creates conflicts also between non-timber benefits (here, carbon storage and biodiversity).

The forestry activities with the most substantial effects on ecosystem services are those that most severely affect the structure of the forest ecosystem. These include, for example, intensive and/or extensive wood harvesting, management of stand structure and tree species composition, and mechanical soil preparation. However, the natural processes that generate ecosystem services may be very complex, and much information on forestry's effects on them is still lacking. The effects of timber production on ecosystem services may be highly site-dependent and varied – simultaneously harmful for some services and beneficial for others.

One of the most extensively studied ecosystem services with respect to its responses to forest management is climate regulation via carbon dynamics. Carbon storage and sequestration by boreal forests is hugely important for global climate change mitigation (Pan et al. 2011), and forestry may have substantial impacts on these functions. However, these impacts are not always straightforward. Timber production and climate regulation via carbon sequestration are in conflict if forestry operations decrease the system's ability to fix carbon or if they result in releases of carbon into the atmosphere from long-term storages in the forest ecosystem, e.g., from forest soils. Generally, management choices increasing tree growth have a positive effect on the balance, whereas increasing the harvesting level results in negative effects (Liski et al. 2001). Management strategies to increase forest carbon stocks include extending rotation lengths (Cooper 1983; Liski et al. 2001; Kaipainen et al. 2004; Triviño et al. 2015), changes in initial stand density and thinning strategies (e.g., Niinimäki et al. 2013; Pihlainen et al. 2014), and forest fertilization (e.g., Boyland 2006). Extending forest rotation period allows trees to grow larger and forests to accumulate more litter and soil organic matter, whereas forest fertilization increases tree growth and litter input to the soil from living biomass and forest thinnings. Forest management choices can also reduce the carbon stocks and the carbon sink capacity through the intensification of biomass harvests in the form of timber or forest harvest residues (Repo et al. 2011; Kallio et al. 2013; Sievänen et al. 2013) and the shortening of forest rotation period (Kaipainen et al. 2004). Forest management minimizing the disturbances in the stand structure and soil reduces the risk of unintended carbon losses (Jandl et al. 2007). However, there may be trade-offs between short-term and long-term carbon sequestration. On top of the direct effects of forest management on carbon-related ecosystem functions, forestry's contribution to carbon sequestration and/or emissions may crucially depend on the entire life cycle of the forest product. Moreover, the role of forests in climate regulation is not limited to carbon dynamics, but includes processes such as water and energy fluxes (Naudts et al. 2016), surface albedo (Lutz and Howarth 2014), and production of aerosols that contribute to cloud formation (Spracklen et al. 2008).

Another extensively studied forest ecosystem service is regulation of surface water quality. Clear-cut harvesting and mechanical site preparation, which are common practices in boreal forestry, may have negative impacts on runoff water quality by increasing nutrient and organic matter load (Kreutzweiser et al. 2008). Also fertilization may increase the nutrient load from forests to surface waters (Laudon et al. 2011), while also affecting ground vegetation (Strengbom and Nordin 2008).

Alterations to the structure and composition of the stand and the physical properties of the soil may affect the site's suitability as habitat for beneficial species such as collectable forest products, pollinators, natural pest control agents, and decomposers. Here, it is particularly typical that forestry has contrasting impacts on the various benefits, because different species have differing habitat and resource requirements. For example, the recreationally and economically important wild bilberry (*Vaccinium myrtillus*) has declined in abundance in Fennoscandia due to extensive clear-cut harvesting and soil preparation activities. Conversely, species that thrive in young stands or benefit from increased canopy openness may benefit from forestry activities (Clason et al. 2008). Pest outbreaks may be controlled by removing naturally felled trees and thus minimizing the availability of food and breeding resources of pest species (Jactel et al. 2009). Then again, the lack of nesting resources created by natural disturbances has been also suggested to negatively affect pollinators (Rodríguez and Kouki 2015).

Some of the ecosystem services provided by production forests directly benefit forestry itself. These include ecosystem functions that maintain the productivity of the forest ecosystem or protect the forest property, such as nutrient cycling, erosion prevention, mitigation of storm damages, and regulation of pest outbreaks. Losses in these ecosystem services would eventually risk also timber yields. The impacts of forestry operations on the physical and chemical properties of soils that maintain productivity are highly site-dependent, but in many cases they have been estimated to be small in effect (Kreutzweiser et al. 2008). Boreal forests are typically able to recover from nutrient losses caused by biomass removal and increased leaching due to comparatively long rotation cycles as well as atmospheric deposition (Kreutzweiser et al. 2008). However, this ability may be threatened if productivity is artificially increased or biomass harvesting is intensified (Laudon et al. 2011). This would mean there is a conflict between these services and timber production, but only if the services are evaluated in a short time perspective.

The importance of the prevention and mitigation of natural disturbances is increasingly recognized as these disturbances are predicted to become more common in response to global change. The storm resistance of a stand may potentially be reinforced by choices regarding the structure of the stand and the surrounding landscape, that is, by stand diversification and minimization of height differences, gaps, and stand edges (Zeng et al. 2009; Zeng et al. 2010). Resistance to biotic disturbances may be more complicated to promote, because different pests may respond to stand management differently and because actions planned to control pest populations may also harm their natural enemies (Jactel et al. 2009).

Overall, it can be concluded that there is great variation between different ecosystem services in the extent to which the impacts of forestry on them are known. Changes in processes that involve interactions within species communities over various temporal and spatial scales are difficult to quantify and predict. The negative effects of forestry on numerous forest species and species groups have been recognized, but the long-term implications for the forest functions that rely on these communities are still poorly understood. Moreover, even when changes in the supply of ecosystem services are predicted, the consequences of these changes to human well-being are rarely evaluated. However, it is clear that intensive forestry has the potential to lead to substantial losses in several crucial forest ecosystem services.

Because of the linkages between management, biodiversity, and ecosystem services, it is obvious that economically, ecologically, and socially sustainable forest management planning must simultaneously consider all these aspects. Management solely focusing on intensive timber production at the landscape scale will (and has) incur(red) severe ecological and societal costs in terms of loss of ecosystem structures and functions as well as species and, consequently, may put many ecosystem services at risk. Over the past 20–30 years, forest management has been in transition from mere exploitation to sustainability and has adopted several methods to allow for more balanced use of the goods and benefits forests provide. In the next section, we provide an overview of these methods. While debate on the most beneficial forest management methods is ongoing and important information is still lacking (Kuuluvainen et al. 2012), one general message can already be derived from the past research: we need more variation in management regimes. In the following section, we also provide insights how to combine alternative management regimes at the landscape level to figure out desirable management plans that will meet the potentially conflicting objectives in an optimal way.

We can also conclude from the overview we provided earlier that even if biodiversity is the basis for ecosystem functioning and services and there tends to be a positive correlation between them, management for a full set of ecosystem services does not necessarily align with the management that most effectively supports biodiversity. For example, prevention and mitigation of natural disturbances is increasingly recognized as an important regulating ecosystem service, but many studies have shown that natural disturbances are crucial to the maintenance of biodiversity. Thus, while management planning is increasingly incorporating measures to ensure multiple ecosystem services simultaneously, we cannot assume that this will be good for biodiversity as well.

5 Management for Solving the Conflicts

5.1 *Management Approaches for Multiuse Forestry*

The currently dominant practice of even-aged management with clear felling was originally targeted to solely maximize the volume of harvested timber. However, there is a gradual recognition of shortcomings on the economic effectiveness of clear-cutting (Hyytiäinen and Tahvonen 2002; Hyytiäinen et al. 2004; Kuuluvainen et al. 2012) and of the conflict of such approach with other services from forest habitats beyond wood extraction. To address these, there has been a range of alternatives proposed. The common denominator in the deviations from the general practice is the preservation of key structural and biological elements necessary for certain ecological processes of forest ecosystems (e.g., seedling, groundwater regulation, connectivity between forest stands).

Monocultures are largely preferred in current intensive forestry (Clark and Covey 2012). For instance, in Finland and Sweden, only 14% and 10% of the total forest land, respectively, are composed of mixed forests (Christiansen 2014; Peltola 2014) with the proportion of mixed forest in Finland seeing a reduction by half since the 1950s in favor of monocultures of coniferous species (Parviainen and Västilä 2011). The model of monoculture is attractive in intensive forestry for its conceptual and management simplicity while providing large amounts of timber by focusing on the most productive, commercially interesting tree species. Nonetheless, mixed woods have a series of advantages that can make them appealing to stakeholders. First, mixed stands are usually not so badly affected than monocultures by extreme conditions linked to climate change such as wind throws, pest outbreaks, or drought (Kelty 1992; Felton et al. 2010; Felton et al. 2016). Second, mixed stands still enjoy relatively high timber productivity (Kelty 1992; Gamfeldt et al. 2013). This is due to broad-leaved trees reducing soil acidification typical in monocultures of preferred coniferous species (Jönsson et al. 2003) and to niche partition (i.e., use of partly different resources) between species (Kelty 1992; Amoroso and Turnblom 2006). Eventually this can translate into higher carbon sequestration. For instance, Gamfeldt et al. (2013) calculated that forest stands with five species incorporated 11% more carbon to the soil than monocultures. Third, the number of tree species in a stand is directly related to general species diversity (Felton et al. 2010; Gamfeldt et al. 2013; Felton et al. 2016). This is not surprising as several specialist forest species use resources from one or just a few species, while other species benefit from heterogeneity of resources found in mixed forests.

Another fundamental approach to reconcile wood production with other functions is through retaining trees in the harvesting interventions (so-called green tree retention) (Rosenvald and Lõhmus 2008; Gustafsson et al. 2012; Fedrowitz et al. 2014). From the forestry perspective, leaving trees in the stand can be a cost-efficient way to facilitate forest regrowth. Trees of the commercially interesting species are sometimes left as seeding trees or as shading trees for sun-intolerant species of interest. Also, given that a significant portion of trees retained, it is possible to prevent the rise of the water table in the stand that can be detrimental for forest regrowth. Additionally, some commercially less interesting tree species (e.g., Aspen) may be best left untouched to avoid sucker regeneration (Frey et al. 2003). Beyond timber production, tree retention can have important implications for other ecosystem services and biodiversity by preserving biological resources and promoting stand heterogeneity and connectivity. In this sense one can expect that the output of the tree retention option is directly related to what percentage of the forest is retained and to what tree species are spared. For instance, although stands with retained trees contain more threatened species than clear-cut forests, the relevance of this practice is related to how much is left (Vanha-Majamaa and Jalonen 2001; Hyvärinen et al. 2006; Aubry et al. 2009; Santangeli et al. 2013). Aggregating retained trees also seems to be more successful in retaining biodiversity value

(Carlén et al. 1999; Aubry et al. 2009) as opposed to the preferred practice in silviculture of random tree retention when seeking seeding and shadow trees. It has also been recommended that green tree aggregations should be situated in areas rich in threatened species, typically moist sites (Vanha-Majamaa and Jalonen 2001), and in different woodland key habitats (Timonen et al. 2010) if they are present in the stand. Given that a large portion of species, especially invertebrates and fungi, are dependent on deadwood, it is also advisable to leave dead or decaying trees also because the economic value of those is rather small, and this may even have positive economic effects via better seedling establishment (Alaspää et al. 2015).

A special case of green tree retention is the continuous cover or uneven-aged management, where the upper stratum of the forest is removed (also called thinning from above) inducing faster growth of the lower strata (Pommerening and Murphy 2004). Mounting evidence shows that this approach can be equally or more worthy for forestry purposes as the conventional model with clear-cuts (Laiho et al. 2011; Kuuluvainen et al. 2012; Rämö and Tahvonen 2014; Tahvonen 2015). Additionally, it is considered more resilient than the even-aged models to natural disturbances like windthrows, insect pests, or fires partly because of the different age classes and more alternatives to recovery after disturbances (O'Hara and Ramage 2013; Felton et al. 2016; Pukkala 2016). It can also outperform the conventional approach when considering some other ecosystem services, e.g., climate regulation (Pukkala 2016), but not all, e.g., collectable goods (Peura et al. 2016). From a biodiversity point of view, the largest contribution of continuous cover forestry is on the grounds of habitat connectivity, a key threat to intensively managed landscapes (Fahrig 2003; Nordén et al. 2015). One has to note that despite the concept of continuous cover forestry has received more interest for restoring biodiversity, it is obvious that a landscape covered by intensively managed uneven-aged forests cannot fulfill the demands of all species, particularly specialist species that rely on large amounts of deadwood or require closed forest structures. Uneven-aged management can, however, improve connectivity between more suitable patches for most of those more selective species. Still, applying permanent tree retention where some trees are allowed to grow, die, and decay within the context of continuous cover forestry is likely to greatly improve potential of this management model.

The clear-cut approach in intensively managed forests often includes a number of interventions with thinning from below to select the tree species of commercial interest, to promote faster growth of the remaining trees, and to prevent natural mortality (Daniel et al. 1979; Bamsey 1995). Despite the fact that it takes longer for a non-thinned stand to reach equally large trees than a thinned stand, the additional deadwood generated in non-thinned stands can render non-thinning a cost-efficient strategy for promoting those species dependent on deadwood (Tikkanen et al. 2007; Tikkanen et al. 2012; Mönkkönen et al. 2014).

Under the risk of potential hazards to forest, managers might be moved to shorten the rotation times on the clear-cut model. However, while reducing rotation cycles may be good against some risk agents like windthrow, cambium feeders, or root rot, at the same time this strategy increases the risks of regeneration pests and fire (Björkman and Nimelä 2015; Roberge et al. 2016). At the same time, shortening rotation typically has negative effects (vice versa for extending rotation) on relevant attributes for forest biodiversity (e.g., connectivity of old forest, reduction of dead wood, reduction in understory complexity) and regulating and supporting services (e.g., carbon storage, soil quality, and hydrologic integrity) (Jandl et al. 2007; Pawson et al. 2013; Pihlainen et al. 2014; Triviño et al. 2015; Felton et al. 2016; Roberge et al. 2016).

5.2 Policy Tools to Enforce and Promote Management for Multi-Objective Forestry

There is an interlinked array of policy tools designed to promote sustainable forestry. National forest legislation sets minimum standards for forest management. In addition, international forest certification standards are developed to promote sustainable forest management. The forest certification standards use the compliance with national laws as a starting point but have some elements that require more than the national legislation.

In Nordic countries two voluntary forest certification standards prevail: the Forest Stewardship Council (FSC) and the Programme for the Endorsement of Forest Certification schemes (PEFC). Both systems include certificates for forest management and for the chain of custody. The standard for forest management sets requirements for responsible forest management, whereas the chain of custody is a mechanism for tracking certified material from the forest to the final product to ensure that the wood, wood fiber, or non-wood forest produce contained in the product or product line can be traced back to certified forests. In Finland, 17,6 million hectares of the 20,3 million hectares of productive forest land are certified with the PEFC and approximately a million hectares with the FSC. In Sweden 11,5 million hectares are certified with the PEFC, and FSC certificates are given to 12 million hectares, which corresponds to almost half of the productive forest land in Sweden. Globally 300 million hectares of forest are certified with the PEFC and 190 million with the FSC (FCS International 2017,[2] PEFC International 2017[3]).

The FSC and the PEFC schemes consider ecosystem services and biodiversity. The standards do not use the concept of ecosystem services but explicitly account for multiple benefits of forests and their multifunctionality (FSC Finland 2010). A forest owner is required to acquire information on the occurrences of threatened species and plan the management activities safeguarding protection of rare, endangered species and their habitats. For instance, fellings are forbidden during the bird nesting season in both standards. Forest owners are also encouraged to ensure adequate resources for the protection of biological diversity and soil and water resources. The standards also require consideration of recreation values, and forest

---

[2] https://ic.fsc.org/en (Accessed 3.4.2017)

[3] www.pefc.org (Accessed 3.4.2017)

owners are required to take into consideration routes and structures for ecotourism and recreational use when planning forest management activities. The PEFC standard also recognizes the carbon sequestration in forests and requires that forests should be preserved as carbon sinks (e.g., PEFC Finland 2014). In addition, the Nordic everyman's rights or the freedom roam is safeguarded in the standards. In environmental considerations the FSC is more demanding than the PEFC (Gulbrandsen 2005). The most noticeable difference is that FSC requires at least 5% if the productive forest is permanently set aside (FSC Finland 2010; FSC Sweden 2010).

Forest certification can contribute to biodiversity conservation, but the level and the effect depend on the requirements of the forest certification scheme and its implementation (Elbakidze et al. 2011). According to Gullison (2003), forest certification can generally benefit biodiversity conservation in three ways by (1) reducing logging pressure on high conservation value, (2) preventing land-use change, and (3) improving ecological value of certified forests for biodiversity. In Nordic countries, national Forest Acts together with other environmental legislation set minimum environmental requirements for forest management. Nieminen (2006) concludes that direct, additional ecological benefits from forest certification in Finland are small compared to the environmental requirements in national legislation. However, the Forest Acts in Finland and Sweden only require conservation of biodiversity values already present, whereas the FSC standard demands the creation of new values by creating snags, leaving retention trees, and prescribed burning. These restoration measures are important in creating new structures such as old trees, deadwood, and deciduous trees and disturbances like fire that have decreased in managed forests (Johansson et al. 2013).

Although a direct cause-effect relationship between forest certification and environmental benefits is difficult to show (Rametsteiner and Simula 2003; Gulbrandsen 2005; Johansson and Lidestav 2011), studies conducted in Finland and Sweden indicate that forest certification can bring additional environmental benefits compared to the requirements of the legislation. These benefits result mainly from criteria for retention trees, prescribed burning, areas set aside from management, and restrictions in management operations in valuable habitats (Nieminen 2006; Johansson et al. 2013). Tree retention aims to reduce the intensity of harvest in clear-cutting by leaving single trees or tree groups and has several important functions: it (1) can "lifeboat" species over the regeneration phase, (2) increases structural diversity in young production forests, (3) enchases the connectivity in forest landscape, (4) promotes species dependent on deadwood and living trees in early successional environments, and (5) sustains ecosystem functions such as nitrogen retention (Gustafsson et al. 2010; Kruys et al. 2013). The Finnish legislation has no specific requirements for retention trees, whereas as according to the Swedish legislation, 2%–10% of timber value should be left after harvesting, prioritizing rare and broad-leaved species as well as old trees and nesting trees (Johansson et al. 2013; Finnish Forest Act 2014). The FSC scheme gives specific quantitative and qualitative requirements for retention trees. Neither the Finnish nor the Swedish legislation has requirements for prescribed burning. The Finnish and Swedish Forest

Acts list key habitats and require that these habitats are managed and used in such a manner that the general conditions for the preservation of these habitats important for the biological diversity of forests are safeguarded (Finland) or that damage from forestry is minimized or avoided (Sweden), whereas the FSC standard does not allow any commercial forest measures in these key habitats (FSC Finland 2010; FSC Sweden 2010; Johansson et al. 2013).

Forest certification has also indirect environmental benefits. Forest certification has harmonized forest management practices, improved communication among stakeholders, and clarified practical instructions for forest management (Nieminen 2006). In addition, the certification schemes have increased environment awareness and consideration of environmental aspect among forest owners (Johansson and Lidestav 2011) and employees of the forest sector (Nieminen 2006). Studies on corrective action requirements issued by certification bodies indicate that the auditing process improves the management of existing forests because the corrective action requirements must be addressed to obtain the certificate (Gullison 2003; Auld et al. 2008).

Forest certification has positive but limited impact on sustainable forest management, maintenance of ecosystem services, and biodiversity conservation. For example, tree retention may reduce harmful consequences of clear-cutting on biota, but it cannot maintain the characteristics of intact mature forests (Gustafsson et al. 2010). In addition, the long-term benefits of tree retention to red-listed species are questioned (Johansson et al. 2013). The forest certification integrates conservation measures into production forests, which complies with sustainable land-use strategy of the Millennium Ecosystem Assessment (2005) promoting different ecosystem services with the same land use. However, according to Johansson et al. (2013), forest certification together with environmental legislation does not guarantee biodiversity conservation in Sweden. This is because the levels of deadwood, the share of deciduous trees, areas for set aside, and other environmental consideration in the standards do not meet the thresholds identified in the scientific literature (Johansson et al. 2013). In conclusion, forest certification can alleviate the negative effects of forest management and be seen as a complementary, but not substitutive, measure to formal forest protection (Rametsteiner and Simula 2003; Elbakidze et al. 2011; Johansson et al. 2013).

In addition to forest certification, new payment schemes have been proposed to guide forest management with economic incentives (e.g., Farley & Costanza (2010)). According to Engel et al. (2008), payments for ecosystem services (PES) are voluntary transactions for a well-defined environmental service that a service buyer acquires from a service provider, who secures service provision of this service. Besides traditional tax and subsidy instruments, these new PES instruments can promote the provision of public goods in forestry (see review by Ollikainen (2016)). For example, subsidize-and-tax model and carbon rent models have been proposed as alternatives to implement payments for carbon sequestration and storage for forest owners (van Kooten et al. 1995; Lintunen et al. 2016). An interest to participate to carbon offset mechanisms has been shown, for example, among Norwegian forest owners (Håbesland et al. 2016). Previous studies on offset mechanism for ecosystem services and biodiversity have introduced new, promising instruments for policy-makers. However, the development of operational payment systems in forestry is in its infancy (Ollikainen 2016). One example of existing payment system in boreal forests is the Southern Finland Forest Biodiversity Programme (METSO). The METSO program is a voluntary-based conservation program aiming to halt the ongoing decline in the biodiversity of forest habitats and species and establish stable favorable trends especially in Southern Finland's forest ecosystems. The program offers monetary compensation for forest owners for permanent or temporary protection of forest, or support for nature management in forest habitats (METSO 2015).

5.3   Evaluating Conflicts

Conflicts exist when there is a need to balance between the desired outcomes of different objectives. Conflicts between different functions and services of the forest ecosystem can be evaluated in a variety of fashions. One common way to evaluate conflicts between two services or functions is by evaluating their trade-offs (King et al. 2015). The key idea behind the trade-off analysis is to gain an understanding of loss in one objective that must be endured for the benefit of a second objective. The trade-off analysis can be done through an optimization framework, which seeks to find the maximum of one objective, while a second objective is constrained to a specific proportion of the specific objective's maximum. Though an iterative method of adjusting the proportion is required, a frontier can be established, and the trade-offs which occur between the objectives can be examined.

With a two-dimensional problem, the conflict between each objective is easy to graph and describe to policy-makers. When the problem involves the consideration of conflicts between multiple ecosystem services, describing and portraying the conflicts become complicated. By analyzing the potential change in management prescriptions, the conflicts can be examined through a multi-objective optimization problem (Miettinen 1999). Through an interactive process (such as Nimbus; Ojalehto et al. 2007), it is possible for stakeholders to define their preferences between the selected set of criteria. These kinds of interactive processes work well when there is a clear decision-maker who can accurately define his/her preferences. When considering the potential stakeholders in landscape level planning, each stakeholder may provide a different perspective of how to evaluate the environment. Some of the stakeholders may not be able or willing to accurately define their preferences (i.e., Nordström et al. 2009). In these group decision-making processes, the focus is to promote understanding between stakeholders and improve the acceptability of the management plan (Hjortsø 2004).

Rather than simply evaluating alternative management plans, it is also possible to quantify the conflicts between objectives (Mazziotta et al. 2017). When evaluating ecological objectives, the case may exist where some objectives are compatible and maximizing one objective causes only a small decrease in the possible maximum of the other objective. This is, for example, the case with the habitat availability of the capercaillie (*Tetrao urogallus*) in Finland which can be maximized with fairly small reductions in timber production (Mönkkönen et al. 2014). Alternatively, when the objectives are not compatible, maximizing one objective will cause a large decrease in the possible maximum of the other objective; this is the case, e.g., between timber production and carbon storage. Mazziotta et al. (2017) developed an index of compatibility, which is essentially the percentage of an objective (x) possible when maximizing a second objective (y). These compatibility indices are not symmetrical, so the compatibility index of x when maximizing y does not necessarily equal the compatibility index of y when we maximize x.

From decision-making and forest management planning perspectives, trade-off analysis can be divided into two classes (Felton et al. 2017). The problem can be formulated as a "how to" question focusing on identifying a single or limited set of management alternatives from a larger set, which meet a specific set of objectives and desired constraints. Often this approach provides a mean of determining the optimal values for a set of objectives. In contrast, a "what-if" approach aims at understanding the implications of different scenarios regarding "what" will be the implications for the objectives, "if" this policy or management intervention takes place.

Along the lines of how-to approach, it is possible to develop an objective function, which finds a suitable compromise solution for a set of objectives. Mazziotta et al. (2017) suggest the use of a compromise programming formulation (Yu 1973; Zeleny 1982), where the idea is to minimize the maximum deviations from the ideal value for all objectives under consideration, i.e., to minimize total losses due to a decision. The solution produced demonstrates how maximizing a set of objectives requires that no objective obtains their maximum potential values. One must consider the preferential interpretation of the objective function (Diaz-Balteiro et al. 2012).

Quantification of conflict between ecological objectives can enhance our understanding of trade-offs between various conservation objectives. However, the key conflict is between economic and ecological objectives. Mazziotta et al. (2017) highlighted how much the ecological objectives would suffer by requiring 95% of the maximum net present value be extracted from the forest. As expected, the achievement of all objectives decreased substantially. On average, the values of the ecological objectives decreased about 20% from the case when only ecological objectives were considered. From a conservation perspective, this is a substantial decline. To resolve the conflicts caused by the economic requirement, a reduction to two-thirds of the maximum NPV would be required (Mazziotta et al. 2017). From an economic perspective, this may not be an acceptable reduction. From this point, the trade-off between ecological objectives and economic demands is evident and will require some compromise.

Alternatively, rather than comparing specific management alternatives to evaluate the conflicts between criteria, we can view the issue through shifts in policy. Policy suggestions can be evaluated by analyzing the statements, objectives, and goals using the "what-if" approach. Optimized forest management plans can be made by making specific assumptions which correspond to the policy statement. The conflict between criteria of interest can then be evaluated through the systematic relaxation of the assumptions made, i.e., adjusting the "if" part. So rather than finding a specific solution which best fits the policy, we can find a set of solutions (which are Pareto optimal; any positive adjustment in one objective must be offset by a negative adjustment in another objective) which highlights the potential conflicts between criteria.

5.4   Consequences of Increasing Timber Harvesting Rate: A Case Study

The Finnish government is currently promoting the growth of the bioeconomy, where the use of renewable resources is encouraged (Ministry of Employment and the Economy 2014). For forestry in Finland, this is being implemented through an increase of the maximum annual allowable harvest. Currently, forest operations in Finland are not near the maximum annual allowable harvest (Peltola 2014), but because of conflicts, increases in harvests will most likely cause harm to other ecosystem services such as climate regulation, recreational use, and biodiversity. Through modeling and simulation by use of forest management software, it is possible to predict future forest resources according to different management regimes and then evaluate the ecological performance of the forest. By adjusting the management alternatives for different portions of the forest, we can predict how the increased use of forest resources will impact the forest's other uses. To conduct the analysis in a more systematic fashion, optimization can be used to determine the optimal spatial allocation of the specific management regimes.

For this example, we evaluate the trade-offs and potential conflicts for a variety of criteria at a watershed level when there is a requirement for even flow of timber harvested. We apply a what-if approach with the policy statement of maximizing the even flow of timber over the duration of the planning horizon. We compare this policy shift to the case where only a proportion of the maximum even flow of timber will be harvested. In principle, the amount of timber provided by the forest will be constant for all periods during the planning horizon. However, the required timber harvested will be less than the theoretical maximum.

For this example, we analyze a watershed located in central Finland. The region consists of slightly over 5060 ha (composed of 3356 stands) of managed boreal forests. This region is composed of primarily three tree species: Norway spruce (*Picea abies (L.) Karst.*) composes 57% of the total volume, Scots pine (*Pinus sylvestris L.*) is 26%, and silver birch (*Betula pendula*) is 16.5%. The remaining (0.5%) component consists of other deciduous trees (*Betula pubescens* and *Populus tremula*). Figure 6 describes the age distribution of the forest, which is rather even – with a slight bulge in the age classes of 60–80.

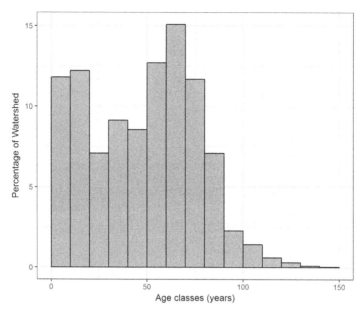

Fig. 6 Age distribution of the forest within the watershed

To predict the possible future states of the forest holding, we used the forest management software SIMO (Rasinmäki et al. 2009). For each stand, 19 alternative management regimes were simulated. All stands included the option of not conducting any actions in the forest, in other words to set the stand aside. Other regimes followed the recommendations from the forest management guide (Äijälä et al. 2014), including slight variations on these recommendations (such as to delay or speed up the timing of the management actions; Table 2). Additionally, a management regime corresponding to a method of conducting continuous cover forestry (CCF) was included (Pukkala et al. 2013).

Once the alternative management alternatives were simulated, the next step is to find the combination of management actions which best fulfills the objectives of the stated policy. For this case, the policy is to increase harvests to the maximum allowable sustainable harvestable yield. At the country level, this value is computed by the authority responsible for natural resources and takes into account the growth rate of the forest, protected forested areas to evaluate what is the maximum sustainable harvestable yield. For a watershed level, the maximum sustainable harvestable yield could be comparable to the maximum even flow of timber for the area under consideration. Finding the maximum even flow of timber is an optimization problem which maximizes the first period harvested yield, subject to a constraint that for all other periods under consideration, the yield cannot be less than the first period yield. The object function is [Model 1]

$$\max z = \sum_{K}^{k=1} \sum_{Jk}^{j=1} c_{kj1} x_{kj} \tag{1}$$

Table 2 Alternative management regimes simulated for stands in the landscape

Management regime	Description
Tapio (BAU)	Conventional regeneration harvest regime (see section Forestry). Simulated sped up (−5) and delayed (5, 10, 15, 30) final fellings
Tapio harvesting w/o thinnings (BAU w/o thin)	Otherwise similar to Tapio but no thinnings applied before final felling by clear-cutting. Simulated sped up (−5) and delayed (5, 10, 15, 30) final fellings
Tapio harvesting w thinnings (BAU w thin)	Otherwise similar to Tapio but thinnings are always applied before final harvest by clear-cutting. Simulated sped up (−5) and delayed (5, 10, 15, 30) final fellings
Tapio seed tree (GTR, GTR w thin)	Otherwise similar to Tapio but regeneration after harvest is through seed trees, rather than planting (with and without thinning). Simulated sped up (−10) and delayed (20) final fellings
Continuous cover forestry (CCF)	Rather than conducting harvests, only thinnings are conducted, depending on the specific stand properties. Follows suggestions by Pukkala et al. (2013)
Set-aside (SA)	No management

The brackets after the abbreviation indicate the number of years final felling is sped up (−) or delayed (+)

subject to

$$\sum_{K}^{k=1}\sum_{J_k}^{j=1} c_{kj1} x_{kj} \leq \sum_{K}^{k=1}\sum_{J_k}^{j} c_{kji} x_{kj}, i = 2,\ldots,I \tag{2}$$

$$\sum_{J_k}^{j=1} x_{kj} = 1, k = 1,\ldots,J \tag{3}$$

$$x_{kj} \geq 0 \forall k = 1,\ldots,K, j = 1,\ldots,J_k \tag{4}$$

where K is the total number of stands of the forest holding, J_k is the number of management regimes for stand k, c_{kji} is the amount of timber which is available by managing stand k according to management regime j at time period i, x_{kj} is the decision to manage a specific proportion of stand k according to management regime j, I is the total number of time periods under consideration, and z will represent the maximum amount of even-flow timber for the duration of the planning horizon. Constraint [2] represents the requirement for even flow; constraint [3] is an area constraint, requiring that the sum of the decisions must equal to 1; and constraint [4] requires that the decision variable is not a negative number. In forest management, this type of problem is referred to as a model I problem (Johnson and Scheurman 1977), and one of the key features is that the spatial integrity of stands is maintained. This allows for easy mapping of which management regimes are proposed for which stand.

Once we have evaluated the maximum amount of even-flow of timber, we can then analyze the possible impacts of relaxing the specific constraints. In this specific case, we will reduce the required annual timber harvested and simultaneously maximize a set of normalized ecosystem services. Other than the timber provided, we are interested in promoting the production of bilberries (*Vaccinium myrtillus*), increasing the amount of carbon stored in the forest, increasing the amount of deadwood in the forest, and increasing the habitat suitability for a set of species. We included six vertebrate species representing a wide spectrum of habitat associations and also conservation and social values ranging from game birds (capercaillie, hazel grouse) to red-listed (Siberian flying squirrel) and indicator species (three-toed woodpecker, lesser-spotted woodpecker, and long-tailed tit) (Mönkkönen et al. 2014).

To do this, we employ a different optimization model. This model optimizes the normalized set of ecosystem services, while introducing an additional constraint to ensure that the first period timber harvest meets a proportion of the maximum even-flow of timber (z). The objective function of this model is [Model 2]

$$\max \sum_{K}^{k=1} \sum_{J_k}^{j=1} \sum_{I}^{i=1} \sum_{L}^{l=1} \frac{d_{kjil} x_{kj}}{y_l} \qquad (5)$$

subject to

$$\sum_{K}^{k=1} \sum_{J_k}^{j=1} c_{kj1} x_{kj} \geq z * p \qquad (6)$$

and constraints 2, 3, and 4, where d_{kjil} is the value of the specific ecosystem service (l) of interest from the set of all ecosystem services under consideration (L), y_l is the normalization constant for the ecosystem service l, and p is a parameter which represents the desired proportion of the maximum even-flow of harvest. The objective function [5] maximizes the normalized set of ecosystem services, while constraint [6] requires that the first period timber is at a specific proportion of the maximum even-flow. Constraint [2] ensures that all future timber harvest is not lower than the first period harvests.

By running Model 2 iteratively, it is possible to evaluate how the relaxation of the even-flow constraints will impact the other ecosystem services. For those ecosystem services which are negatively impacted by increased harvests, by relaxing the even-flow constraint, it is expected that those ecosystem services will increase. Alternatively, for those ecosystem services which are positively impacted by increased harvests, relaxing the even-flow constraint will reduce those ecosystem services. To provide an example, we have conducted this analysis by adjusting p from 0.6 to 0.95 in increments of 0.05 and with $p = 0.975$.

To provide an elaborative description of how tightening the timber requirement impacts the selected ecosystem services, we set the case where $p = 0.6$ to be the starting point, and all other iterations are compared to that case. The current harvest rate in Finland varies annually between 60% and 75%, and thus the starting point roughly corresponds with the current situation.

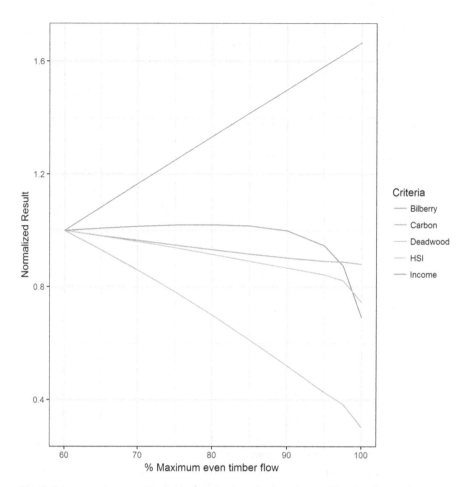

Fig. 7 Represents the normalized change in the five criteria under consideration. Income increases linearly (due to the flow constraint), while the remaining criteria vary depending on the optimization

Figure 7 highlights the trade-off between increasing timber harvests and the biodiversity and ecosystem service indicators of interest. With an initial small increase in required timber harvests (i.e., from $p = 0.6$ to $p = 0.65$), the decrease in the ecosystem services other than timber is rather small; however, with a continued increase in timber harvested, the impact on the ecosystem services becomes rather severe. Alternative indicators of ecosystem services and biodiversity, nonetheless, show different patterns of decline. Carbon storage linearly declines monotonically with increasing harvest rate (proportion of the maximum even-flow of harvest), whereas bilberry yield shows first slight increase and very steep decline at very high levels of timber harvesting. Thus, each increment in harvest rate results in equal reduction in carbon storage, but bilberry yields can be maintained at the initial levels

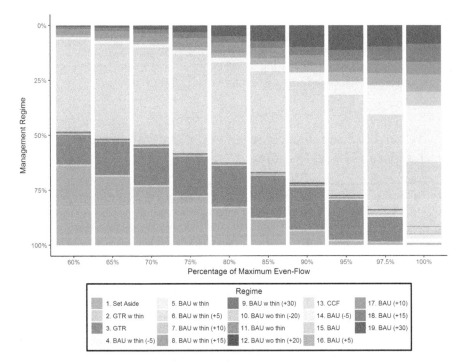

Fig. 8 Proportion of management regimes as the proportion of timber harvest is increased. BAU refers to alternative clear-cut-based management regimes with variable thinning intensities and rotation lengths (GTR = green tree retention, w thin = with thinnings before clear felling, w/o thin = without thinnings). CCF refers to continuous cover forestry with no final felling by clear-cut, and set-aside denotes permanent protection (no management). For a description of management regimes, see Table 2

until timber harvesting exceeds 95% of the maximum. Bilberry benefits from open forest structures, e.g., after thinning, but suffers from clear-cutting (Peura et al. 2016) explaining the nonlinear response to increased timber harvesting.

The shift of increased sustainable harvest implies an intensification of forest use. When we move toward the maximum sustainable harvest, a greater proportion of forest area moves from being "set aside" toward other management regimes (Fig. 8). When the required sustainable harvest is relatively low, the majority of the area that is harvested is done using the management regime of CCF. As the required level of harvest increases, all management regimes which conduct harvests increase, and forest managed according to CCF also increases. Only at the maximum sustainable harvest level does the traditional management regime become the dominant method of managing the forest. However, there is still a large component of the forest being managed with CCF and still some forest being managed without the conduct of thinnings prior to final felling.

With optimization processes, finding an optimal solution to a particular case is rather straightforward; the difficulty remains in being able to implement the solution. In Scandinavia, being able to implement these kinds of solutions requires motivation from the policy-makers, but more importantly requires acceptance and a majority of compliance to be able to achieve these kinds of results.

6 Toward Sustainable Management of Boreal Forests: Landscape Level Planning

Natural boreal forests are disturbance-driven ecosystems. Some consequences of industrial forest harvesting resemble the effects of natural disturbances such as forest fire. Although the analogy between forest management and fire disturbance in boreal ecosystems has some merit, it is important to recognize that natural disturbance and human-induced disturbances differ considerably (Bergeron et al. 2002; Table 1).

Research has shown that many ecosystem services and biodiversity are in conflict with intensive timber production in boreal forests. The conflicts stem from the changes forest management causes to the structure and dynamics of forests. The conventional regeneration forest management including site preparation, planting trees, and 1–3 thinning operations before final felling by clear-cutting (see section Forestry), if applied consistently on the entire landscape, causes ecological (Mönkkönen et al. 2014) and social costs (Triviño et al. 2015). Biodiversity losses arise because a proportion of species do not have adaptations to cope with changes in resource availability, habitat structures, and their spatial configuration. Also ecosystem functioning is altered. There are some conflicts among non-timber ecosystem services and, moreover, some between biodiversity and ecosystem service indicators, but these conflicts are generally weaker than for timber provision (Pohjanmies et al. 2017b, Triviño et al. 2017). Because alternative management regimes such as continuous cover forestry (Pukkala et al. 2011) or refraining from thinnings (Mönkkönen et al. 2014) are more beneficial for some objectives but worse for some others than the intensive Fennoscandian forest management regime, no management regime alone is optimal. Therefore, the best option for multiple objectives would be to diversify management, but finding an efficient balance among alternative management options requires careful planning.

Forest management planning can be conducted for a wide range of interests and for varying spatial scales. For forest owners who own small parcels of land, their interests may be purely financial or they may be interested in managing their forests with an aim to enhance the ecological functions of the forest. A key issue in forestry planning is the spatial scale. At the stand scale, reconciling alternative economic, ecological, and social objectives is difficult because only one management regime can be applied at a time. But at the landscape scale, it may be possible to find plans, i.e., combinations of management regimes, that provide acceptable compromise solutions because of flexibility provided by increasing combinations of stand level management decisions. Large-scale planning would be desired also because some ecological functions operate at scales larger than a single holding (i.e., habitat requirements of species; Mönkkönen et al. 1997; Kurki et al. 2000). The costs of increasing the scale are obviously the increasing complicatedness of the problems and the difficulty of putting plans into practice when they encompass several forest holdings. Because of the dynamism of the forest ecosystems, forest planning should also consider long-time perspectives and future generations: the consequences of today's management decisions may be realized only after several decades.

Resolving the conflicts necessitates applying a mixture of management regimes on a landscape, i.e., applying the conventional regeneration harvest regime on a proportion of stands and alternative regimes on others to better achieve multiple objectives. The desired combination of alternative management regimes depends on the decision-maker's preferences and objectives. Even when the decision-maker aims at maximizing timber revenues, she should not apply the recommended management consistently but only on a fraction of stands (Mönkkönen et al. 2014; Fig. 8). With decreasing emphasis on timber production and increasing emphasis on non-timber benefits and biodiversity, the utility of the conventional regeneration harvest management further decreases. Refraining from intermediate thinnings, extending stand rotations, and increasing the amount of area set aside from forestry seem necessary to safeguard biodiversity and non-timber ecosystem services (see, e.g., Triviño et al. 2017).

The good news is that typically pairwise conflicts are solvable at relatively low cost if land-use planning is done at the landscape scale (Pohjanmies et al. 2017b). For example, giving up 5% of the maximum timber revenues enabled maintaining up to 278% more habitats for species (Mönkkönen et al. 2014) or increasing the landscape's capacity to store carbon by 9% and to sequester carbon by 23% (Triviño et al. 2015).

The bad news is that attaining to high values for more than two objectives at the same time seems very difficult. In fact, the objective of having high timber harvest rates aggravates the conflicts among non-timber objectives as shown by Triviño et al. (2017). Triviño et al. (2017) aimed at reconciling timber production with provisioning of other ecosystem services (i.e., store carbon for climate regulation) while maintaining suitable habitat for forest biodiversity. They applied seven alternative forest management regimes, ranging from the current recommended regime to set aside, using a forest growth simulator in a large boreal forest production landscape. With multi-objective optimization, they identified the optimal combination of forest management regimes to minimize the trade-offs between timber harvest revenues, carbon storage, and biodiversity maintenance. Results indicate that it was not possible to achieve high levels of carbon storage or biodiversity if the objective of forest management was to maximize timber harvest revenues. However, with small reductions of timber revenues (1%–5%), it was possible to greatly increase the multifunctionality of the landscape, especially the biodiversity indicators (47%–90% of the maximum deadwood and 65%–88% of the habitat availability) (see Fig. 9). For more severe reductions in timber harvest revenues, e.g., 80%–95%, it

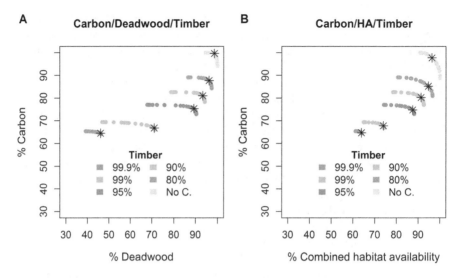

Fig. 9 Multi-objective optimization results: curves representing the trade-offs between carbon storage and two biodiversity indicators (deadwood index and combined habitat availability of six vertebrate species) for different levels of timber harvest revenues. The black star in each Pareto optimal set indicates the compromise management plan (Figure adapted from Triviño et al. 2017)

was possible to almost achieve the maximum levels for both carbon storage and biodiversity indicators (see Fig. 9). Even with modest levels of timber objective, there was a strong trade-off between carbon storage and biodiversity objectives, and both objectives remained far from their maximum values. When timber objectives were relaxed, close to maximum levels for both carbon storage and biodiversity objectives could be achieved.

The results also showed that no management regime alone is able to maximize timber revenues, carbon storage, and biodiversity individually or simultaneously and that a combination of different regimes is needed to resolve the conflicts among these objectives (see Fig. 10). Forest management actions, alternative to the conventional regeneration harvest, such as reducing thinning intensity, extending the rotation period, and increasing the amount of area set aside from forestry may be necessary to safeguard biodiversity and non-timber ecosystem services in Fennoscandia. They concluded that it is possible to reduce the trade-offs between different objectives by applying diversified forest management planning at the landscape level.

The example above suggests that strong emphasis on timber production at the landscape scale makes it impossible to simultaneously achieve high levels in more than one other objective no matter how landscape is managed. Intensifying timber production results in increasingly strong conflict between biodiversity protection and climate regulation via carbon storing, even though initially biodiversity and carbon-related ecosystem services were not in conflict. We can therefore conclude that the current objective of bioeconomy policies to considerably increase timber

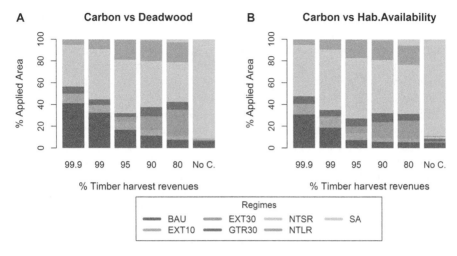

Fig. 10 Changes in percentage of area in the landscape allocated for the different management regimes for the compromise outcome in the Pareto optimal set (the black stars from Fig. 9) at decreasing levels of timber harvest revenues (from 99.9% to "no constraints"). The acronyms of the management regimes are BAU (business as usual), EXT10 (extended rotation by 10 years), EXT30 (extended rotation by 30 years), GTR30 (green tree retention), NTSR (no thinning short rotation), NTLR (no thinning long rotation, and SA (set-aside) (Figure adapted from Triviño et al. 2017)

flow from forest is not sustainable in Fennoscandian forests (see also section Consequences of Increasing Timber Harvesting Rate: A Case Study) because already at current timber harvest rates biodiversity is at risk (Hanski 2000; Mönkkönen et al. 2014), many ecosystem services have declined (Pohjanmies et al. 2017a), and even with very careful landscape level planning and management, optimization resolving the conflicts among several objectives is not possible.

Two alternative forest harvesting strategies are proposed to meet timber demands with other objectives: land sharing, which combines timber extraction with other objective protection across the entire concession, and land sparing, where high-intensity harvesting in one place is combined with the low intensity or no harvesting somewhere else (Edwards et al. 2014). The discussion above clearly suggests that sustainable forest use via landscape sharing would require rather low overall intensity of timber production. Thus, if timber requirements are large, landscape sparing becomes a desired option for economic and social reasons. Segregating the role of landscapes is justified also from the mere nature conservation point of view. Several ecological processes underpinning both ecosystem services and biodiversity have a minimum threshold that is context dependent. Ecological research has concluded that if a limited area of species habitats can be protected, they should be protected in spatially aggregated clusters rather than as randomly scattered fragments. This will generally increase the conservation benefits for a given total area protected (Rybicki and Hanski 2013). The Nagoya agreement recognizes the need to focus on "especially areas of particular importance for biodiversity and ecosystem services,

... ecologically representative and well-connected systems of protected areas ... integrated into the wider landscapes" (Strategic Plan for Biodiversity 2011–2020 and the Aichi Targets Aichi Targets, strategic goal C, target 11). Thus, we conclude that regional forest resource management planning should start differentiating landscapes where environmental and social objectives have priority over timber production landscapes.

Finding a balance between timber production landscapes and multiuse landscapes is yet another challenging objective for natural resource management. Hanski (2011) suggested the third-of-third rule of thumb, which implies that conservation landscapes would cover one-third of the total region, and within conservation landscapes one-third of the total area be protected resulting in roughly 10% level of set-asides. This is less than the target set in Nagoya, but this 10% would be additional to the existing protected areas. Further, this 10% figure is only double the amount of set-asides required by FSC certification standards. Within conservation landscapes, biodiversity and people coexist, and the ecosystem services provided by biodiversity and natural habitats play an integral part in providing direct benefits to local communities and to the society at large.

In boreal forest settings, Hanski's (2011) suggestion would mean concentrating timber production on two-thirds of the total area within a region. Also in timber production landscapes, applying multiple management regimes, including extended rotations, refraining from thinnings, and continuous cover forestry, is necessary for maximum timber values. Multiuse landscapes would cover the remaining one-third. Here set-asides comprise a prominent proportion of area but still managed forest for multiple purposes dominate. There is a growing support for management in production forest to recreate conditions found within a given region by natural disturbances with the rationale that the process and species are adapted to such conditions (Attiwill 1994; Burton et al. 2003; Kuuluvainen 2009; Kuuluvainen and Aakala 2011; Larocque 2016). However, vast forest areas of the boreal biome have much lower natural disturbance rate than that required by the forestry sectors. Therefore, it has been suggested to conduct functional zoning in which one part of the region focuses on protection, another part is managed according to natural disturbance dynamics, and the rest is devoted to intensive forestry (Seymour and Hunter 1992; Côté et al. 2010; Strengbom et al. 2011; Lindenmayer et al. 2012; Tittler et al. 2016).

According to mitigation hierarchy (Biodiversity Consultancy Ltd.), avoidance is often the easiest, cheapest, and most effective way of reducing potential negative impacts of any development. Therefore, development, e.g., forestry, should be concentrated on areas where negative impacts can be avoided. Kareksela et al. (2013) developed the method of inverse spatial prioritization and applied it to land-use allocation for peat-land mining. This approach can also be used to identify multiuse landscape vs. timber production landscapes within a region. In practice, this means finding landscapes with the lowest timber but highest biodiversity and non-timber ecosystem service values (multiuse landscapes) and, conversely, landscapes with the highest timber production potential and lowest biodiversity and non-timber ESS values (timber production landscapes). We suggest that a regional approach where timber production landscapes are separated from multiuse landscapes using systematic zoning tools, such as inverse spatial prioritization, would be a promising way of reconciling multiple conflicting objectives for boreal forests and their management.

7   Concluding Remarks

In this chapter, we have provided a general overview of boreal forest ecosystems and their management for timber production, maintaining biodiversity and ensuring the flow of non-timber ecosystem services. We showed evidence that in boreal production forests the conflicts between the primary provisioning service of timber and other benefits are real, severe, and challenging to solve. This in line with the more general finding that unbalanced focus on one or few provisioning ecosystem services typically results in severe trade-offs (Howe et al. 2014). Research into the processes affecting the supply of different forest ecosystem services may assist in designing forestry practices and planning management regimes that protect diverse forest benefits. We need to understand the mechanisms causing trade-offs and recognize situations where they likely occur when we want to generate solutions to these trade-offs.

Forestry's effects on ecosystem services may be generated at various spatial scales (e.g., a single stand, a landscape) that are relevant for different forms of forest ownership and management (e.g., a private forest holding, a state-owned forest). We concluded that long-term landscape level planning that simultaneously takes into account alternative objectives and the capacity of each land parcel (stand) to meet these objectives is a necessary, but not sufficient, condition for sustainable forest use. In addition, regional level planning where the roles of landscapes are differentiated is needed. Maintaining a diverse set of forest services requires coordination of activities among forest owners. Therefore, we need incentives for landowners to make decisions that reflect the value of ecosystem services and biodiversity conservation in general. This is, however, particularly challenging in several parts of the boreal forests like Fennoscandia where ownership of forest land propriety is heavily divided.

An extra challenge in regional or landscape level forest planning stems from the fact that biodiversity or alternative ecosystem services provide goods and benefits to different stakeholders. Some commodities such as timber are considered private property, benefiting primarily the landowner while others are considered public goods. For example, climate change mitigation provides a global benefit by reducing atmospheric CO_2 levels, while water quality regulation, recreational use and natural collectable products (e.g., berries and mushrooms) profit mostly the local community. Private landowners typically lack the incentive to manage land to provide ecosystem services and biodiversity conservation benefits in cases where the benefits produced on their land accrue to others. A recent review (Howe et al. 2014) showed that trade-offs among ecosystem services are especially likely to occur when one of the services is a provisioning service and one of the stakeholders involved has a private interest in the services. In summary, besides new management practices and planning tools, new regulations and/or incentives are required to improve the protection of public interests in boreal production forests.

References

Ahti T, Hämet-Ahti L, Jalas J (1968) Vegetation zones and their sections in northwestern Europe. Ann Bot Fenn 5:169–211

Äijälä O, Koistinen A, Sved J, Vanhatalo K, Väisänen P (2014) Metsänhoidon suositukset. [the good practice guidance to forestry, (in Finnish)]. Metsäkustannus oy, forestry development Centre Tapio, Helsinki

Alaspää K, Muukkonen P, Mäkipää R (2015) Lahopuun merkitys kasvualustana eteläboreaalisen vanhan luonnon tilaisen kuusimetsän uudistumisessa. Metsätieteen Aikakauskirja 2015:237–245

Amoroso MM, Turnblom EC (2006) Comparing productivity of pure and mixed Douglas-fir and western hemlock plantations in the Pacific northwest. Can J For Res 36:1484–1149

Angelstam P, Kuuluvainen T (2004) Boreal forest disturbance regimes, successional dynamics and landscape structures – a European perspective. Ecol Bull 51:117–136

Angelstam P, Pettersson B (1997) Principles of present Swedish forest biodiversity management. Ecol Bull 46:191–203

Angelstam PK (1998) Maintaining and restoring biodiversity in European boreal forests by developing natural disturbance regimes. J Veg Sci 9:593–602. https://doi.org/10.2307/3237275

Arnborg T (1990) Forest types of northern Sweden. Vegetatio 90:1–13

Attiwill PM (1994) The disturbance of forest ecosystems: the ecological basis for conservative management. For Ecol Manag 63:247–300

Aubry KB, Halpern CB, Peterson CE (2009) Variable-retention harvests in the Pacific northwest: a review of short-term findings from the DEMO study. For Ecol Manag 258:398–408

Auld G, Gulbrandsen LH, McDermott CL (2008) Certification schemes and the impacts on forests and forestry. Annu Rev Environ Resour 33:187–211

Bamsey CR (ed) (1995) Innovative silviculture systems in boreal forests. Proceedings, Symposium, 2–8 October 1994, Edmonton, Alberta. Clear Lake Ltd., Edmonton, Alberta

Berg A, Ehnstrom B, Gustafsson L, Hallingback T, Jonsell M, Weslien J (1994) Threatened plant, animal, and fungus species in Swedish forests: distribution and habitat associations. Conserv Biol 8:718–731

Bergeron Y, Leduc A, Harvey BD, Gauthier S (2002) Natural fire regime: a guide for sustainable management of the Canadian boreal forest. Silva Fenn 36:81–95

Biodiversity Consultancy Ltd. Mitigation hierarchy. Available from http://www.thebiodiversity-consultancy.com/approaches/mitigation-hierarchy

Björkman C, Nimelä P (eds) (2015) Climate change and insect pests. CABI, Wallingford

Bonan GB (2008) Forests and climate change: Forcings, feedbacks, and the climate benefits of forests. Science 320(5882):1444–1449

Boyland M (2006) The economics of using forests to increase carbon storage. Can J For Res 36:2223–2234

Bradshaw CJA, Warkentin IG, Sodhi NS (2009) Urgent preservation of boreal carbon stocks and biodiversity. Trends Ecol Evol 24:541–548

Brandt JP, Flannigan MD, Maynard DG, Thompson ID, Volney WJA (2013) An introduction to Canada's boreal zone: Ecosystem processes, health, sustainability, and environmental issues. Environ Rev 21:207–226

Bryant D, Nielsen D, Tangley L (1997) The last frontier forests: ecosystems and economies on the edge. World Resource Institute, Washington

Burton PJ, Bergeron Y, Bogdanski BEC, Juday GP, Kuuluvainen T, McAfee BJ, Ogden A, Teplyakov VK, Alfaro RI, Francis DA, Gauthier S, Hantula J (2010) Sustainability of boreal forests and forestry in a changing environment. In: Mery G, Katila P, Galloway G, Alfaro RI, Kanninen M, Lobovikov M, Varjo J (eds) Forests and society – responding to global drivers of change, vol 25. International Union of Forest Research Organizations, IUFRO World Series, Vienna, Austria, pp 249–282

Burton PJ, Messier C, Smith DW, Adamowicz WL (eds) (2003) Towards sustainable Management of the Boreal Forest. NRC Research Press, Ottawa

Cajander AK (1949) Forest types and their significance. Acta For Fenn 56:1–71

Cardinale BJ, Duffy JE, Gonzalez A, Hooper DU, Perrings C, Venail P, Narwani A, Mace GM, Tilman D, Wardle DA, Kinzig AP, Daily GC, Loreau M, Grace JB, Larigauderie A, Srivastava DS, Naeem S (2012) Biodiversity loss and its impact on humanity. Nature 486:59–67

Carlén O, Mattsson L, Atlegrim O, Sjöberg K (1999) Cost efficiency in pursuing environmental objectives in forestry. J Environ Manag 55:111–125

Chapin FS, Mcguire AD, Randerson J, Pielke R, Baldocchi D, Hobbie SE, Roulet N, Eugster W, Kasischke E, Rastetter EB, Zimov SA, Running SW (2000) Arctic and boreal ecosystems of western North America as components of the climate system. Glob Chang Biol 6:211–223

Christiansen L (ed) (2014) Swedish statistical yearbook of forestry 2014. Swedish Forest Agency, Jönköping, Sweden

Clark JA, Covey KR (2012) Tree species richness and the logging of natural forests: a meta-analysis. For Ecol Manag 276:146–153

Clason AJ, Lindgren PMF, Sullivan TP (2008) Comparison of potential non-timber forest products in intensively managed young stands and mature/old-growth forests in south-central British Columbia. For Ecol Manag 256:1897–1909

Cogbill C (1985) Dynamics of the boreal forests of the Laurentian highlands, Canada. Can J For Res 15:252–261

Cooper CF (1983) Carbon storage in managed forests. Can J For Res 13:155–166

Costanza R, Fisher B, Mulder K, Liu S, Christopher T (2007) Biodiversity and ecosystem services: a multi-scale empirical study of the relationship between species richness and net primary production. Ecol Econ 61:478–491

Côté P, Tittler R, Messier C, Kneeshaw DD, Fall A, Fortin M-J (2010) Comparing different forest zoning options for landscape-scale management of the boreal forest: possible benefits of the TRIAD. For Ecol Manag 259:418–427

Cox CB, Moore PD (2010) Biogeography. An ecological and evolutionary approach, 8th edn. Wiley, New York

Daniel TW, Helms JA, Baker FS (1979) Principles of silviculture. McGraw-Hill Book Company, New York

De Jong J, Dahlberg A, Stokland JN (2004) Död ved i skogen: Hur mycket behövs för att bevara den biologiska mångfalden? Sven Bot Tidskr 98:278–297

Díaz S, Lavorel S, de Bello F, Quétier F, Grigulis K, Robson TM (2007) Incorporating plant functional diversity effects in ecosystem service assessments. Proc Natl Acad Sci 104:20684–20689

Diaz-Balteiro L, González-Pachón J, Romero C (2012) Goal programming in forest management: customising models for the decision-maker's preferences. Scand J For Res 7581:1–8

Duncan C, Thompson JR, Pettorelli N (2015) The quest for a mechanistic understanding of biodiversity–ecosystem services relationships. Proc R Soc B Biol Sci 282:20151348

EC (2012) Commission staff working document accompanying the document communication on innovating for sustainable growth: a bioeconomy for Europe. European Commission, Brussels

Edwards DP, Gilroy JJ, Woodcock P, Edwards FA, Larsen TH, Andrews DJR, Derhé MA, Docherty TDS, Hsu WW, Mitchell SL, Ota T, Williams LJ, Laurance WF, Hamer KC, Wilcove DS (2014) Land-sharing versus land-sparing logging: reconciling timber extraction with biodiversity conservation. Glob Change Biol 20:183–191

Eid T, Hobbelstad K (2000) AVVIRK-2000: a large-scale forestry scenario model for long-term investment, income and harvest analyses. Scand J For Res 15:472–482

Elbakidze M, Andersson K, Angelstam P, Armstrong GW, Axelsson R, Doyon F, Hermansson M, Jacobsson J, Pautov Y (2013) Sustained yield forestry in Sweden and Russia: how does it correspond to sustainable forest management policy? Ambio 42:160–173

Elbakidze M, Angelstam P, Andersson K, Nordberg M, Pautov Y (2011) How does forest certification contribute to boreal biodiversity conservation? Standards and outcomes in Sweden and NW Russia. For Ecol Manag 262:1983–1995

Engel S, Pagiola S, Wunder S (2008) Designing payments for environmental services in theory and practice: an overview of the issues. Ecol Econ 65:663–674

Esseen P-A, Ehnström B, Ericson L, Sjöberg K (1997) Boreal forests. Ecol Bull 46:16–47

Fahrig L (2003) Effects of habitat fragmentation on biodiversity. Annu Rev Ecol Evol Syst 34:487–515

FAO (2010) Global Forest Resources Assessment 2010. FAO For Pap 378

Farley J, Costanza R (2010) Payments for ecosystem services: from local to global. Ecol Econ 69:2060–2068

Fedrowitz K, Koricheva J, Baker SC, Lindenmayer DB, Palik B, Rosenvald R, Beese W, Franklin JF, Kouki J, Macdonald E, Messier C, Sverdrup-Thygeson A, Gustafsson L (2014) Can retention forestry help conserve biodiversity? A meta-analysis. J Appl Ecol 51:1669–1679

Felton A, Ranius T, Roberge J-M, Öhman K, Lämås T, Hynynen J, Juutinen A, Mönkkönen M, Nilsson U, Lundmark T, Nordin A (2017) Projecting biodiversity and wood production in future forest landscapes: 15 key modeling considerations. J Environ Manag 197:404–414

Felton A, Gustafsson L, Roberge J-M, Ranius T, Hjältén J, Rudolphi J, Lindbladh M, Weslien J, Rist L, Brunet J, Felton AM (2016) How climate change adaptation and mitigation strategies can threaten or enhance the biodiversity of production forests: insights from Sweden. Biol Conserv 194:11–20

Felton A, Lindbladh M, Brunet J, Fritz Ö (2010) Replacing coniferous monocultures with mixed-species production stands: an assessment of the potential benefits for forest biodiversity in northern Europe. For Ecol Manag 260:939–947

Finnish Forest Act (2014) (1093/1996; amendments up to 567/2014 included)

Franklin JF, Spies TA, Van PR, Carey AB, Thornburgh DA, Berg DR, Lindenmayer DB, Harmon ME, Keeton WS, Shaw DC, Bible K, Chen J (2002) Disturbances and structural development of natural forest ecosystems with silvicultural implications, using Douglas-fir forests as an example. For Ecol Manag 155:399–423

Frey BR, Lieffers VJ, Landhäusser SM, Comeau PG, Greenway KJ (2003) An analysis of sucker regeneration of trembling aspen. Can J For Res 33:1169–1179

FSC Finland (2010) Standard for Finland. FSC-STD-FIN-(Ver1–1)-2006 Finland natural forests. FSC Finland, Helsinki, p 66

FSC Sweden (2010) Swedish FSC standard for Forest certification including SLIMF indicators FSC-STD-SWE-02-02-2010 Sweden natural, plantations and SLIMF EN. FSC Sweden, Uppsala, p 99

Gamfeldt L, Snäll T, Bagchi R, Jonsson M, Gustafsson L, Kjellander P, Ruiz-Jaen MC, Fröberg M, Stendahl J, Philipson CD, Mikusiński G, Andersson E, Westerlund B, Andrén H, Moberg F, Moen J, Bengtsson J (2013) Higher levels of multiple ecosystem services are found in forests with more tree species. Nat Commun 4:1340

Gauthier S, Bernier P, Kuuluvainen T, Shvidenko AZ, Schepaschenko DG (2015) Boreal forest health and global change. Science 349:819–822

Gentry AH (1988) Tree species richness of upper Amazonian forests. Proc Natl Acad Sci U S A 85:156–159

Global Land Cover Facility Tree Canopy Cover. 2010. Available from www.landcover.org

Gromtsev A (2002) Natural disturbance dynamics in the boreal forests of European Russia: a review. Silva Fenn 36:41–55

Gulbrandsen LH (2005) The effectiveness of non-state governance schemes: a comparative study of Forest certification in Norway and Sweden. Int Environ Agreements Polit Law Econ 5:125–149

Gullison RE (2003) Does forest certification conserve biodiversity? Oryx 37:153–165

Gustafsson L, Baker SC, Bauhus J, Beese WJ, Brodie A, Kouki J, Lindenmayer DB, Lõhmus A, Pastur GM, Messier C, Neyland M, Palik B, Sverdrup-Thygeson A, Volney WJA, Wayne A, Franklin JF (2012) Retention forestry to maintain multifunctional forests: a world perspective. Bioscience 62:633–645

Gustafsson L, Kouki J, Sverdrup-Thygeson A (2010) Tree retention as a conservation measure in clear-cut forests of northern Europe: a review of ecological consequences. Scand J For Res 25:295–308

Haines-Young R, Potschin M (2013) Common international classification of ecosystem services (CICES): consultation on version 4, August–December 2012

Haines-Young R, Potschin M (2011) Common international classification of ecosystem services (CICES): 2011 update. Expert meet Ecosyst accounts 1–17

Haines-Young RH, Potschin MB (2010) The links between biodiversity, ecosystem services and human well-being. In: Raffaelli DG, Frid CLJ (eds) Ecosystems ecology: a new synthesis. Cambridge University Press

Hansen AJ, Spies TA, Swanson FJ, Ohmann JL (1991) Conserving biodiversity in managed forests. Bioscience 41:382–392

Hanski I (2000) Extinction debt and species credit in boreal forests: modelling the consequences of different approaches to biodiversity conservation. Ann Zool Fennici 37:271–280

Hanski I (2011) Habitat loss, the dynamics of biodiversity, and a perspective on conservation. Ambio 40:248–255

Hanski I, Hammond P (1995) Biodiversity in boreal forests. Trends Ecol Evol 10:5–6

Harrison PA, Vandewalle M, Sykes MT, Berry PM, Bugter R, de Bello F, Feld CK, Grandin U, Harrington R, Haslett JR, Jongman RHG, Luck GW, da Silva PM, Moora M, Settele J, Sousa JP, Zobel M (2010) Identifying and prioritising services in European terrestrial and freshwater ecosystems. Biodivers Conserv 19:2791–2821

Hjortsø CN (2004) Enhancing public participation in natural resource management using soft OR — an application of strategic option development and analysis in tactical forest planning. Eur J Oper Res 152:667–683

Howe C, Suich H, Vira B, Mace GM (2014) Creating win-wins from trade-offs? Ecosystem services for human well-being: a meta-analysis of ecosystem service trade-offs and synergies in the real world. Glob Environ Chang 28:263–275

Huntley B (1993) Species-richness in north-temperate zone forests. J Biogeogr 20:163–180

Huntley B, Birks HJB, Harry JB (1983) An atlas of past and present pollen maps for Europe, 0–13,000 years ago. Cambridge University Press, Cambridge

Hynynen J, Ahtikoski A, Siitonen J, Sievänen R, Liski J (2005) Applying the MOTTI simulator to analyse the effects of alternative management schedules on timber and non-timber production. For Ecol Manag 207:5–18

Hyvärinen E, Kouki J, Martikainen P (2006) Fire and green-tree retention in conservation of red-listed and rare deadwood-dependent beetles in Finnish boreal forests. Conserv Biol 20:1710–1719

Hyytiäinen K, Hari P, Kokkila T, Mäkelä A, Tahvonen O, Taipale J (2004) Connecting a process-based forest growth model to stand-level economic optimization. Can J For Res 34:2060–2073

Hyytiäinen K, Tahvonen O (2002) Economics of forest thinnings and rotation periods for Finnish conifer cultures. Scand J For Res 17:274–288

Håbesland DE, Kilgore MA, Becker DR, Snyder SA, Solberg B, Sjølie HK, Lindstad BH (2016) Norwegian family forest owners' willingness to participate in carbon offset programs. For Policy Econ 70:30–38

Imbeau L, Mönkkönen M, Desrochers A (2001) Long-term effects of forestry on birds of the eastern Canadian boreal forests: a comparison with Fennoscandia. Conserv Biol 15:1151–1162. https://doi.org/10.1046/j.1523-1739.2001.0150041151.x

Jactel H, Nicoll BC, Branco M, Gonzalez-Olabarria JR, Grodzki W, Långström B, Moreira F, Netherer S, Orazio C, Piou D, Santos H, Schelhaas MJ, Tojic K, Vodd VF, Hervé J, Bruce CN, Manuela B, Ramon J, Wojciech G, Bo L, Francisco M, Sigrid N, Christophe O, Dominique P, Helena S, MJ S, Karl T, Floor V, Jactel H, Nicoll BC, Branco M, Gonzalez-Olabarria JR,

Grodzki W, Långström B, Moreira F, Netherer S, Orazio C, Piou D, Santos H, Schelhaas MJ, Tojic K, Vodde F, Hervé J, Bruce CN, Manuela B, Ramon J, Wojciech G, Bo L, Francisco M, Sigrid N, Christophe O, Dominique P, Helena S, MJ S, Karl T, Floor V, Jactel H, Nicoll BC, Branco M, Gonzalez-Olabarria JR, Grodzki W, Långström B, Moreira F, Netherer S, Orazio C, Piou D, Santos H, Schelhaas MJ, Tojic K, Vodde F (2009) The influences of forest stand management on biotic and abiotic risks of damage. Ann For Sci 66:701

Jandl R, Lindner M, Vesterdal L, Bauwens B, Baritz R, Hagedorn F, Johnson DW, Minkkinen K, Byrne KA (2007) How strongly can forest management influence soil carbon sequestration? Geoderma 137:253–268

Johansson J, Lidestav G (2011) Can voluntary standards regulate forestry? — assessing the environmental impacts of forest certification in Sweden. For Policy Econ 13:191–198

Johansson T, Hjältén J, de Jong J, von Stedingk H (2013) Environmental considerations from legislation and certification in managed forest stands: a review of their importance for biodiversity. For Ecol Manag 303:98–112

Johnson KN, Scheurman H (1977) Techniques for prescribing optimal timber harvest and investment under different objectives - discussion and synthesis. For Sci (Monogr) 18

Jönsson U, Rosengren U, Thelin G, Nihlgård B (2003) Acidification-induced chemical changes in coniferous forest soils in southern Sweden 1988–1999. Environ Pollut 123:75–83

Kaipainen T, Liski J, Pussinen A, Karjalainen T (2004) Managing carbon sinks by changing rotation length in European forests. Environ Sci Pol 7:205–219

Kallio AMI, Salminen O, Sievänen R (2013) Sequester or substitute—consequences of increased production of wood based energy on the carbon balance in Finland. J For Econ 19:402–415

Kangas A, Kurttila M, Hujala T, Eyvindson K, Kangas J (2015) Decision support for Forest management. Springer International Publishing, Cham

Kareksela S, Moilanen A, Tuominen S, Kotiaho JS (2013) Use of inverse spatial conservation prioritization to avoid biological diversity loss outside protected areas. Conserv Biol 27:1294–1303

Karjalainen E, Sarjala T, Raitio H (2010) Promoting human health through forests: overview and major challenges. Environ Health Prev Med 15:1–8

Kelty MJ (1992) Comparative productivity of monocultures and mixed-species stands. In: Kelty MJ, Larson BC, Oliver CD (eds) the ecology and Silviculture of mixed-species forests. Springer, pp 125–141

Kettunen M, Vihervaara P, Kinnunen S, D'Amato D, Badura T, Argimon M, Ten Brink P (2012) Socio-economic importance of ecosystem services in the Nordic countries. Synthesis in the context of the economics of ecosystems and biodiversity (TEEB). TemaNord 2012:559. Available form https://doi.org/10.6027/TN2012-559

King E, Cavender-Bares J, Balvanera P, Mwampamba TH, Polasky S (2015) Trade-offs in ecosystem services and varying stakeholder preferences: evaluating conflicts, obstacles, and opportunities. Ecol Soc 20. art25

Kreutzweiser DP, Hazlett PW, Gunn JM (2008) Logging impacts on the biogeochemistry of boreal forest soils and nutrient export to aquatic systems: a review. Environ Rev 16:157–179

Kruys N, Fridman J, Götmark F, Simonsson P, Gustafsson L (2013) Retaining trees for conservation at clearcutting has increased structural diversity in young Swedish production forests. For Ecol Manag 304:312–321

Kurki S, Nikula A, Helle P, Lindén H (2000) Landscape fragmentation and forest composition effects on grouse breeding success in boreal forests. Ecology 81:1985–1997

Kuuluvainen T (2002) Natural variability of forests as a reference for restoring and managing biological diversity in boreal Fennoscandia. Silva Fenn 36:97–125

Kuuluvainen T (2009) Forest management and biodiversity conservation based on natural ecosystem dynamics in northern Europe: the complexity challenge. Ambio 38:309–315

Kuuluvainen T, Aakala T (2011) Natural forest dynamics in boreal Fennoscandia: a review and classification. Silva Fenn. 45:823–841

Kuuluvainen T, Grenfell R (2012) Natural disturbance emulation in boreal forest ecosystem management — theories, strategies, and a comparison with conventional even-aged management. Can J For Res 42:1185–1203

Kuuluvainen T, Saaristo L, Keto-Tokoi P, Kostamo J, Kuuluvainen J, Kuusinen M, Ollikainen M, Salpakivi-Salomaa P, Hallanaro E-L, Jäppinen J-P (2004) Metsän kätköissä : Suomen metsäluonnon monimuotoisuus. Helsinki, Edita

Kuuluvainen T, Tahvonen O, Aakala T (2012) Even-aged and uneven-aged forest management in boreal Fennoscandia: a review. Ambio 41:720–737

Laiho O, Lahde E, Pukkala T (2011) Uneven- vs even-aged management in Finnish boreal forests. Forestry 84:547–556

Larocque GR (2016) Ecological forest management handbook. CRC Press. Taylor & Francis Group, Boca Raton

Latham R, Ricklefs R (1993) Continental comparisons of temperate-zone tree species diversity. In: Species diversity in ecological communities. Historical and geographical perspectives. University of Chicago Press, Chicago, pp 291–314

Laudon H, Sponseller R, Lucas R, Futter M, Egnell G, Bishop K, Ågren A, Ring E, Högberg P (2011) Consequences of more intensive forestry for the sustainable Management of Forest Soils and Waters. Forests 2:243–260

Lindenmayer DB, Franklin JF, Lõhmus A, Baker SC, Bauhus J, Beese W, Brodie A, Kiehl B, Kouki J, Pastur GM, Messier C, Neyland M, Palik B, Sverdrup-Thygeson A, Volney J, Wayne A, Gustafsson L (2012) A major shift to the retention approach for forestry can help resolve some global forest sustainability issues. Conserv Lett 5:421–431

Lintunen J, Laturi J, Uusivuori J (2016) How should a forest carbon rent policy be implemented? For Policy Econ 69:31–39

Liski J, Pussinen A, Pingoud K, Mäkipää R, Karjalainen T (2001) Which rotation length is favourable to carbon sequestration? Can J For Res 31:2004–2013

Lundmark T, Bergh J, Nordin A, Fahlvik N, Poudel BC (2016) Comparison of carbon balances between continuous-cover and clear-cut forestry in Sweden. Ambio 45:203–213

Lutz DA, Howarth RB (2014) Valuing albedo as an ecosystem service: implications for forest management. Clim Chang 124:53–63

Luyssaert S, Schulze E-D, Börner A, Knohl A, Hessenmöller D, Law BE, Ciais P, Grace J (2008) Old-growth forests as global carbon sinks. Nature 455:213–215

Mace GM, Norris K, Fitter AH (2012) Biodiversity and ecosystem services: a multilayered relationship. Trends Ecol Evol 27:19–26

Maes J, Paracchini ML, Zulian G, Dunbar MB, Alkemade R (2012) Synergies and trade-offs between ecosystem service supply, biodiversity, and habitat conservation status in Europe. Biol Conserv 155:1–12

MARSI (2014) Luonnonmarjojen ja – sienten kauppaantulomäärät vuonna 2014. Raportteja ja selvityksiä 3/2015

Maynard DG, Paré D, Thiffault E, Lafleur B, Hogg KE, Kishchuk B (2014) How do natural disturbances and human activities affect soils and tree nutrition and growth in the Canadian boreal forest? Environ Rev 22(2):161–178

Mazziotta A, Podkopaev D, Triviño M, Miettinen K, Pohjanmies T, Mönkkönen M (2017) Quantifying and resolving conservation conflicts in forest landscapes via multiobjective optimization. Silva Fenn 51(1):1778. https://doi.org/10.14214/sf.1778

MEA (2005) Millennium ecosystem assessment. Ecosystems and human well-being: synthesis. Island Press, Washington, DC

METSO 2015. http://www.metsonpolku.fi/en-US Accessed 6.4.2017

Miettinen K (1999) Nonlinear multiobjective optimization. Kluwer Academic Publishers, Boston

Ministry of Employment and the Economy (2014) Sustainable growth from bioeconomy. The Finnish Bioeconomy Strategy

Moen A (1999) National Atlas of Norway – vegetation. Norwegian Mapping Authority, Hønefoss

Myllyntaus T, Hares M, Kunnas J (2002) Sustainability in danger?: slash-and-burn cultivation in nineteenth-century Finland and twentieth-century Southeast Asia. Environ Hist Durh N C 7:267

Mönkkönen M (1999) Managing Nordic boreal forest landscapes for biodiversity: ecological and economic perspectives. Biodivers Conserv 8:85–99

Mönkkönen M, Juutinen A, Mazziotta A, Miettinen K, Podkopaev D, Reunanen P, Salminen H, Tikkanen O-P (2014) Spatially dynamic forest management to sustain biodiversity and economic returns. J Environ Manag 134:80–89

Mönkkönen M, Reunanen P, Nikula A, Inkeroinen J, Forsman J (1997) Landscape characteristics associated with the occurrence of the flying squirrel Pteromys Volans in old-growth forests of northern Finland. Ecography (Cop) 20:634–642

Mönkkönen M, Viro P (1997) Taxonomic diversity of the terrestrial bird and mammal fauna in temperate and boreal biomes of the northern hemisphere. J Biogeogr 24:603–612

Natural resources of Canada (2016) Non-timber forest products. Available from http://www.nrcan. gc.ca/forests/industry/products-applications/13203 Accessed 26 May 2016

Naudts K, Chen Y, McGrath MJ, Ryder J, Valade A, Otto J, Luyssaert S (2016) Europe's forest management did not mitigate climate warming. Science 351:597–600

Nieminen A (2006) Metsäsertifioinnin ekotehokkuus. Metlan työraportteja (Working Papers of the Finnish Forest Research Institute) 39: 1–85

Niinimäki S, Tahvonen O, Mäkelä A, Linkosalo T (2013) On the economics of Norway spruce stands and carbon storage. Can J For Res 43:637–648

Nilsson S (1997) Forests in the temperate-boreal transition: natural and man-made features. Ecol Bull:61–71

Nordén B, Dahlberg A, Brandrud TE, Fritz Ö, Ejrnaes R, Ovaskainen O (2015) Effects of ecological continuity on species richness and composition in forests and woodlands: a review. Ecoscience 21:34–45

Nordström E-M, Romero C, Eriksson LO, Öhman K (2009) Aggregation of preferences in participatory forest planning with multiple criteria: an application to the urban forest in Lycksele, Sweden. Can J For Res 39:1979–1992

O'Hara KL, Ramage BS (2013) Silviculture in an uncertain world: utilizing multi-aged management systems to integrate disturbance. Forestry 86:401–410

Ojalehto V, Miettinen K, Mäkelä M (2007) Interactive software for multiobjective optimization: IND-NIMBUS, WSEAS. Trans Comput 6:87–94

Ollikainen M (2016) Forest management, public goods, and optimal policies. Annu Rev Resour Economics 8:207–226

Olson DM, Dinerstein E, Wikramanayake ED, Burgess ND, Powell GVN, Underwood EC, D'amico JA, Itoua I, Strand HE, Morrison JC, Loucks CJ, Allnutt TF, Ricketts TH, Kura Y, Lamoreux JF, Wettengel WW, Hedao P, Kassem KR (2001) Terrestrial ecoregions of the world: a new map of life on earth. Bioscience 51:933

Östlund L, Roturier S (2010) Forestry historical studies in the province of Västerbotten, northern Sweden: a review of Lars Tirén (1937). Scand J For Res 26:91–99

Östlund L, Zackrisson O, Axelsson a-L (1997) The history and transformation of a Scandinavian boreal forest landscape since the 19th century. Can J For Res 27:1198–1206

Paassilta M, Moisio S, Jaakola L, Häggman H, Oulu University press (2009) Voice of the Nordic wild berry industry. A survey among the companies 84

Pan Y, Birdsey RA, Fang J, Houghton R, Kauppi PE, Kurz WA, Phillips OL, Shvidenko A, Lewis SL, Canadell JG, Ciais P, Jackson RB, Pacala SW, McGuire AD, Piao S, Rautiainen A, Sitch S, Hayes D (2011) A large and persistent carbon sink in the world's forests. Science 333:988–993

Paquette A, Messier C (2011) The effect of biodiversity on tree productivity: from temperate to boreal forests. Glob Ecol Biogeogr 20:170–180

Parviainen J, Västilä S (eds) (2011) State of Finland's forests 2011. Ministry of Agriculture and Forestry & Finnish Forest Research Institute (Metla)

Pawson SM, Brin A, Brockerhoff EG, Lamb D, Payn TW, Paquette A, Parrotta JA (2013) Plantation forests, climate change and biodiversity. Biodivers Conserv 22:1203–1227

PEFC Finland (2014) Criteria for PEFC Forest certification PEFC FI 1002:2014. PEFC Finland, Helsinki

Peltola A (ed) (2014) Finnish statistical yearbook of forestry 2014. Forest Research Institute, Tampere, Finland

Pennanen J (2002) Forest age distribution under mixed-severity fire regimes – a simulation-based analysis for middle boreal Fennoscandia. Silva Fenn 36:213–231. https://doi.org/10.14214/sf.559

Peura M, Triviño M, Mazziotta A, Podkopaev D, Juutinen A, Mönkkönen M (2016) Managing boreal forests for the simultaneous production of collectable goods and timber revenues. Silva Fenn 50:1672. https://doi.org/10.14214/sf.1672

Pihlainen S, Tahvonen O, Niinimäki S (2014) The economics of timber and bioenergy production and carbon storage in scots pine stands. Can J For Res 44:1091–1102

Pohjanmies T, Triviño M, Le Tortorec E, Mazziotta A, Snäll T, Mönkkönen M (2017a) Impacts of forestry on boreal forests: an ecosystem services perspective. Ambio 46:743–755. https://doi.org/10.1007/s13280-017-0919-5

Pohjanmies T, Triviño T, Le Tortorec E, Salminen H, Mönkkönen M (2017b) Conflicting objectives in production forests pose a challenge for forest management. Ecosyst Serv 28:298–310

Pommerening A, Murphy S (2004) A review of the history, definitions and methods of continuous cover forestry with special attention to afforestation and restocking. Forestry 77:27–44

Potapov P, Yaroshenko A, Turubanova S, Dubinin M, Laestadius L, Thies C, Aksenov D, Egorov A, Yesipova Y, Glushkov I, Karpachevskiy M, Kostikova A, Manisha A, Tsybikova E, Zhuravleva I (2008) Mapping the world's intact forest landscapes by remote sensing. Ecol Soc 13.:Artn 51

Pukkala T (2016) Which type of forest management provides most ecosystem services? For Ecosyst 3:9

Pukkala T, Lähde E, Laiho O (2013) Species interactions in the dynamics of even- and uneven-aged boreal forests. J Sustain For 32:371–403

Pukkala T, Lähde E, Laiho O, Salo K, Hotanen J (2011) A multifunctional comparison of even-aged and uneven-aged forest management in a boreal. Region 862:851–862

Rametsteiner E, Simula M (2003) Forest certification—an instrument to promote sustainable forest management? J Environ Manag 67:87–98

Rasinmäki J, Mäkinen A, Kalliovirta J (2009) SIMO: an adaptable simulation framework for multiscale forest resource data. Comput Electron Agric 66:76–84

Rassi P, Hyvärinen E, Juslén A, Mannerkoski I (eds) (2010) The 2010 red list of Finnish species. Helsinki, Ympäristöministeriö & Suomen ympäristökeskus

Redsven V, Hirvelä H, Härkönen K, Salminen O, Siitonen M (2012) MELA2002 reference manual, 2nd edn. The Finnish Forest Research Institute

Repo A, Tuomi M, Liski J (2011) Indirect carbon dioxide emissions from producing bioenergy from forest harvest residues. GCB Bioenergy 3:107–115

Roberge J-M, Laudon H, Björkman C, Ranius T, Sandström C, Felton A, Sténs A, Nordin A, Granström A, Widemo F, Bergh J, Sonesson J, Stenlid J, Lundmark T (2016) Socio-ecological implications of modifying rotation lengths in forestry. Ambio 45(Suppl 2):109–123

Rodríguez A, Kouki J (2015) Emulating natural disturbance in forest management enhances pollination services for dominant Vaccinium shrubs in boreal pine-dominated forests. For Ecol Manag 350:1–12

Rosenvald R, Lõhmus A (2008) For what, when, and where is green-tree retention better than clear-cutting? A review of the biodiversity aspects. For Ecol Manag 255:1–15

Rosenzweig M (1995) Species diversity in space and time. Cambridge University Press, Cambridge, MA

Rybicki J, Hanski I (2013) Species-area relationships and extinctions caused by habitat loss and fragmentation. Ecol Lett 16:27–38

Rämö J, Tahvonen O (2014) Economics of harvesting uneven-aged forest stands in Fennoscandia. Scand J For Res 29:777–792

Saastamoinen O, Matero J, Horne P, Kniivilä M, Haltia E, Vaara M, Mannerkoski H (2014) Classification of boreal forest ecosystem goods and services in Finland. Publications of the University of Eastern Finland. Reports and studies in forestry and natural sciences Number 11, University of Eastern Finland, Faculty of Science and Forestry, School of Forest Sciences

Santangeli A, Wistbacka R, Hanski IK, Laaksonen T (2013) Ineffective enforced legislation for nature conservation: a case study with Siberian flying squirrel and forestry in a boreal landscape. Biol Conserv 157:237–244

Schindler D (1998) Sustaining aquatic ecosystems in boreal regions. Conserv Ecol 2:18

Seidl R, Schelhaas M-J, Rammer W, Verkerk PJ (2014) Increasing forest disturbances in Europe and their impact on carbon storage. Nat Clim Chang 4:806–810

Seymour RS, Hunter MLJ (1992) New forestry in eastern spruce-fir forests: principles and applications to Maine. Miscellaneous Publication 716. Maine Agricultural Forest Experimental Station

Seymour RS, White AS (2002) Natural disturbance regimes in northeastern North America—evaluating silvicultural systems using natural scales and frequencies. For Ecol Manage 55(1–3):357–367

Shorohova E, Kuuluvainen T, Kangur A, Jogiste K (2009) Natural stand structures , disturbance regimes and successional dynamics in the Eurasian boreal forests: a review with special reference to Russian studies. Ann For Sci 66:1–20

Sievänen R, Salminen O, Lehtonen A, Ojanen P, Liski J, Ruosteenoja K, Tuomi M (2013) Carbon stock changes of forest land in Finland under different levels of wood use and climate change. Ann For Sci 71:255–265

Siitonen J (2001) Forest management, coarse woody debris and saproxylic organisms: Fennoscandian boreal forests as an example. Ecol Bull 49:11–41

Skovsgaard JP, Vanclay JK (2008) Forest site productivity: a review of the evolution of dendrometric concepts for even-aged stands. Forestry 81:13–32

Spracklen DV, Bonn B, Carslaw KS (2008) Boreal forests, aerosols and the impacts on clouds and climate. Philos Trans A Math Phys Eng Sci 366:4613–4626

Snyder PK, Delire C, Foley JA (2004) Evaluating the influence of different vegetation biomes on the global climate. Clim Dyn 23(3–4):279–302

Strassburg BBN, Kelly A, Balmford A, Davies RG, Gibbs HK, Lovett A, Miles L, Orme CDL, Price J, Turner RK, Rodrigues ASL (2010) Global congruence of carbon storage and biodiversity in terrestrial ecosystems. Conserv Lett 3:98–105

Strategic Plan for Biodiversity 2011–2020 and the Aichi Targets Aichi Targets. Available from http://www.cbd.int/doc/strategic-plan/2011-2020/Aichi-Targets-EN.pdf

Strengbom J, Dahlberg A, Larsson A, Lindelöw Å, Sandström J, Widenfalk O, Gustafsson L (2011) Introducing intensively managed spruce plantations in Swedish forest landscapes will impair biodiversity decline. Forests 2:610–630

Strengbom J, Nordin A (2008) Commercial forest fertilization causes long-term residual effects in ground vegetation of boreal forests. For Ecol Manag 256:2175–2181

Stupak I, Asikainen A, Jonsell M, Karltun E, Mizaraitė D, Pasanen K, Pärn H, Raulund-Rasmussen K, Röser D, Schroeder M, Varnagirytė I, Vilkriste L, Callesen I, Gaitnieks T, Ingerslev M, Mandre M, Ozolincius R, Saarsalmi A, Armolaitis K, Helmisaari H-S, Indriksons A, Kairiukstis L, Katzensteiner K, Kukkola M, Ots K, Ravn HP, Tamminen P (2007) Sustainable utilisation of forest biomass for energy—possibilities and problems: policy, legislation, certification, and recommendations and guidelines in the Nordic, Baltic, and other European countries. Biomass Bioenergy 31:666–684

Syrjänen K, Kalliola R, Puolasmaa A (1994) Landscape structure and forest dynamics in subcontinental Russian European taiga. Ann Zool Fenn 31:19–34

Tahvonen O (2015) Economics of rotation and thinning revisited: the optimality of clearcuts versus continuous cover forestry. For Policy Econ 62:88–94

Thom D, Seidl R (2015) Natural disturbance impacts on ecosystem services and biodiversity in temperate and boreal forests. Biol Rev 91:760–781

Thomas CD, Anderson BJ, Moilanen A, Eigenbrod F, Heinemeyer A, Quaife T, Roy DB, Gillings S, Armsworth PR, Gaston KJ (2013) Reconciling biodiversity and carbon conservation. Ecol Lett 16:39–47

Tikkanen O-P, Heinonen T, Kouki J, Matero J (2007) Habitat suitability models of saproxylic red-listed boreal forest species in long-term matrix management: cost-effective measures for multi-species conservation. Biol Conserv 140:359–372

Tikkanen O-P, Matero J, Mönkkönen M, Juutinen A, Kouki J (2012) To thin or not to thin: bio-economic analysis of two alternative practices to increase amount of coarse woody debris in managed forests. Eur J For Res 131:1411–1422

Tilman D, Reich PB, Isbell F (2012) Biodiversity impacts ecosystem productivity as much as resources, disturbance, or herbivory. Proc Natl Acad Sci 109:10394–10397

Timonen J, Siitonen J, Gustafsson L, Kotiaho JS, Stokland JN, Sverdrup-Thygeson A, Mönkkönen M (2010) Woodland key habitats in northern Europe: concepts, inventory and protection. Scand J For Res 25:309–324

Tittler R, Messier C, Goodman RC (2016) Triad Forest management: local fix or global solution. In: Laroque GR (ed) Ecological forest management handbook. CRC Press, Boca Raton, pp 33–45

Toivanen T, Kotiaho JS (2007) Mimicking natural disturbances of boreal forests: the effects of controlled burning and creating dead wood on beetle diversity. Biodivers Conserv 16:3193–3211

Triviño M, Juutinen A, Mazziotta A, Miettinen K, Podkopaev D, Reunanen P, Mönkkönen M (2015) Managing a boreal forest landscape for providing timber, storing and sequestering carbon. Ecosyst Serv 14:179–189

Triviño M, Pohjanmies T, Mazziotta A, Juutinen A, Podkopaev D, Le Tortorec E, Mönkkönen M (2017) Optimizing management to enhance multifunctionality in a boreal forest landscape. J Appl Ecol 54:61–70

Tuhkanen S (1980) Climatic parameters and indices in plant geography. Svenska växtgeografiska sällskapet

Turner MG (2010) Disturbance and landscape dynamics in a changing world. Ecology 91:2833–2849

Turtiainen M, Saastamoinen O, Kangas K, Vaara M (2012) Picking of wild edible mushrooms in 1997-1999 and 2011. Silva Fenn 46(4):569–581

UNEP, FAO, UNFF (2009) Vital Forest graphics. UNEP/GRID-Arendal

Uprety Y, Asselin H, Dhakal A, Julien N, Shrestha K, Haddad P, Haddad P, Arnason J, Steffany A, Bennett S, Prentki M, Bennett S, Arnason J, Haddad P, Arnason J, Haddad P (2012) Traditional use of medicinal plants in the boreal forest of Canada: review and perspectives. J Ethnobiol Ethnomed 8:7

van Kooten GC, Binkley C, Delcourt G (1995) Effect of carbon taxes and subsidies on optimal forest rotation age and supply of carbon services. Am J Agr Econ 77:365–374

Vanha-Majamaa I, Jalonen J (2001) Green tree retention in Fennoscandian forestry. Scand J For Res 16:79–90

Vanha-Majamaa I, Lilja S, Ryömä R, Kotiaho JS, Laaka-Lindberg S, Lindberg H, Puttonen P, Tamminen P, Toivanen T, Kuuluvainen T (2007) Rehabilitating boreal forest structure and species composition in Finland through logging, dead wood creation and fire: the EVO experiment. For Ecol Manag 250:77–88

Vanhanen H, Jonsson R, Gerasimov Y, Krankina ON, Messier C (2012) Making Boreal Forests Work for People and Nature:15

Westling A (ed) (2015) Rödlistade arter i Sverige 2015. Artdatabanken, Uppsala

Wikström P, Edenius L, Elfving B, Eriksson LO, LäMåS T, Sonesson J, ÖHMAN K, Wallerman J, Waller C, Klintebäck F (2011) The Heureka forestry decision support system: an overview. Math Comput For Nat Sci 3:87–94

Vilà M, Vayreda J, Comas L, Ibáñez JJ, Mata T, Obón B (2007) Species richness and wood production: a positive association in Mediterranean forests. Ecol Lett 10:241–250

Yu PL (1973) A class of solutions for group decision problems. Manag Sci 19:936–946

Zackrisson O (1977) Influence of forest fires on the north Swedish boreal forest. Oikos 29:22

Zeleny M (1982) Multiple criteria decision making. McGraw-Hill, New York

Zeng H, Garcia-Gonzalo J, Peltola H, Kellomäki S (2010) The effects of forest structure on the risk of wind damage at a landscape level in a boreal forest ecosystem. Ann For Sci 67:111

Zeng H, Peltola H, Väisänen H, Kellomäki S (2009) The effects of fragmentation on the susceptibility of a boreal forest ecosystem to wind damage. For Ecol Manag 257:1165–1173

Natural Disturbances and Forest Management: Interacting Patterns on the Landscape

Lee E. Frelich, Kalev Jõgiste, John A. Stanturf, Kristi Parro, and Endijs Baders

1 Introduction

Forests provide for much of the biodiversity and ecosystem services that many scientists say are needed for a sustainable world. These ecological functions of forests are resilient to certain rates and severities of disturbance, since forests have evolved under the influence of natural disturbance regimes (Turton and Alamgir 2015). Although forest harvesting is relatively benign compared to intensive agriculture—many native species can continue to live in commercial forests—we also need to ensure that harvesting, when added to natural disturbance, does not exceed the resilience of forests, possibly leading to degradation in the condition of forests over time (Woodcock et al. 2015). In some regions the condition of forests is already degraded, and a trajectory of restoration of biodiversity and ecosystem services is needed while allowing for timber harvesting and the occurrence of natural disturbances. Thus, interactions between harvesting and natural disturbances are an important topic of ecological research, and here we provide a landscape ecology perspective on human and natural disturbance.

L. E. Frelich (✉)
University of Minnesota, Center for Forest Ecology, St. Paul, MN, USA
e-mail: freli001@umn.edu

K. Jõgiste · K. Parro
Institute of Forestry & Rural Engineering, Estonian University of Life Sciences, Tartu, Estonia

J. A. Stanturf
Center for Forest Disturbance Science, US Forest Service Southern Research Station, Athens, GA, USA

E. Baders
Latvian State Forest Research Institute "Silava", Salaspils, Latvia

© Springer International Publishing AG, part of Springer Nature 2018
A. H. Perera et al. (eds.), *Ecosystem Services from Forest Landscapes*,
https://doi.org/10.1007/978-3-319-74515-2_8

Forest health can be most simply defined as the ability to maintain productivity and all native species reasonably expected to be present given the climate and physiographic setting over time (authors of this chapter). This implies adequate habitat for species that depend on trees. This in turn implies that rates and types of disturbance fluctuate within certain bounds, so that a variety of tree species, and tree and stand ages, are always present on the landscape. In addition, these disturbance processes must create stable conditions across spatial scales, with tree (i.e., within stand, also including microscale habitat features like deadwood and tip-up mounds), stand, and landscape scales commonly employed as a hierarchical scheme (Seidl et al. 2011). Another way to put it is that connectivity exists in time and space for all species; the special habitats which some species require are never completely absent from a given landscape and are never too isolated for adequate gene flow to prevent inbreeding. Harvesting adds to the total rate of disturbance (although it is not always a completely additive situation, see below), simplifying structure and making the average age of trees younger than with natural disturbances only, possibly making it more difficult to maintain special habitats that occur in older trees and stands.

Harvesting regimes designed by those knowledgeable about forest dynamics are intended to capture impending tree mortality before it happens and send those trees to the mill before they turn into snags or fall on the ground. However, the natural forest ecosystem evolved over millions of years prior to arrival of humans, to function with all of that material falling onto the ground. Thus, harvesting preempts natural disturbance by diverting organic material that would otherwise fall on the ground during and after natural disturbances. This preemption leads to fewer biological legacies, defined as the total of all living and dead organic materials and patterns that persist through disturbance and are incorporated into the recovering ecosystem (Franklin et al. 2000; Jõgiste et al. 2017). Another way to put it is that harvesting short-circuits the forest's nutrient recycling system, as well as the production of microhabitats for species that depend on the detrital pool (coarse woody debris, CWD, and other organic matter). Therefore, the decay-based trophic pyramid becomes smaller, and it is possible that fewer species may be accommodated at the top (Grove 2002). Note that natural disturbance can also preempt harvest; the desired stable supply of material can be interrupted (for example, by insect outbreaks or hurricanes) or the quality greatly reduced due to charring in fires or twisting and breakage in windstorms.

Attempts to leave some materials representing a biological legacy while harvesting will partially mitigate the situation with regard to interrupting temporal and spatial connectivity of habitats that species need and short-circuiting natural cycles to some extent. However, the forest manager's ability to optimize the natural disturbance and harvest interaction may be limited. Despite the best intentions and prescriptions, there is no perfect situation—over time natural disturbances will inevitably take down some timber, and salvaged timber will not be as high in quality as the timber harvested from live standing trees. Also, timber production will lead to less woody material going into the detrital pool, and differing spatial patterns of tree species and age classes, than the natural system would have.

Close-to-nature forestry is a concept that attempts to emulate natural disturbance regimes with respect to size, frequency, and severity of disturbances, using unlogged reference forests as a blueprint or by mimicking processes thought to occur within natural forests in regions where no reference forests are available (Woodcock et al. 2015). Retention forestry attempts to maintain structures (biological legacies) that are necessary for maintenance of biodiversity and that may take a long time to create (i.e., large trees and large CWD). Woodland key habitat (WKH) is an enhancement of retention forestry that specifies retention in a variety of habitats present on the landscape (e.g., riparian corridors, late-successional forests, forest ponds or swamps). Any of these concepts could be employed to retain biological legacies within large harvests, such as standing dead snags, coarse woody debris, older trees with defects, buffer strips along streams, and individual live trees or patches of forest that provide continuous habitat for species that cannot survive in open conditions (Woodcock et al. 2015).

A variety of governmental and private frameworks have been developed to implement sustainable forest management practices. The Montreal Process and Pan-European Forest Process encompass a large majority of the world's temperate and boreal forests and specify the two parts of the definition of forest health given by the authors of this chapter—conservation of species diversity and maintenance of productive capacity—as criteria for sustainable forest management, although it has been pointed out that empirical evidence for the success of these programs is often lacking (Siry et al. 2005; Chandran and Innes 2014). China has the Natural Forest Conservation Program that addresses similar issues (Wang et al. 2007). Best management practices (BMPs) are specific actions developed to benefit water quality or wildlife species of interest and specify types and quantities of biological legacies (e.g., standing dead snags, logs, and old trees) to leave (on a per ha basis) for the benefit of wildlife, as well as landscape components such as buffer strips along waterways (Aust and Blinn 2004). Certification (e.g., Forest Stewardship Council, FSC, and the Programme for the Endorsement of Forest Certification (PEFC; http://pefc.org/) specifies BMP-like guidelines for retention of biological legacies within stands, and for large land owners, also what proportion of the landscape can be harvested each year, creating a sustainable mosaic of forest stands of varying ages across the landscape. This is tied into some concepts of forest health where the age distribution of stands, as well as that of trees within each species, must be stable over time at the landscape and regional scales (Castello and Teale 2011). Very large acreages of temperate-zone forestland covering ca 50% of all forests are certified in Europe and North America as of 2016 (FSC 2016; PEFC 2016).

The scientific literature on interactions between natural disturbance and harvesting has concentrated on two aspects of the issue that are taken up in more detail below: first, salvage of wind and fire disturbed areas, including ecological impacts on regeneration and quality of the materials salvaged, and second, the similarities and differences between the silvicultural system used on a given landscape and the natural disturbance regime. Natural disturbance regimes are considered to be a type of "gold standard" against which to compare harvesting, since it is widely presumed that the natural regime was able to maintain productivity and biodiversity of forests over centuries or millennia. Forest ecosystems under natural disturbance regimes are (structurally) quite complex, even right after a stand-leveling wind event or crown fire. It can be said that such forests are "born complex" (complex structure of combined living and dead stand), and they are likely to be on alternative successional trajectories compared to harvested forests (Donato et al. 2011). Tree- and stand-age distributions; amount of carbon and nutrients in various pools (especially CWD); successional patterns; species richness of taxonomic groups including plants, wildlife, insects, fungi, and lichens at stand and landscape scales; as well as the number and distribution of microhabitats that are present in reference forests influenced primarily by natural disturbances are common types of information used to compare natural disturbance regimes with harvesting regimes (Hale et al. 1999; Lindenmayer et al. 2000; Frelich and Reich 2003).

The big question to address throughout this chapter is can managed forests produce timber, maintain the same level of productivity and ecosystem services, and maintain all species that would be present in natural forests? If so, then the large array of ecosystem services provided by forests can continue. In addition to timber-based forest products, these services include non-timber products such as forest-grown food and biological products (i.e., healthcare products and pharmaceuticals), clean water, maintenance of soil quality, wildlife habitat, and recreational opportunities; all of these ecosystem services depend on forest health, which is in turn greatly affected by natural and human disturbance. We address these issues at the landscape scale. However, considerations at smaller (stand) and larger (regional to global) scales are also brought in where necessary to provide context for landscapes.

2   Forest Resilience to Human and Natural Disturbance

Resilience in forests is the ability to recover to a pre-disturbance state; this is related to the maximum severity of disturbance that a forest can absorb and still return to its pre-disturbance state (Holling 1973; Paine et al. 1998). Resilience must also be considered in terms of what aspect of the pre-disturbance state (i.e., structure or composition, stand or landscape) and which specific disturbance (DeRose and Long 2014). Prior to human intervention, the historic natural disturbance regime on most forested landscapes created a variety of successional states (some exceptions occur such as landslides and volcanic eruptions that restart primary succession, but those are beyond the scope of this chapter). It also allowed for connectivity across the landscape and continuous presence of adequate habitat for all species over time and for replacement of nutrients lost at the time of disturbance, all with some unknown level of redundancy. In other words more disturbances could have occurred without impairing forest health as defined above (Standish et al. 2014) or without exceeding the resilience limits of the forest. Now that harvesting activities dominate most forested landscapes, the question is how much human disturbance can be added to natural disturbance before a given landscape gets close to or exceeds the limit for continued resilience? Reserved areas (protected forests with no harvesting), will-ingness to limit harvest levels if needed (through BMPs and certification), and how natural and human disturbances interact all play into this equation.

There is much debate about the International Union for Conservation of Nature target that by 2020 at least 17% of terrestrial habitat should be conserved in pro-tected areas and other effective, area-based conservation measures (IUCN 2014). At this point about 11.5% of forests worldwide are protected (Dudley and Phillips 2006). It would be relatively simple to leave 11.5 (or 17%) of the landscape unhar-vested, in the form of parks, wilderness areas, and other reserves and then not worry about how the rest of the landscape is managed. However, at these percentages, these reserves are likely too isolated for metapopulation dynamics to occur among them to maintain most or all species and probably do not provide enough area for all species to survive, as informed by the theory and models of island biogeography and species-area curves (Arrhenius 1921; Gleason 1925). Also, many regions of the world have much less than the average of 11.5% protected forest. Furthermore, it also seems unlikely that there is so much redundancy in natural systems that all spe-cies can survive with only 11.5% of the original area, and certainly goals for forest-dependent ecosystem services such as water quality and carbon sequestration cannot be met on these small percentages of the landscape. Therefore, species habitats and ecological function must also be provided within the managed forest matrix (Lindenmayer and Franklin 2002), that covers the majority of the landscape. The previously discussed certification programs help, since impairment of ecological health of forests can be slowed, stopped, or reversed by certification, allowing the surrounding matrix to augment some functions of protected areas. If reserves with only natural disturbances are available, they will be useful parts of the landscape mosaic, contribute their share of ecosystem services, and provide reference condi-tions for the managed landscape; however, for most of the landscape over most of the planet, forest managers will always be stuck with the inherent conflict between commercial products and necessary biological legacies.

The situation may not be quite as bad as it seems; although harvests do add to the total rate of disturbance—tree- and stand-age distributions are skewed toward younger age classes when harvests occur—natural and human disturbances are only partially additive, due to some level of preemption effects between human and natu-ral disturbances. The nonadditive preemption effects of natural and harvest distur-bance are illustrated by the following examples.

- Example 1. Even-aged management by clear-cutting in forest ecosystems with wind-dominated natural disturbance regimes (Fig. 1). This would include forests dominated by shade-tolerant species where severe fires occur less often than the maximum lifespan of the dominant tree species, e.g., temperate forests domi-nated by various combinations of hemlock (*Tsuga*), maple (*Acer*), beech (*Fagus*), and basswood or linden (*Tilia*) in eastern North America, central and northern Europe, Japan, and eastern China and temperate rainforests of Douglas fir (*Pseudotsuga*), hemlock, and Sitka spruce (*Picea sitchensis*) in western North America (Frelich 2002). Harvesting at a certain stand-age threshold can prevent

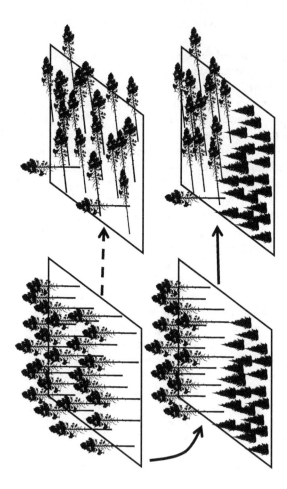

Fig. 1 Preemption effect of logging and nonadditive impacts of logging plus natural wind disturbance. Two scenarios with different outcomes are shown. Scenario 1 (upper two diagrams), the forest is almost totally blown down by a windstorm (dotted arrow). Scenario 2 (lower two diagrams), half of the stand is logged, and 10 years later the same windstorm as in scenario 1 occurs, and only the unlogged portion is affected by the storm (solid arrows). In scenario 2, half of the timber went to the mill rather than into the dead biomass pool, and also the area impacted by the windstorm was half that in scenario 1, due to the young, non-susceptible age class of the trees in the area that was harvested. The simple assumption that, over the long term, all stands will be logged and all stands will also blow down does not hold up

development of older stands that are more susceptible to wind damage, thus pre-empting future stand-leveling wind disturbance. The same future windstorms will occur, but due to harvesting, fewer susceptible stands will be present; the frequency of stand-leveling wind events on the landscape will be lower, and therefore over the long term, the natural + harvesting impacts will not add up to twice the area of the landscape (i.e., all acreage would not be disturbed twice, both logged and blown down). Therefore, harvesting has a less than additive impact on total (natural + harvesting) disturbance rate. The main consequences of interest from a biological legacy point of view are (1) the even-aged stands created by clear-cutting will have much less CWD on the ground and standing snags than even-aged stands after windthrow (Fig. 2) and (2) there may be fewer old stands on the landscape. Note that these effects on legacies could be partly or mostly mitigated by BMP guidelines that require leaving snags and older stands.

- Example 2. Uneven-aged management by selection cutting (not high grading), in the same forest ecosystems as in example 1. Effects and consequences similar to example 1 occur but on a finer-grained spatial mosaic. Within a given stand, harvesting removes older cohorts of trees that would have blown down or died from disease later on. In addition, gap sizes from selection can have much less variability than natural gap sizes.
- Example 3. Clear-cutting in forests with fire-dominated disturbance regimes. Many temperate conifer and southern boreal forests start out with birch (*Betula*) and aspen (*Populus*) species after fire and succeed to coniferous pine (*Pinus*), spruce (*Picea*), and fir (*Abies*) species which then burn, restarting the successional sequence. This includes boreal and hemi-boreal "mixed-wood" ecosystems with varied mixtures of birch, aspen, spruce, and fir and high-severity fire regimes of the northern USA and southern Canada, northern parts of Europe, and Asia, (Frelich 2002). Preemption of fire effects by clear-cutting stands at relatively young ages of 50–80 years could be quite large, since succession to conifers would be prevented, along with susceptibility to fire that comes with conifer dominance. The consequences would be absence of old stands, absence of conifer stands on the landscape, and diminished fire occurrence.
- Example 4. Clear-cutting in conifer or conifer-hardwood forests with mixed-severity fire regimes. This includes red and white pine (*P. resinosa* and *P. strobus*) in the northeastern USA, Scots pine (*P. sylvestris*) in northern Europe, and Korean pine (*P. koraiensis*) in northeastern Asia (Ishikawa et al. 1999; Frelich 2002). These forests are characterized by exceptionally large spatial heterogeneity in disturbance severity and mixture of species with varied successional status, at within-stand, stand, and landscape scales. Harvesting regimes can homogenize the variability in spatial and temporal distribution of disturbance size, severity, and trees of varying successional status. Possibilities include maintaining the forest in almost entirely early-successional species or late-successional species, depending on details of the relationship between harvesting and life history characteristics of the tree species. For example, large clear-cuts accompanied by scarification can maintain birch and aspen via resprouting in mixture with the pines. Conversely, in regions where most harvests occur during winter and regen-

Fig. 2 Large volume of CWD on the forest floor of a young post-blowdown forest in Estonia (upper) and post-fire pine and spruce forest in Minnesota, USA (lower). Note the huge volume of both standing dead snags and downed CWD in both cases (Photos by Kalev Jõgiste (upper) and Dave Hansen (lower))

eration survives under the snow, harvesting may allow advance regeneration of maple, beech, spruce, and fir to dominate postharvest stands. In either of these cases, the impacts of harvesting will decrease future occurrence of mixed-severity fires by maintaining tree species with fuel characteristics not conducive to fire (Nowacki and Abrams 2008).

Although these preemption effects mean that forest characteristics such as stand-age distribution are not altered as much by harvesting as one might initially suspect, they still do not mitigate the short-circuiting of material flow into the deadwood detrital pool (CWD) caused by harvesting; volumes of CWD after wind and fire are very large (Fig. 2), nor do they mitigate the absence of certain species habitats at within-stand, stand, and landscape scales, such as tip-up mounds and old trees (Fig. 3). Thus, knowing the level of redundancy natural systems have, with regard to flow of materials and nutrients through the detrital pool and presence of special habitats, is extremely important. If we knew the exact level of redundancy and resilience natural forest systems have with respect to these biological features (e.g., only 10% needed to maintain forest health and 90% redundant and available for harvest or 50% needed and 50% redundant, etc.), then we could harvest that exact amount of timber without the fear of negative impacts. However, that would still assume that future natural mortality caused by disturbances could be perfectly captured by harvests. Stands and trees would have to be harvested before they were about to be hit by a disturbance, or we would have to live with lower-quality salvaged forest products. And of course, we really do not know the exact level of redundancy in natural forest ecosystems. We know that "more is better" for retention of biological legacies, but do not know what proportion of production can routinely be removed to allow long-lasting (i.e., centuries to millennia) maintenance of forest health.

Two additional concepts are needed to understand the interactions of human and natural disturbances on the landscape. First, large infrequent disturbances (LIDS, Turner et al. 1998) remind us that we do not always control landscape dynamics. Generally, we take it for granted that the representation of successional stages of forests depends on disturbance regime and that at the landscape level, all stages of post-disturbance forest are likely to exist (Turner 2010). However, infrequent large disturbances punctuate (override) the dynamics created by lesser severity agents (including management disturbances). A preexisting steady-state shifting mosaic may be wiped away and the future landscape pattern henceforth be predetermined by the variety of resilience mechanisms of ecosystems before the big event (Foster et al. 1998a). Secondly, novel disturbance regimes may shape the future landscape due to human effects and climate change. Old guidelines of successional response to disturbance may need to be discarded, and instead we may need to use basic principles of ecosystem response such as resilience, legacies of land use, and hierarchy of disturbances, to predict forest response. However, note that novel disturbance regimes (which also include management outcomes) can create a highly stochastic component leading to non-equilibrium dynamics of forest ecosystems (Foster et al. 1998b; Mori 2011). For example, climate change could easily push forests outside of their "safe operating space" with respect to disturbance severity, size, and frequency, thereby leading to an unexpected response to future disturbance (Johnstone et al. 2016). Another example is historical; large herbivores probably shaped the structure of European temperate forests over centuries (Vera 2003). Domesticated herbivores kept the landscape partially open, albeit the initial clearing was made by humans. In addition, browsing and grazing become interconnected when wild and domesticated herbivores are present. Although novel from the

Fig. 3 Complexity of post-disturbance stand structure including living and deadwood legacies. Upper, post-wind tip-up root structures in temperate maple-hemlock forest in the Porcupine Mountains, Michigan, USA. Lower, post-fire legacy of live and dead standing and downed trees in boreal jack pine, fir, black spruce, and cedar (*Thuja occidentalis*) forest in Minnesota, USA (Photos by Kristi Parro (upper) and Dave Hansen (lower))

perspective of natural disturbances, this regime that includes domestic grazing as a disturbance has become well known, with predictable effects, and the ecosystem has adapted to it. The problem with future climate change and the accompanying invasive tree diseases is that forest managers need to accumulate experience with the response of the forest to novel disturbance agents to create a sustainable system. They will not have that body of experiences to draw on until long after the changes occur.

2.1   To Salvage or Not to Salvage?

The issues around salvaging of windthrown or burned forests are complex. Salvaging can affect future successional trajectories by altering the probability of establishment of certain species, and it can also change the likelihood and effects of future disturbances (Parro et al. 2015). Post-fire salvage has been shown to be damaging to forest ecosystems in some cases (Lindenmayer et al. 2004; Donato et al. 2006), by adding to the severity of natural disturbance so that the total disturbance severity exceeds the threshold that pre-disturbance species can tolerate (Frelich 2002). The ecosystem is at a vulnerable state; microsites for germination, as well as any seedlings that are germinating, can be eliminated by movement of equipment. Revegetation that prevents erosion can be delayed (Donato et al. 2006; Wagenbrenner et al. 2015), and depending on the timing of the salvage operations, this may add to delays already occurring in cases where the fire occurred between good seed years for the main tree species. Slash and downed logs after fires or windstorms can provide critical shade that allows seedlings to avoid desiccation and shelters them from browsing ungulates (Frelich 2002).

While fire provides bare soil to shade-intolerant pioneer species, salvage logging as an additional disturbance may cause negative impacts on tree regeneration (Lindenmayer and Noss 2006; Vanha-Majamaa et al. 2007). For example, in Estonia, salvage logging may increase the negative impact of ground vegetation species like *Calluna vulgaris* and grasses (Parro et al. 2009), reducing available nitrogen of already nutrient-stressed habitat (Mallik 2003) and hampering forest regeneration. On wet sites the effects of fire combined with salvage logging are less significant, as the proportion of soil patches untouched by fire is higher and the soil is able to maintain its moisture. Salvage logging removing woody materials from vast areas of vegetation supports the establishment of *Calluna* fields within 2–3 years (Sedlakova and Chytry 1999). This together with temperature fluctuations and erosion prolongs the regeneration of forest (Hannerz and Hanell 1997). Additionally, the abundance of regeneration on salvage-logged areas may be 60% lower than on burned-only areas and lead to pure pine stands with advanced height growth by 15 years after fire, skipping the successional stage with deciduous trees (Parro et al. 2015).

Windthrow salvage generally has few long-term effects on abundance and diversity of tree regeneration, although if salvage involves soil scarification, then it may promote admixture of shade-intolerant species into the regeneration by removing understory herb vegetation and creating a window of opportunity (Royo et al. 2016). Saproxylic beetles of varied species require different types and diameters of dead-wood, and therefore whether structural features such as tree crowns are left intact, in addition to larger diameter logs, will affect species richness after salvage (Thorn et al. 2014). Windthrow salvage can change the diversity and types of microsites available within stands, including the creation of compacted soil on skid trails (Peterson and Leach 2008a, b).

Salvage may have other effects than those on biodiversity—it may change the chances and intensities of future disturbance (Thompson et al. 2007). Salvage of windthrow may or may not change susceptibility to subsequent fire. However, lack of salvage can make it more difficult to use prescribed fire due to increased fire intensity, smoke production, and safety risks. If slash is left (from windthrow or logging), then subsequent fires likely can be equally intense (Stephens 1998), since fires get most of their energy from 100-h (2.5–7.5 cm diameter) fuels and spread based on contiguity of 10- and 100-h fuels, including the branches of trees that are commonly left after salvaging. Only small proportions of bolewood are burned in forest fires, and bolewood is the main material that is removed in harvests. Exceptions occur in cases where so-called logging residue (10- and 100-h fuels) is removed to be used for biomass burning, which may indeed reduce the likelihood and intensity of subsequent fires. Windfall slash is concentrated near the forest floor, and there-fore soils are heated more during post-wind fires, and consumption of duff is usually more complete, leaving a different type of seedbed than after windfall alone or after fires in standing forests. Salvage after intense stand-killing fire in conifer forests may have little impact on subsequent fire probability or intensity, because consump-tion of 10- and 100-h fuels leaves little for the next fire; new fuel must accumulate before an intense fire can occur. However, the situation is different after low- to moderate-intensity fires, since the needles could be killed by crown scorch (note that scorch is defined as tissue death resulting from radiant heat and rising convec-tion of hot air into the canopy; do not confuse this with char due to direct contact with flames that occurs in crown fires), resulting in a standing dead forest where salvaging would likely remove bolewood and leave a lot of recently killed fine fuels and branch wood (10- and 100-h fuels) on the ground, which could actually increase the risk of subsequent high-intensity fires (Donato et al. 2006).

2.2 Reference Forest Ecosystems

Reference ecosystems (also known as baselines) allow comparisons between areas that are similar except for the presence/absence of harvesting, and there are two types. The first type compares two stands or landscapes with common disturbance history, but that will differ henceforth and that the references will not experience harvesting. These could be primary or secondary forests (depending on the current condition of the landscape matrix) and provide straightforward comparison of the effects of harvesting from a given common starting point. The second type is areas previously under the influence of natural disturbances only—typically natural areas that can serve as blueprints for forest restoration and maintenance of biodiversity at stand and landscape scales. This type of reference represents spatial and temporal patterns of disturbance severity and interactions with landforms and the resulting patterns of biological legacies that historically regulated and maintained ecosystem biodiversity and productivity. Examples include large natural areas which are now the only places to get good information on landscape distributions of stand ages and successional stages created by natural disturbances (Frelich 2002). The two types of reference ecosystems provide different takes on one of the basic tenets of science: a baseline by which to judge the sustainability of managed forests. Such baselines are required by some organizations (e.g., FSC) for certification of large landowners.

Naturalness is a debatable concept, and questions about human impact in its many forms can make reference conditions hard to define (Colak et al. 2003). There is a gradient in naturalness (inverse to a gradient in management intensity) which in many parts of the world is truncated, so that what is at the most "natural" end of the gradient in one region would be considered to be significantly disturbed by human activities in another region. There is no choice for forest managers other than to work with what they have. In some cases synthetic reference conditions based on historical reconstruction, paleoecological analyses of fossil pollen in sediments and other evidence, can be constructed in places where humans have had a pervasive influence for many centuries (parts of Europe and Asia), while in North America, there are vast tracts of forest that were never harvested (old-growth reserves, parks and wilderness areas) available to use as reference ecosystems, although they may not be devoid of all anthropogenic disturbances, for example, grazing or firewood collection (Frelich et al. 2015).

Regardless of one's definition of naturalness and ability to access information about natural disturbance impacts, clearly, comparing processes after natural disturbances and human disturbances can provide insight into forest sustainability. Forest managers also need to recognize that certain changes in forest health can only be brought about by fire or wind. Harvesting cannot do a perfect job of maintaining forest health, even with the best intentions. For example, coppicing does not disturb the soil (e.g., aspen and oak harvest in North America, Europe, and Asia), but it cannot be repeated too many times since rootstocks will eventually build up diseases and productivity will go down (Stanosz and Patton 1987). Perhaps only the occurrence of fire (and occasional regeneration by seed) once every several harvest rotations as it occurs in natural aspen and oak forests can correct this situation. There is much left to learn from reference ecosystems.

3 Combined Natural and Human-Dominated Landscape Mosaics

What types of stand and landscape dynamics occur in regions with different natural disturbance regimes, different levels of human influence, and different cultural outlooks?

3.1 Stand-Scale Considerations

Spatial patterns tend to have more variability in natural systems, which could be important for many species that contribute to forest productivity. It is probably not essential to duplicate the natural age structure and spatial patterns of stands exactly, but rather to assure that stands with a range of structural characteristics are present in sufficient area to provide habitat for all species. Several types of combined harvest and natural disturbance occur under different disturbance regimes, each with a unique set of challenges for managers who would like to mimic natural disturbance for purposes of maintaining biodiversity:

Crown fire systems These systems have large high-severity disturbances, typical of pine-, spruce-, and fir-dominated temperate and boreal systems. Large clear-cuts best mimic landscape patch dynamics, but even with that silvicultural system in place, there would still be a number of differences with a natural fire-generated landscape:

- CWD: after severe fires in conifer forests, there are large numbers of standing snags that case harden and slowly fall over several decades, creating a steady supply of newly downed CWD on the forest floor and CWD in many stages of decay in the newly developing forest (Heinselman 1996, Fig. 2). This is surprisingly similar to the CWD deposition process in old, uneven-aged forests with episodic windthrow.
- Successional status of regeneration: severe fires kill shade-tolerant advance regeneration and reset succession to early-successional birch, aspen, and species with serotinous cones (e.g., jack pine (*Pinus banksiana*) or black spruce (*Picea mariana*) in the North American boreal forest). However, winter clear-cutting that is commonly practiced is less severe than the crown fires and can lead to release of advanced regeneration and a push toward late-successional conditions.
- Seedbed conditions: the organic horizon is often consumed in wildfires but would still be present after clear-cutting.
- Seed dispersal and germination: the spatial pattern of seed trees left after clear-cutting (leave trees) differs from that after fire (Turner et al. 2003). Seeds of fire-dependent serotinous species may not be dispersed if branches are left on the ground in the shade of advance regeneration, and seeds from shade-intolerant

serotinous or non-serotinous leave trees may not be successful if advance regeneration and shrub layer are still intact and compete with germinating tree seedlings.

Scarification or a prescribed fire would be needed to more closely mimic the effects of crown fires on successional status, seedbed, and seed dispersal and germination. Since all of the branches and twigs would be present after clear-cutting, any prescribed fire might be too severe and consume tree seeds that are on the ground. The supply of large CWD to the forest floor over time that occurs after fire would be almost impossible to mimic.

Mixed fire regimes These ecosystems have crown and surface fires, with red pine, white pine, Scots pine, or Korean pine. Uneven-aged silviculture with occasional small clear-cuts (patch cuts) would mimic the natural disturbance regime. The same problems with advance regeneration of shade-tolerant species (and in this case, also shrubs) as with substituting harvest for fire in Crown Fire Systems may occur. For example, the dense understory of beaked hazel (*Corylus cornuta*) and balsam fir (*Abies balsamea*) in Minnesota red and white pine forests would need to be intentionally removed during harvest but would be killed by severe fire (Heinselman 1996).

Surface fire systems Larch (*Larix* spp.), pine, and oak (*Quercus* spp.) woodlands and savannas in central North America, central Europe, and eastern Asia fall in this group. Although fires cannot be suppressed in most ecosystems with crown fires, they can often be suppressed or otherwise excluded in surface fire systems. This can be a major issue leading to filling in of the understory by shade-tolerant tree species and shrubs and reduction of understory diversity (Peterson and Reich 2008). Generally, these forests have uneven-aged tree populations with infrequent or low density of reproduction. It is hard for harvesting alone to mimic the effects of fire; partial harvests followed by prescribed burns can be the most effective way to mimic a natural surface fire regime, maintain forest health, and allow reproduction of trees after harvest. Fire-scarred trees are a natural part of these ecosystems that provide unique microhabitats not found in ecosystems managed by harvesting alone.

Wind-dominated systems These have multi-aged stands with variable gap sizes and rare stand-leveling disturbance. Uneven-aged silviculture with occasional small clear-cuts could be a good mimic. The following considerations would apply:

- Gap-size distributions and ranges of sizes vary considerably in natural forests (Runkle 1982; Kathke and Bruelheide 2010), but gaps tend to be more uniform in size in commercial forests unless managers make a deliberate attempt to vary gap size. Variable gap sizes can lead to variable species composition, since species with different levels of shade tolerance can take advantage of different sized gaps, although other factors like forest floor characteristics and browsing by deer can create complex effects related to gap size (Kern et al. 2013; Willis et al. 2015).

- Keeping trees smaller by harvesting at younger ages will always create a forest less susceptible to wind damage, as will favoring certain species that are less susceptible to wind. By combining these two strategies, it is possible to largely negate wind damage (Stanturf et al. 2007). However, this could be counter to resilience and biodiversity strategies—needed to meet goals of society and to deal with climate change—to increase the diversity of species and ages of trees.
- Trees injured by wind, but that persist for decades to centuries afterwards, often provide structural complexity that includes nesting habitat for cavity-nesting birds. Snapped off dead snags after windstorms can be left because they do not take growing space from surrounding live trees.
- Windthrow creates a mixture of snapped-off snags and uprooted trees with associated tip-up mounds (Vodde et al. 2011), with a lot of fine branch and twig materials from the tree crowns. Thus, there is a mosaic of wet and well-drained microsites on the forest floor (Beatty and Stone 1986) and places protected from deer browsing by the branches of tree crowns (de Chantal and Granström 2007), or by location on tip-up mounds (Krueger and Peterson 2006), and a variety of microsites for regeneration of diverse tree species (Fig. 4). An example is that the pioneer tree species black cherry (*Prunus pensylvanica*) only occurred on post-blowdown tip-up mounds in Pennsylvania, USA (Peterson and Pickett 1995).
- Prior to advent of harvesting, all dead trees became CWD. With removal of timber, the base of the CWD-based food pyramid or food web is smaller, and there-

Fig. 4 Diverse microsites and resulting diversity in regeneration of tree species (*Picea, Acer, Betula, Sorbus*) in a post-windstorm forest in Estonia, as described by Vodde et al. (2011) (Photo by Kalev Jõgiste)

fore, it is reasonable to hypothesize that fewer species may be supported at higher trophic levels. In addition, there is less area covered by CWD and other microsites, thus causing a reduction in species supported due to species-area curve considerations, which may affect even the number of species that can be sustained at low trophic levels.

- Isolation of CWD and large, old trees. This would be an island biogeography effect but at a finer scale than most people usually think about such effects. Microsites like CWD, old trees, and tip-up mounds are further apart when fewer are left in the forest. This may affect the metapopulation dynamics of mosses, lichens, fungi, and insects (Jönsson et al. 2008; Morrissey et al. 2014).

Purposeful creation of a variety of gap sizes and microsites, leaving standing dead snags, old live trees, and both CWD and branches during harvests would be needed to mimic these natural features. Although BMPs mentioned earlier address some of these issues, most were developed for habitat of certain wildlife species, and do not specify the range of sizes needed to replicate the variability that occurs after natural windthrow.

3.2   Landscape-Scale Considerations

Stand-age distributions can be changed by management but only slowly. Stand-age distributions under natural disturbance regimes are either negative exponential in cases where probability of disturbance is equal across stand ages (as in boreal forest with crown fires) or uniform (flat) in young age classes followed by a negative exponential decay after a threshold age is reached at which stands become susceptible to disturbance (Fig. 5, Frelich 2002). In contrast, landscapes dominated by harvesting tend to have uniform stand-age distributions, with deficits of old stands compared to natural systems (Kuuluvainen 2009). Sometimes they also have a unimodal peak due to an episode of settlement and land clearing followed by reforestation (Fig. 6), which exaggerates the imbalance because fewer young and old stands are present (relative to natural landscapes) than on regulated landscapes with uniform stand-age distributions. In some cases young and old stands are completely missing.

Truncation of the landscape stand-age distribution so that maximum stand age is relatively young is the most common situation on managed landscapes, since clearcuts replace old multi-aged or old even-aged stands of long-lived species. Some modern landscapes are headed toward a split between young stands (in the portion of the landscape subject to clear-cut harvest) and old stands (in the portion of the landscape that is reserved). There are a few cases where stands are harvested at ages similar to the average age at which natural disturbances occur, particularly in places with coniferous forests with crown fires (e.g., boreal forest in North America where crown-fire rotations can be 50–100 years), although there can still be truncation of maximum age. With fires occurring at random with respect to stand age, 36.8% of stands survive one rotation period, 13.5% survive two rotation periods, and 5.0%

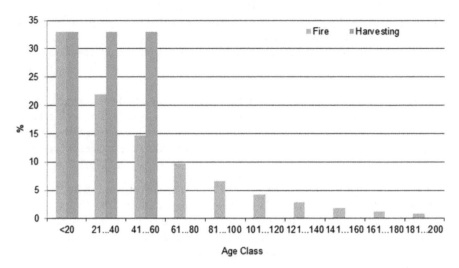

Fig. 5 Landscape age distribution for a boreal forest with randomly timed fires and for a well-regulated landscape with clear-cut harvesting. The rotation period is about 50 years in both cases. Note the shortage of old stands for the regulated landscape

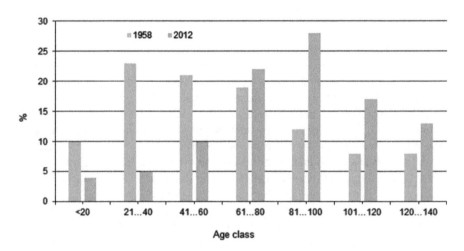

Fig. 6 Landscape age distribution in Estonia in 1958 and 2012, note that the age distribution is not balanced and the peak has moved. It takes a long time to restore such imbalances in landscape age distribution

survive three (Frelich 2002). Thus, even if the natural disturbance rotation on average is 50 years, there would still be 5% of all stands that are 150+ years old, and for a 100-year natural fire rotation, there would be a similar percentage 300+ years old; this is a consequence of the negative exponential stand-age distribution. These old stands can have tremendously different environments and species composition than younger stands, with uneven-aged canopies, deep moss cover, and high variability in light and nutrient environments, fostering high levels of species richness in plants (Reich et al. 2012), fungi, and insects within a single hectare, and they can have indicator species that only do well in such stands.

In systems with mostly uneven-aged stands under the natural disturbance regime, there can also be truncation of the stand-age distribution. To add a cross-scale (between landscape, stand, and tree) element here, in addition, the ages of cohorts within stands can also be truncated either by clear-cuts or by selection cutting that takes place at shorter intervals than natural gap-forming disturbances. In natural stands there are a number of old trees that can have canopy residence times 2–3 times the average, so that if the average is 150 years, there will be a substantial number of trees 300+ years old and a few that are 450 years old. These trees contribute unique microhabitats to the forest, such as hollow trunks and crevices in the bark that harbor certain species of mosses, lichens, and insects. These habitats provide nesting for woodpeckers, bats, and other animals and rotting wood that supports many species of fungi, and they produce extraordinarily large tip-ups and pieces of coarse woody debris when they die that contribute to the microhabitat diversity and detritus-based food web of the forest for decades, or in the case of slowly decaying conifers, even centuries after tree death.

Note that to really replicate the landscape-scale patterns, stands would have to be harvested at a range of ages, instead of at a single preferred age. Ignoring this results in substantial changes in age distribution (e.g., unbalanced age distribution in Fig. 6). This would contribute to the variability in the amount and quality of wood obtained per hectare, since older stands might have less wood, or might have more wood but in fewer larger pieces, or could have more defects in the wood, while simultaneously having parts of the same tree with higher quality (fine-grained) wood. Younger stands may have more, smaller pieces of wood. These days utilization of wood has advanced to use a variety of sizes, and this variability in size and quality of material going to the mill might not be the problem as it was a few decades ago.

3.3   Reserved or Natural Forest

Here we will mention some additional considerations regarding reference ecosystems that were not considered above. Natural area reserves (remnants in a human-dominated landscape) may not be representative of natural landscape age class structure, ecosystem function, and spatial patterns that would result when only natural disturbance occurs. This is due to remnants being left in areas that were unsuitable for logging, which often have unique topographic and soil features that precluded harvest. At the same time, remnant forests on flat sites with high productivity tend to be replaced with agriculture or clear-cut forests and are often underrepresented in natural area reserves or reference ecosystems (Lõhmus and Kraut 2010; Kraut 2016). In some political jurisdictions, efforts have been made to create a system of reference ecosystems and natural areas that are representative of the natural landscape. In such a system, there will usually be an excess of available reference stands in some regions (with lesser degrees of human settlement), so that only very high-quality sites are selected to be reserved, while in other regions (with greater degrees of human settlement) very few choices exist and lower quality stands may be chosen (Margules and Pressey 2000).

Forest managers tend to have a static view of forested natural area remnants and view them as "lost" when hit by natural disturbances, when in fact such disturbances often create an exhibit of natural mechanisms of resilience to disturbance. This resilience may even carry these stands back to natural conditions faster than harvested second growth or post-agriculture forests. We need to recognize natural resilience and learn from it to make the managed matrix of the forest landscape more resilient. Also, in human-dominated landscapes, reserved stands with old growth tend to be fixed in location, whereas old stands have a tendency to occur in different locations over time in a natural system, due to the mostly random patterns of natural disturbances. Some exceptions occur, such as stands in sheltered valleys being protected from wind and possibly fire for long periods of time. The natural landscape may have a mixture of fixed old-growth locations as well as moving ones. In any case, the large degree of biological legacies left after natural disturbance insure temporal continuity for ecological processes and native species, and spatial connectivity is also likely given the number and spatial dispersion of older stands usually present on the landscape (Frelich 2002).

The moving pattern also accompanies human settlement: for example, old pastures and fields are converted into forest through long history in the Baltic states of Estonia, Latvia, and Lithuania. However, recovery of such forests to natural condition depends on the spatial proximity to stands that have been continuously forested for a long time—continuous forest cover over centuries can be important for many species—and the proportion of the landscape with no land-use change is crucial so that these species can move into afforesting stands. Even if the stands with continuous forest cover are not natural stands, the disturbance legacies allowed by various silvicultural systems may keep biodiversity alive and ready to move into afforested areas. How long it takes to restore natural conditions after land-use change depends on many factors, and distance to artificial and natural legacies may have great effects.

3.4 Disturbance-Harvest Interactions

Large, infrequent disturbances (LIDS, Turner et al. 1998), including severe storms and fires, can cause problems in the flow of timber over time, namely, a surplus of salvage timber followed by low flow of forest products for a few decades while the post-disturbance forest develops. These effects on flow of harvestable products are dependent on the spatial extent of ownerships or management units that are affected, throughout which the loss can be averaged out, compared to the size of disturbance. At the stand scale of small ownerships, one storm or fire can create a very unbalanced age distribution, with a large surplus of salvage timber that cannot be used before it decays, followed by no possibility for stands of harvestable age for several decades, whereas for large ownerships of millions of hectares, one storm or fire is unlikely to affect more than a small percentage. However, in fire-dependent boreal systems, big events happen more frequently than in gap-dominated systems with multi-aged stands, and million ha fires can occur that disrupt timber flow even at the scale of a national forest.

Unfortunately, introduced tree diseases and pests can make the dynamics of a given genus or species of tree unstable over its entire range, and although exotic tree diseases are found throughout the temperate zone, North America seems to be especially susceptible to this phenomenon (Roy et al. 2014); examples currently in play in the USA include Dutch elm disease (*Ophiostoma ulmi*), hemlock woolly adelgid (*Adelges tsugae*), emerald ash borer (*Agrilus planipennis*), and many others, historically including chestnut blight (*Cryphonectria parasitica*). This problem can only be eventually solved by planting resistant varieties or by evolution of resistant trees by natural selection, followed by slow restoration of populations across the landscape, processes that will likely take decades at the very least but centuries in most cases.

A diversity of tree species and ages at stand, landscape, and regional scales is likely to help mitigate the magnitude of effects of all disturbances on the flow of timber products. Younger stands or multi-aged stands are less likely to be leveled by windstorms than old even-aged stands. Oak stands are less likely to be leveled than aspen stands or pine stands. Fuel breaks with roads, fields, and nonflammable species of trees can limit the spatial extent of fire damage to the economic system. However, there is always the possibility of extreme disturbance events that cannot be planned for or mitigated.

3.5 Role of Culture

In the USA, forests have abundant CWD and other detritus with complex microtopography, which is seen as normal and even attractive. In contrast, a more manicured forest aesthetic is normal in Western Europe, although places like Estonia with close-to-nature forestry have conditions intermediate between the USA and Western Europe. The region of the eastern Baltic Sea is probably the transition zone between intensive and extensive forest management (Liira and Kohv 2010). Compared to Western Europe, the Baltic states have more seminatural forests of high biodiversity value, due to a high diversity in approaches to management because of large variation of soil and terrain conditions within small areas and variation in historical forest management traditions. In Latvia and Lithuania the traditional silvicultural system is dominated by clear-cuts followed by soil preparation and regeneration with native species—mainly spruce, pine, and birch. The regeneration method can either be planting, natural regeneration, or a combination of both. Priority is given to natural regeneration, mostly from advance regeneration, while artificial regeneration is used only in cases where there is no possibility of natural regeneration. However, to an increasing extent, it has been found that forest owners select close-to-nature silvicultural practices (notably in Estonia) such as selective logging, shelterwood regeneration, and regeneration in small groups of varied sizes. Due to financial investment in post-clear-cut forest regeneration, these alternative silvicultural systems usually have been practiced by the private forest sector, especially for smaller forest owners (Rendenieks et al. 2015), and they also have smaller exposure for neighboring stands, thereby decreasing the risks for damage by wind, frost, and insect pests.

China presents a great contrast to European and North American temperate forest zones. Here, some lands have been deforested for centuries or millennia, the few natural forests that still exist (although perhaps protected by the Natural Forest Conservation Program) are very far away from most of the landscape, and establishing tree cover of some type to prevent flooding and soil erosion is a significant goal (Wang et al. 2007).

3.6   Role of Restoration

Restoration forestry could restore resilience to natural and human disturbance, and ecosystem retrogression that has occurred in some places (e.g., with plantation conifers on hardwood sites) could be reversed (Stanturf et al. 2014). These are take-home lessons for restoration in the context of interacting human and natural disturbances: (1) Restoration can proceed only so far in the lifetime of one individual; forests change slowly. (2) Restoration can go in a series of steps—each management agency has to know where they are in the process. (3) People have to be patient and be content to be part of a larger and longer-lasting (then their career or lifetime) process. (4) Restoration "distance" will vary hugely depending on the starting condition and the distance "traveled" across restoration space: going from post-agriculture to natural forest is much longer than from post-clear-cut to natural forest in a continuously forested system (Frelich and Puettmann 1999). In addition, note that ecosystems generally have hysteresis in their response to disturbance—the time and processes that need to happen to get back to a natural state are generally different than those that created the degraded system to begin with. Particularly, the legacy of the previous ecosystem state affects the future dynamics. The key question for forest managers is can the ecosystem produce the legacy components needed to allow continuation of forest cover in the desired manner, or should measures be planned to assist the course of restoration (Jõgiste et al. 2017)?

4   Conclusions and Synthesis

Forest managers need to recognize the balance between materials going into detrital pools in forests and materials being removed for human products. Forests did not evolve to produce economic products; therefore, the degree to which seminatural forests without intensive management, such as adding fertilizers, can sustain removals in long term (over centuries to millennia) is not known. The comparison of commercial forests with reference forests that experience only natural disturbance can show whether the health of ecosystems, including the productivity of fiber, viability of native species, and the associated ecosystem services, is being maintained. It could be that there is enough redundancy in forest ecosystems so that substantial harvests can be maintained, but this is a hypothesis that forest managers and researchers need to test. Research on the level of redundancy of natural systems is needed. We need better answers, for a variety of forest ecosystems, to the question: what is the real level of sustainable harvest? Experiments designed to answer this question need to take into account the maintenance of productivity and all taxonomic groups of species as standards and, when possible, controls consisting of reference ecosystems that have never experienced harvesting and previously harvested stands that are currently similar to stands that will be harvested but that will be reserved from harvest. Such a "double control" system would show how far away current forests are from natural conditions and whether forests subject to restoration harvesting, certification standards, and traditional forest management practices progress toward natural conditions faster or slower than forests at the same starting point that experience only natural disturbances. Such research should be long term—at least several decades—to get meaningful results, although this can be difficult given the constantly changing political and management environment.

Within this broad framework of research, details are needed in more regions of the world on topics such as the ecological function of coarse woody debris, old stands of trees, spatial patterns of human disturbance and how it differs from natural disturbance, resilience to changing climate combined with harvesting, and restoration of natural function in areas undergoing afforestation. Knowledge of these topics is extremely variable from region to region and seems to depend partly on the presence or absence of small groups of researchers who happen to be interested, but also partly on opportunities presented by funding agencies and the regional landscape. A few examples include research on CWD in the Pacific Northwest, USA; metapopulation dynamics of mosses, lichens, and wood-dwelling insects on old trees in Scandinavia; landscape patterns of large-scale natural disturbance as compared to harvested landscapes in Pacific Northwest and Great Lakes region of the USA; land-use legacies in New England, the USA, and Europe; and afforested landscapes in northern Europe and China (e.g., Harmon et al. 1986; Heinselman 1996; Foster et al. 1998a; Wenhua 2004; Snäll et al. 2005; Donato et al. 2011).

Protected, somewhat natural forests tend to occupy very small portions of the landscape in regions where human needs are greatest and a small (11.5%) portion for the world as a whole. Therefore, there is a great need to use the managed forest matrix as a base for biodiversity conservation (Wallington et al. 2005; Lindenmayer and Franklin 2002). Multiple uses and commodities obtained from forests along a management/forest condition gradient from natural to intensively managed create a multidimensional space for options and decisions for forest resource managers. A central question for human society is how much of the biodiversity and ecological function created by natural disturbances can and should be accommodated in the matrix? Part of the answer is hopefully more and more over time, rather than less and less. Restoration and conservation tools proved appropriate by practice to achieve the desired state of the forest ecosystem will also have to be adapted to take into account the effects of climate change. Still, somehow we need to cope with the need for restoration and uncertainty of the future, using the range of realistic management actions and outcomes.

There is a gradient from nature-dominated (wilderness in the USA) to totally human-dominated forests in many regions of the world, while other places have extensive tree plantations of afforested abandoned agricultural lands, for example, most of Europe and China. Thus, there is a variable need for restoration of the natural functions of forests. Restoration forestry is needed in afforested areas after agricultural abandonment or in areas where plantation forestry with off-site tree species was extensively practiced. There is a limit to how fast such areas can have attributes such as tree composition, landscape age structure, tree-age distributions within stands, microsite variability, and gap-size variability restored.

Harvesting stands and trees over a range of ages in the forest matrix of the landscape, thus minimizing the truncation of stand and tree-age distributions due to harvest, combined with a system of reserved/reference forests to serve as a baseline for the long-term experiment known as forest management, is probably the best way to create resilient forests, ensure adequate habitat for native species at all spatial extents, and maintain productivity over time. The reserved stands (1) help create the long tail of old stands of the landscape stand-age distribution that is present in natural landscapes, (2) show how forests respond to major disturbances that do occur, (3) at the same time also show how forests are changing due to invasive disease species and climate change in the absence of harvesting, and (4) serve as natural areas for the benefit of the human population. Therefore these reference forests should capture all significant forest or ecosystem types and be distributed widely across a region, so that not all are affected by large disturbances at once and so that their several roles continue after unexpectedly large events occur. Creating a range of stand ages that are harvested on the rest of the landscape, combined with the reserved forests, allows for attainment of a somewhat natural age distributions at the landscape scale, which is essential for stable flow of materials and maintenance of biodiversity, the main tenets of forest health. This in turn insures that ecosystem services such as carbon sequestration, genetic resources, clean water and air, and human recreational needs will continue to be provided.

References

Arrhenius O (1921) Species and area. J Ecol 9:95–99

Aust WM, Blinn CR (2004) Forestry best management practices for timber harvesting and site preparation in the eastern United States: an overview of water quality and productivity research during the past 20 years (1982-2002). Water Air Soil Pol 4:5–36

Beatty SW, Stone EL (1986) The variety of soil microsites created by tree falls. Can J For Res 16:539–548

Castello JD, Teale SA (eds) (2011) Forest health: an integrated perspective. Cambridge University Press, Cambridge

Chandran A, Innes JL (2014) The state of the forest: reporting and communicating the state of forests by Montreal Process countries. Int For Rev 16:103–111

Çolak AH, Rotherham ID, Çalikoglu M (2003) Combining 'naturalness concepts' with close-to-nature silviculture. Forstwissenschaftliches Centralblatt 122:421–431

de Chantal M, Granström A (2007) Aggregations of dead wood after wildfire act as browsing refugia for seedlings of Populus Tremula and Salix Caprea. Fort Ecol Manag 250:3–8

DeRose RJ, Long JN (2014) Resistance and resilience: a conceptual framework for silviculture. For Sci 60:1205–1212

Donato DC, Fontaine JB, Campbell JL et al (2006) Post-wildfire logging hinders regeneration and increases fire risk. Sci 311:352

Donato DC, Campbell JL, Franklin JF (2011) Multiple successional pathways and precocity in forest development: can some forest be born complex? J Veg Sci 23:576–584

Dudley N, Phillips A (2006) Forests and protected areas, guidance on the use of IUCN protected area management categories. IUCN, Gland, Switzerland/Cambridge, UK

Foster DR, Knight DH, Franklin JF (1998a) Landscape patterns and legacies resulting from large, infrequent forest disturbances. Ecosystems 1:497–510

Foster DR, Motzkin G, Slater B (1998b) Broad-scale disturbance: regional forest dynamics in Central New England. Ecosystems 1:96–119

Franklin JF, Lindenmayer DB, MacMahon JA et al (2000) Threads of continuity: ecosystem disturbances, biological legacies and ecosystem recovery. Cons Biol Pract 1:8–16

Frelich LE (2002) Forest dynamics and disturbance regimes. Cambridge University Press, Cambridge

Frelich LE, Puettmann K (1999) Restoration ecology. In: Hunter ML Jr (ed) Maintaining biodiversity in forest ecosystems. Cambridge University Press, Cambridge, pp 498–524

Frelich LE, Reich PB (2003) Perspectives on development of definitions and values related to old-growth forests. Env Rev 11:S9–S22

Frelich LE, Montgomery R, Oleksyn J (2015) Northern temperate forest. In: Peh K, Corlett R, Bergeron Y (eds) Handbook of forest ecology. Routledge Press, London, pp 30–45

FSC (2016) FSC facts and figures December 5, 2016. https://ic.fsc.org/en/facts-figures. Accessed 7 Jan 2017

Gleason HA (1925) Species and area. Ecology 6:66–74

Grove SJ (2002) Saproxylic insect ecology and the sustainable management of forests. Ann Rev Ecol Syst 33:1–23

Hale CM, Pastor J, Rusterholz KA (1999) Comparison of structural and compositional characteristics in old-growth and mature, managed hardwood forests of Minnesota, USA. Can J For Res 29:1479–1489

Hannerz M, Hanell B (1997) Effects on the flora in Norway spruce forests following clearcutting and shelterwood cutting. For Ecol Manag 90: 29–49

Harmon ME, Franklin JF, Swanson FJ, Sollins P, Gregory SV, Lattin DJ et al (1986) Ecology of coarse woody debris in temperate ecosystems. Adv Ecol Res 15:133–302

Heinselman ML (1996) The boundary waters wilderness ecosystem. University of Minnesota Press, Minneapolis

Holling CS (1973) Resilience and stability of ecological systems. Ann Rev Ecol Syst 4:1–23

Ishikawa Y, Krestov PV, Namikawa K (1999) Disturbance history and tree establishment in old-growth Pinus Koraiensis-hardwood forests in the Russian far east. J Veg Sci 10:439–448

IUCN (2014) The green list for protected areas global standard. Available via IUCN. https://www.iucn.org/sites/dev/files/import/downloads/pilot_phase_iucnglpastandard20140515_pdf. Accessed 10 Sep 2016

Jõgiste K, Korjus H, Stanturf JA et al (2017) Hemi-boreal forest: natural disturbances and the importance of ecosystem legacies to management. Ecosphere 8(2):e01706

Johnstone JF, Allen CD, Franklin JF et al (2016) Changing disturbance regimes, ecological memory and forest resilience. Front Ecol Environ 14:369–378

Jönsson MT, Edman M, Jonsson BG (2008) Colonization and extinction patterns of wood-decaying fungi in boreal old-growth Picea Abies forest. J Ecol 96:1065–1075

Kathke S, Bruelheide H (2010) Gap dynamics in a near-natural spruce forest at Mt. Brocken, Germany. For Ecol Manag 259:624–632

Kern CC, D'Amato AW, Strong TF (2013) Diversifying the composition and structure of managed late-successional forests with harvest gaps: what is the optimal gap size? For Ecol Manag 304:110–120

Kraut A (2016) Conservation of wood-inhabiting biodiversity – semi-natural forests as an opportunity. Dissertationes Biologicae Universitatis Tartuensis 287. University of Tartu Press, Tartu

Krueger LM, Peterson CJ (2006) Effects of white-tailed deer on *Tsuga Canadensis* regeneration: evidence of microsites as refugia from browsing. Am Mid Nat 156:353–362

Kuuluvainen T (2009) Forest management and biodiversity conservation based on natural ecosystems dynamics in northern Europe: the complexity challenge. Ambio 38:309–315

Liira J, Kohv K (2010) Stand characteristics and biodiversity indicators along the productivity gradient in boreal forests: defining a critical set of indicators for the monitoring of habitat nature quality. Plant Biosyst 144:211–220

Lindenmayer DB, Noss RF (2006) Salvage logging, ecosystem processes, and biodiversity conservation. Cons Biol 20:949–958

Lindenmayer DB, Margules CR, Botkin DB (2000) Indicators of biodiversity for ecologically sustainable forest management. Cons Biol 14:941–950

Lindenmayer DB, Foster DR, Franklin JF et al (2004) Salvage harvesting policies after natural disturbance. Science 303:1303

Lindenmyer DB, Franklin JF (2002) Conserving forest biodiversity. A Comprehensive multiscale approach. Island Press, Washington DC

Lõhmus A, Kraut A (2010) Stand structure of hemiboreal old-growth forests: characteristic features, variation among site types, and a comparison with FSC-certified mature stands in Estonia. For Ecol Manag 206:155–165

Mallik AU (2003) Conifer regeneration problems in boreal and temperate forests with Ericaceous understory: role of disturbance, seedbed limitation, and keystone species change. Critical Rev Plant Sciences 22: 341–366

Margules CR, Pressey RL (2000) Systematic conservation planning. Nature 405:243–253

Mori AS (2011) Ecosystem management based on natural disturbances: hierarchical context and non-equilibrium paradigm. J Appl Ecol 2011:280–292

Morrissey RC, Jenkins MA, Saunders MR (2014) Accumulation and connectivity of coarse woody debris in partial harvest and unmanaged relict forests. PLoS One 9(11):e113323. https://doi.org/10.1371/journal.pone.0113323

Nowacki GJ, Abrams MD (2008) The demise of fire and "mesophication" of forests in the eastern United States. Bioscience 58:123–138

Paine RT, Tegner MJ, Johnson EA (1998) Compounded perturbations yield ecological surprises. Ecosystems 1:535–545

Parro K, Koster K, Jõgiste K et al (2009) Vegetation dynamics in a fire damaged forest area: the response of major ground vegetation species. Balt For 15:206–215

Parro K, Metslaid M, Renel G et al (2015) Impact of postfire management on forest regeneration in a managed hemiboreal forest, Estonia. Can J For Res 45:1192–1197

PEFC (2016) PEFC global statistics: SFM and CoC certification, Data: September 2016. WWW. pefc.org. Accessed 7 January 2017

Peterson CJ, Leach AD (2008a) Salvage logging after windthrow alters microsite diversity, abundance and environment, but not vegetation. Forestry 81:361–376

Peterson CJ, Leach AD (2008b) Limited salvage logging effects on forest regeneration after moderate-severity windthrow. Ecol Appl 18:407–420

Peterson CJ, Pickett STA (1995) Forest reorganization: a case study in an old-growth forest catastrophic blowdown. Ecology 76:763–774

Peterson DW, Reich PB (2008) Fire frequency and tree canopy structure influence plant species diversity in a forest-grassland ecotone. Plant Ecol 194:5–16

Reich PB, Frelich LE, Voldseth P et al (2012) Understory diversity in boreal forests is regulated by productivity and its indirect impacts on resource availability and heterogeneity. J Ecol 100:539–545

Rendenieks Z, Nikodemus O, Brumelis G (2015) The implications of stand composition, age and spatial patterns of forest regions with different ownership type for management optimisation in northern Latvia. For Ecol Manag 335:216–224

Roy BA, Alexander HM, Davidson J et al (2014) Increasing forest loss wordlwide from invasive pests requires new trade regulations. Front Ecol Environ 12:457–465

Royo AA, Peterson CJ, Stanovick JS et al (2016) Evaluating the ecological impacts of salvage logging: can natural and anthropogenic disturbance promote coexistence? Ecology 97:1566–1582

Runkle JR (1982) Patterns of disturbance in some old-growth mesic forests of eastern North America. Ecology 63:1533–1546

Sedlakova I, Chytry M (1999) Regeneration patterns in a Central European dry heathland: effects of burning, sodcutting and cutting. Plant Ecol 143: 77–87

Seidl R, Fernandes PM, Fonseca TF et al (2011) Modelling natural disturbances in forest ecosystems: a review. Ecol Model 222:903–924

Siry JP, Cubbage FW, Ahmed MR (2005) Sustainable forest management: global trends and opportunities. For Pol Econ 7:551–561

Snäll T, Pennanen J, Kivistö L et al (2005) Modelling epiphyte metapopulation dynamics in a dynamic forest landscape. Oikos 109:209–222

Standish RJ, Hobbs RJ, Mayfield MM et al (2014) Resilience in ecology: abstraction, distraction, or where the action is? Biol Conserv 177:43–51

Stanosz GR, Patton RF (1987) Armillaria root rot in aspen stands after repeated short rotations. Can J For Res 17:1001–1005

Stanturf JA, Goodrick SL, Outcalt KW (2007) Disturbance and coastal forests: a strategic approach to forest management in hurricane impact zones. For Ecol Manag 250:119–135

Stanturf JA, Palik BJ, Dumroese RK (2014) Contemporary forest restoration: a review emphasizing function. For Ecol Manag 331:292–323

Stephens SL (1998) Evaluation of the effects of silvicultural and fuels treatments on potential fire behaviour in Sierra Nevada mixed-conifer forests. For Ecol Manag 105:21–35

Thompson JR, Spies TA, Ganio LM (2007) Reburn severity in managed and unmanaged vegetation in a large wildfire. PNAS 104:10743–10748

Thorn S, Bässler C, Gottschalk T et al (2014) New insights into the consequences of post-windthrow salvage logging revealed by functional structure of saproxylic beetles assemblages. PLoS One 9:e101757. https://doi.org/10.1371/journal.pone.0101757

Turner MG (2010) Disturbance and landscape dynamics in a changing world. Ecology 91:2833–2849

Turner MG, Baker WL, Peterson CJ et al (1998) Factors influencing succession: lessons from large, infrequent natural disturbances. Ecosystems 1:511–523

Turner MG, Romme WH, Tinker DB (2003) Surprises and lessons from the 1988 Yellowstone fires. Front Ecol Env 1:351–358

Turton SM, Alamgir M (2015) Ecological effects of strong winds on forests. In: Peh K, Corlett RT, Bergeron Y (eds) Routledge handbook of forest ecology. Routledge, New York, pp 127–140

Vanha-Majamaa I, Lilja S, Ryömä R, Kotiaho JS, Laaka-Lindberg S, Lindberg H, Puttonen P, Tamminen P, Toivanen T, Kuuluvainen T (2007) Rehabilitating boreal forest structure and species composition in Finland through logging, dead wood creation and fire: the EVO experiment. For Ecol Manag 250: 77–88

Vera FWM (2003) Grazing ecology and forest history. CABI Publishing, Oxon

Vodde F, Jõgiste K, Kubota Y et al (2011) The influence of storm-induced microsites to tree regeneration patterns in boreal and hemiboreal forest. J For Res 16:155–167

Wagenbrenner JW, MacDonald LH, Coats RN et al (2015) Effects of post-fire salvage logging and a skid trail treatment on ground cover, soils, and sediment production in the interior western United States. For Ecol Manag 335:176–193

Wallington TJ, Hobbs RJ, Moore SA (2005) Implications of current ecological thinking for biodiversity conservation: a review of the salient issues. Ecol Soc 10(1):15

Wang G, Innes JL, Lei J et al (2007) China's forestry reforms. Science 318:1556–1557

Wenhua L (2004) Degradation and restoration of forest ecosystems in China. For Ecol Manag 201:33–41

Willis JL, Walters MB, Gottschalk KW (2015) Scarification and gap size have interacting effects on northern temperate seedling establishment. For Ecol Manag 347:237–247

Woodcock P, Halme P, Edwards DP (2015) Ecological effects of logging and approaches to mitigating impacts. In: Peh K, Corlett RT, Bergeron Y (eds) Routledge handbook of fsorest ecology. Routledge, New York, pp 422–435

Ecosystem Services from Forest Landscapes: Where We Are and Where We Go

Louis R. Iverson, Ajith H. Perera, Guillermo Martínez Pastur, and Urmas Peterson

1 Introduction

In this chapter, we aim for three goals: (1) We summarize some key features from the chapters, as they pertain to the overall themes of the book. The chapters themselves provide great resources to bring awareness to some newer and broader aspects of forest ecosystem services as well as a literature-rich, synthetic approach to understanding these advances and future visions for related research and application. We will highlight some of those points here. (2) We then aim to provide some emerging messages resulting from these newer approaches to understanding the complexities of planning for, evaluating, and accentuating the FES. (3) Finally, we provide some insights on science gaps, research priorities, and potentials for knowledge transfer mainly into practitioners and policy makers.

L. R. Iverson (✉)
Northern Institute of Applied Climate Science, Northern Research Station,
US Forest Service, Delaware, OH, USA
e-mail: liverson@fs.fed.us

A. H. Perera
Ontario Forest Research Institute, Ministry of Natural Resources and Forestry,
Sault Ste. Marie, ON, Canada

G. Martínez Pastur
Centro Austral de Investigaciones Científicas (CADIC), Consejo Nacional de Investigaciones
Científicas y Técnicas (CONICET), Ushuaia, Tierra del Fuego, Argentina

U. Peterson
Institute of Forestry and Rural Engineering, Estonian University of Life Sciences,
Tartu, Estonia

© Springer International Publishing AG, part of Springer Nature 2018
A. H. Perera et al. (eds.), *Ecosystem Services from Forest Landscapes*,
https://doi.org/10.1007/978-3-319-74515-2_9

2   Complexity of Forest Ecosystem Services

One immediate realization upon opening the book is that the ecosystem services tied to forests require a broad view, and likely a more complex view, than most people realize. We do not dwell on the "usual suspects" in this book (e.g., timber, biodiversity, carbon sequestration) but emphasize some lesser known aspects of ES. Among them are the large, globally significant contributions forests make to regulating chemical composition of the atmosphere, be they trace gases (chapter "Effects of Climate Change on CH_4 and N_2O Fluxes from Temperate and Boreal Forest Soils", Díaz-Pinés et al.) or plant volatiles (chapter "What Are Plant-Released Biogenic Volatiles and How They Participate in Landscape- to Global-Level Processes?", Niinemets). For example, Díaz-Pinés et al. point out the important role that boreal and temperate forest soils play as regulators of atmospheric gases in the framework of a changing global climate. More than half of global carbon is stored in soils, and dynamic changes in environmental conditions underway now and increasing in the future will likely affect the net atmosphere-forest balance of CH_4 and N_2O fluxes at different temporal and spatial scales. These changes, in turn, will feed back on chemical composition of the atmosphere and, thus, on the global climate. However, knowledge is still rather limited with regard to the relationship between forest composition (and associated microbial processes), interactions with changes brought about by climate, and its importance for the function of forests as climate regulators.

Niinemets (chapter "What Are Plant-Released Biogenic Volatiles and How They Participate in Landscape- to Global-Level Processes?") aptly discusses a vastly understudied and underemphasized ecosystem service derived from plants: plant-released volatile organic compounds (at least 30,000 different compounds identified to date), often specialized for a plethora of biological and ecological functions such as the enhancement of plant stress resistance, or communication among plant organs, with other plants, or even with beneficial insects to slow the spread of herbivory. These services are crucial for stability and performance of ecosystems at local scales but collectively can also modify global climate through multiple feedback loops of stress-induced volatile emissions and cloud condensation nuclei interacting with temperature, solar radiation, and plant productivity. As such, the author argues for a much larger consideration of these trace gases in models intending to predict future climate and outcomes.

2.1   *Planning for Sustainable Ecosystem Services: Integration Across Borders and Across Land Uses*

One of the greatest challenges in the landscape planning is related to maintaining the ES provision in multifunctional managed ecosystems. Research exists in the provision of individual ES; however, only a few studies have considered the trade-offs and synergies among them, and the literature on the associated landscape planning is scarce. Within the book, we present several examples that lead with this topic, trying to integrate the different proposals across boundaries and land-use types.

Elbakidze et al. (chapter "Towards Functional Green Infrastructure in the Baltic Sea Region: Knowledge Production and Learning Across Borders") focus on sustaining natural capital and enhancing multiple ecosystem services through the use of functional green infrastructure, a network of high-quality natural and seminatural areas intended to deliver ecosystem services and protect biodiversity in both rural and urban areas. These green networks are especially important for smaller countries and jurisdictions, which may not be able to provide sufficiently large green areas on their own, and thus benefit from ecosystem services from forest ecosystems that span international and regional boundaries. This process is challenging in any situation but complicated further when coordinating across multiple country boundaries, as the authors present for the Baltic Sea Region. The authors articulate the challenges but also the opportunities with much to gain from increased multilateral, learning-based collaborations regarding all aspects of sustainable forest landscapes.

LaRosa et al. (chapter "Sustainable Planning for Peri-urban Landscapes") review the special considerations when evaluating ES associated with peri-urban landscapes, those areas located partly outside the more compact part of a city with low density and diverse patterns of development spreading into the surrounding rural areas. These areas provide important functions including enhancing biodiversity of urban areas, enhancing proximal recreation opportunities, and reducing heat island effects, pollution, and noise; thus, they play a fundamental role in health, well-being, and social safety. Many of these services depend on the urban forest; thus appropriate spatial planning may be used to enhance certain ES by modifying the size, composition, and structure of the urban forest. On the other hand, peri-urban landscapes also suffer from increased stresses due to the proximity and accessibility of urban activities, and climate change is likely to impact these regions more than rural areas because of the higher concentration of human activities. Peri-urban landscapes are also understudied and provide another fruitful arena for enhancing ecosystem services via research and spatial planning.

Angelstam et al. (chapter "Barriers and Bridges for Landscape Stewardship and Knowledge Production to Sustain Functional Green Infrastructures") use six long-term, place-based case studies throughout Europe to explore their social-ecological systems and the various approaches to enhancing green infrastructure and sustainable forest management. Across the region, landscape histories and governance contexts are very diverse, so that experiences of human and natural scientists, practitioners, and stakeholders from each study were crucial for knowledge production and learning to understand the full process toward functional green infrastructure. They point out the important role that expert knowledge can play to supplement or complement empirical knowledge at least in the short term. The authors provide seven key actions to promote multilevel learning toward enhancing sustainable landscapes and their respective ES.

3 Challenges and Opportunities in Managing for ES in Forested Landscapes

Another great challenge is the balance of ES provision in managed landscapes. In the past, research was focused on maximizing the provisioning ES, despite the other services. However, within the new management paradigm, we must consider the maintenance of other ES inside the areas under intensive management. Several studies have focused on the theoretical framework, but implementation in the field still needs research. In this framework and within the book, the authors present several examples of successful implementation of both theoretical proposals and long-term research of the effects of these new management proposals.

Monkkonen et al. (chapter "Solving Conflicts Among Conservation, Economic and Social Objectives in Boreal Production Forest Landscapes; Fennoscandian Perspectives") focus on preserving certain ecosystem services including biological diversity yet at the same time, maintaining intensive timber extraction in boreal, specifically Fennoscandian, forests. Many ecosystem services and biodiversity are in conflict with intensive timber production in boreal forests; e.g., the proportion of old-growth forests is very small, and natural disturbances such as fires and gap formation are minimized, with negative biodiversity consequences for forest species. The authors present management tools for assessing and finding solutions for conflicts among alternative forest uses, including encouraging more variation in management regimes combined optimally across landscapes, more mixed species stands, more green tree retention, and less thinning. Forest certification, payment schemes, and regional planning differentiating landscapes where environmental and social objectives have priority over timber production landscapes may be policy tools to encourage sustainable landscapes and the broad array of ecosystem services over the long term. This chapter highlights the orthogonality that could occur among ecosystem services (e.g., direct competition between biodiversity conservation and extraction of timber, in this case).

Frelich et al. (chapter "Natural Disturbances and Forest Management: Interacting Patterns on the Landscape") dwell on the question "what is the real level of sustainable harvest" in the context of maximizing ecosystem services and forest health, given the likelihood that natural forests possess some level of redundancy with respect to the amount of dead wood and older trees and stands that are needed to maintain forest health. Though safe levels of harvest are rarely known, practices such as close-to-nature forestry or best management practices with regard to structural features left after harvesting can ensure adequate residuals and help maintain forest resilience to disturbance. However, large infrequent disturbances and novel disturbance regimes in a changing climate may swamp out management practices in shaping the future forest so that additional redundancy, spatially distributed among landscapes, may be needed to sustain ecosystem services across regions. Restoration forestry is also expected to play an increasing role in restoring resilience to forests undergoing increasing natural and human disturbances.

4   Emergent Messages

- *Complexity associated with the knowledge of forest ecosystem services is high.*
  In addition to the vast array of services provided by forest ecosystems, interactions (both positive and negative) among those as well as their numerous feedback loops to influencing composition and function of forest ecosystem
  themselves render information, knowledge, and investigations of forest ecosystem services highly complex.

- *Forest ecosystem services are not limited to localities and fine scales.*
  Services offered by forest ecosystems are broad. Their influence extends
  beyond local benefits (e.g., atmosphere and hydrosphere) and cross ecological,
  geographic, and jurisdictional scales.

- *Much uncertainty exists in scientific/empirical knowledge about services from
  forest ecosystems at broad scales.*
  With regard to atmospheric gas regulation, for example: how do the composition and spatial configuration of forest ecosystems affect the services they provide? How will these services continue as the climate changes and alter the
  atmosphere and ecosystems themselves? These are complex research questions,
  but their answers may divulge vital information about crucial services provided
  by forest ecosystems.

- *Planning for forest ecosystem service provision needs a broad-scale perspective.*
  In many regions where forest cover dwindles due to urbanization or other land
  use, continuation of a sustained provision of ecosystem services is a challenge.
  Spatial configuration of the remaining forest cover, especially their connectivity
  as networks, becomes an important element in planning that requires cross-
  national and regional boundaries. Policy makers and land-use planners must
  broaden their planning horizons and collaborate with neighboring countries and
  regions.

- *Education and awareness of forest ecosystem services is essential.*
  Consideration of forest ecosystem services is limited to rhetoric in land-use
  planning exercises in most countries and regions. In part this is due to a lack of
  awareness and knowledge. Educating and making policy makers and land-use
  planners aware of the range of services that forest ecosystems provide, necessity
  to sustain these services, and how best to design land use for sustained provision
  of the ecosystem services, even with the limited knowledge, are crucial.

- *New approaches and shifts in present management paradigms are necessary.*
  It may be necessary to adopt new approaches (e.g., use of expert knowledge
  and adaptive management in the case of limited knowledge) and tools (e.g., different simulation models for designing and assessing land-use plans). This
  includes changing the culture in land management agencies to embrace new
  approaches in policy development, strategic land-use planning, and optimizing
  land management goals.

5   Applying the Messages

Here we explore the applications of the messages in this book from two viewpoints: their use in land management approaches and decisions and how they can guide future research activities. While these points are not independent, we treat them separately to address the interests of both land managers and researchers.

5.1   *Implementation Approaches to Optimize Forest Ecosystem Services*

The chapters within the book emphasize the complexity and breadth of forest ecosystem services, particularly at broad scales. The many steps involved in the provisioning of ecosystem services involve ascertaining the full suite of ecosystem services desired by the local communities as well as the regional and global community, quantifying and analyzing the capacity of extant forest ecosystems to provide those services, designing an optimal extent and spatial configuration of supplementary forest ecosystems, formulating management plans that both ensure a sustained supply of ecosystem services and minimize conflicts, institutionalizing these plans in land management authorities that may need cultural shifts, educating and transferring knowledge to land management professionals to ensure continuation of these plans, and self-governance. The complexities associated with provisioning ecosystem services are enormous. Their relative magnitude and importance will vary and depend on specific cultural, social, and economic milieus (Fig. 1). Only through innovative and adaptive management approaches can these obstacles may be overcome, and therefore benefits may be achieved.

As such, the approaches to provide a full suite of ecosystem services may be complex as well. First principles of landscape ecology, the study of patterns and processes, e.g., patches, matrix, connectivity, heterogeneity, will be quite useful in designing strategies for sustaining the supply of cross-scale forest ecosystem services. Several general landscape ecology principles can be set forward that are likely valid for many services. First, land planners and managers need to think broadly and spatially as to how various land cover/land-use patches influence each other and how the whole region's ecosystem services can be affected by their decisions. In particular, the concepts of patterns, especially spatial configuration and connectivity of forest ecosystems, matter! An additional challenge for planners and decision-makers is that the ecosystem services that transcend local scales are generally less recognized, especially the regulatory services – they are generally indirect, intangible, and beyond the scale of human perception. Thus, the services that extend beyond individual forest patches, e.g., those which rely on the connected networks of forests, may not be on the radar screen of local agencies responsible for managing forest ecosystem services. This points to the need for (i) education of the local agencies on the regional and global values of the forest patches they manage and (ii) the need for vertical

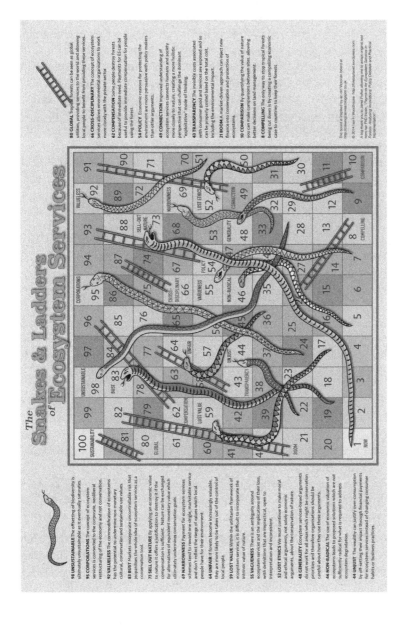

Fig. 1 A graphical depiction of the many benefits and challenges associated with provisioning ecosystem services, the array of both benefits and challenges that face land managers in their decision-making. Reproduced with permission from Janet Fisher (Fisher, J. (2011). Payments for Ecosystem Services in Forests: Analysing Innovations, Policy Debates and Practical Implementation. Doctoral Thesis, University of East Anglia.) and Iain Woodhouse (https://forest-planet.wordpress.com/2014/08/18/the-snakes-and-ladders-of-ecosystem-services/)

coordination (hierarchical nesting) among land management agencies across jurisdictions (e.g., towns-cities-townships-counties-states-regions-provinces-countries).

Second, which follows, is that often planning (and incentives and/or regulation) therefore needs to also occur at higher levels, i.e., at broader scales, even if it means crossing many jurisdictional boundaries. Education and communication among policy makers, including across country boundaries, are therefore essential at this level. In addition, because ecosystem services supply often entails a complex mix of ecology, economics, sociology, and politics in the order of decision-making with broadening scales, cross-jurisdictional assessment and planning to supply forest ecosystem services involve complexities not unlike developing and implementing mitigation strategies for climate change, i.e., most action is local, but planning and regulation needs to be at multiple scales, to achieve maximum impact on enhancing or sustaining the ecosystem services.

Third, the transfer of scientific knowledge to decision-makers and land-use planners needs to be readily accomplished via the tools of the landscape ecologist: maps, GIS, animations of model outputs, visual simulations, remote sensing from a host of sensors, on-ground sensing devices, and, of course, field-obtained plot data. Succinct, synthetic, and visual representations of potential future ecosystem services under various land-use decisions would go a long way toward moving to socially responsible optimization of those services. Extracting knowledge from experts would also be important in this effort to evaluate clear representations of alternative management paths forward; the value of expert and traditional knowledge in supplementing empirical (scientific) knowledge should not be underestimated. Some ecosystem services from forests are traditional and even cultural. Ascertaining the value, demand, and supply of these services may depend on traditional and local knowledge that may not be found in scientific knowledge.

6 Research Opportunities

As we conclude this chapter and this book, the editors reflect on lessons learned and provide the following suggestions for ways forward in research needs and implementation avenues toward successful augmentation and sustenance of the ecosystem services from forests. We leave it to the reader to start with these nuggets, listed in no particular order, and generate appropriate research lines to that end.

1. Develop strategies and methods to simply compile all the information available from the "tools of the trade" into meaningful and digestible packets of information for policy makers to make informed decisions.
2. Research methods that can help determine the "real level of sustainable harvest" as put out by Frelich et al. in chapter "Natural Disturbances and Forest Management: Interacting Patterns on the Landscape".
3. Build up a suite of case studies that cover more and more situations for other jurisdictions to copy. These case studies are often specially funded situations

using a large portion of the landscape ecological toolbox, but the essential tools and methods need to be extracted such that other jurisdictions can replicate the essential elements at low cost. Further, how to best arrive at these essential elements? Included is the need for successful case studies that transcend jurisdictional boundaries with decision-making flowing to and from local, regional, and national levels.

4. Investigate ways to improve understanding of the political dimension to forest ecosystem services/governance that plays, a huge role. Studies like those in chapters "Towards Functional Green Infrastructure in the Baltic Sea Region: Knowledge Production and Learning Across Borders" and "Barriers and Bridges for Landscape Stewardship and Knowledge Production to Sustain Functional Green Infrastructures" can elicit cross-country comparisons such that the best components for optimizing forest ecosystem services can be gleaned among multiple country-level approaches.

5. Investigate ways to better understand the influence of land-use legacies as to what constraints and opportunities exist for forest ecosystem services. Remote sensing from satellites, now with over 40 years of history, can be valuable to trace back at least the short-term legacies. Digitization and analysis of historic maps and aerial photos can extend the information base farther into the past. Learning from the past can help in managing for the future.

6. Investigate further the role of scaling in the fluctuation and provision of ecosystem services. For example, how much does a local land-use decision affect the whole? Or for a more specific example, how much do regional components of trace gases and plant volatiles affect the local conditions for humans and other organisms?

7. Research the specific and general roles of plant-emitted volatile organic compounds in modifying specific ecosystem services. We hope this book has opened the eyes of many readers as to the vast ecological influence of volatiles and the large amount of research still needed to further uncover both the local (e.g., insect resistance or pollinator attraction) and regional/global (e.g., climate regulation or cloud condensation nuclei) impacts across landscapes. Plenty of research questions here!

8. Investigate ways to uncover the net atmosphere-forest balance of CH_4 and N_2O fluxes at different temporal and spatial scales and to evaluate the feedbacks to the global climate system. Again, this is a vastly understudied aspect of forests and their ecosystem services.

9. Investigate the ways in which resilience of a landscape may vary over time, under a changing climate, and the ways in which management actions can enhance resilience. How do we even determine how resilient landscapes are, and which ecosystem services are included when we examine resilience?

10. Develop procedures to evaluate conflicts and trade-offs when opposing ecosystem services need to be optimized across landscapes (e.g., optimal arrangements of land uses within the peri-urban landscape or timber vs. biodiversity emphasis).

11. Develop methods to evaluate, especially within urbanizing landscapes (i.e., peri-urban landscapes), critical limits to the amount and juxtaposition of forested parcels which enable the sustenance of key ecosystem services. For example, how might forest parcel arrangement affect the provision of habitat for various species of birds, pollinating insects, and a biodiverse flora? How do we factor in many of the human influences on these critical limits (e.g., population density, housing/income, traffic patterns/density, gentrification, attitudes toward green space)?

12. Develop new management alternatives for land management which replace traditional (and often one-dimensional) management, and which enhance the provision of ecosystem services, especially nonmonetary services, while also decreasing the conflicts of use among the different social groups that use the ecosystem.

13. Create more studies which analyze synergies and trade-offs with implementation of different management and conservation strategies, especially in the peri-urban regions, the fringes between the urban and natural landscapes.

7   Conclusion

In this book, the authors emphasize, again, that the study of ecosystem services is a very broad and complex topic, which lies within the confluence of environmental, political, and personal values, across scales from very local to global scale. Many variables and strategies come into play, several of which may not have been generally recognized before but are elevated in this book, from the relatively less known molecular-level plant processes (chapters "Effects of Climate Change on CH_4 and N_2O Fluxes from Temperate and Boreal Forest Soils" and "What Are Plant-Released Biogenic Volatiles and How They Participate in Landscape- to Global-Level Processes?") to the widely known forest stand-level processes (chapters "Solving Conflicts Among Conservation, Economic and Social Objectives in Boreal Production Forest Landscapes: Fennoscandian Perspectives" and "Natural Disturbances and Forest Management: Interacting Patterns on the Landscape"), and ecosystem service provisioning strategies that transcend from the local jurisdictional level (chapter "Sustainable Planning for Peri-urban Landscapes") to the international level (chapters "Towards Functional Green Infrastructure in the Baltic Sea Region: Knowledge Production and Learning Across Borders" and "Barriers and Bridges for Landscape Stewardship and Knowledge Production to Sustain Functional Green Infrastructures").This book is not a definitive road map for provisioning ecosystem services from forested landscapes but is a starting point of discussion for future research and applications in a world that must improve their management and conservation strategies, to ensure the sustenance and enhancement of ecosystem services now and into the future.

Index

© Springer International Publishing AG, part of Springer Nature 2018   259
A. H. Perera et al. (eds.), *Ecosystem Services from Forest Landscapes*,
https://doi.org/10.1007/978-3-319-74515-2

CPSIA information can be obtained
at www.ICGtesting.com
Printed in the USA
LVHW06*1756270318
571334LV00002B/22/P